# Creep and Shrinkage in Concrete Structures

# WILEY SERIES IN
# NUMERICAL METHODS IN ENGINEERING

*Consulting Editors*
**R. H. Gallagher,** *College of Engineering,*
*University of Arizona*
*and*
**O. C. Zienkiewicz,** *Department of Civil Engineering,*
*University College Swansea*

---

**Rock Mechanics in Engineering Practice**
*Edited by K. G. Stagg and O. C. Zienkiewicz*

**Optimum Structural Design: Theory and Applications**
*Edited by R. H. Gallagher and O. C. Zienkiewicz*

**Finite Elements in Fluids**
Vol. 1 Viscous Flow and Hydrodynamics
Vol. 2 Mathematical Foundations, Aerodynamics and Lubrication
*Edited by R. H. Gallagher, J. T. Oden, C. Taylor and O. C. Zienkiewicz*

**Finite Elements for Thin Shells and Curved Members**
*Edited by D. G. Ashwell and R. H. Gallagher*

**Finite Elements in Geomechanics**
*Edited by G. Gudehus*

**Numerical Methods in Offshore Engineering**
*Edited by O. C. Zienkiewicz, R. W. Lewis and K. G. Stagg*

**Finite Elements in Fluids**
Vol. 3
*Edited by R. H. Gallagher, O. C. Zienkiewicz, J. T. Oden, M. Morandi
Cecchi and C. Taylor*

**Energy Methods in Finite Element Analysis**
*Edited by R. Glowinski, E. Rodin and O. C. Zienkiewicz*

**Finite Elements in Electrical and Magnetic Field Problems**
*Edited by M. V. K. Chari and P. Silvester*

**Numerical Methods in Heat Transfer**
*Edited by R. W. Lewis, K. Morgan and O. C. Zienkiewicz*

**Finite Elements in Biomechanics**
*Edited by R. H. Gallagher, B. Simon, P. Johnson and J. Gross*

**Soil Mechanics—Transient and Cyclic Loads**
*Edited by G. N. Pande and O. C. Zienkiewicz*

**Finite Elements in Fluids**
Vol. 4
*Edited by R. H. Gallagher, D. Norrie, J. T. Oden and O. C. Zienkiewicz*

**Foundations of Structural Optimization: A Unified Approach**
*Edited by A. J. Morris*

**Creep and Shrinkage in Concrete Structures**
*Edited by Z. P. Bažant and F. H. Wittmann*

# Creep and Shrinkage in Concrete Structures

*Edited by*

**Z. P. Bažant**
*Center for Concrete and Geomaterials*
*Northwestern University.*
*Evanston*

and

**F. H. Wittmann**
*Department of Materials Science*
*Swiss Federal Institute of Technology*
*Lausanne*

*A Wiley-Interscience Publication*

**JOHN WILEY & SONS**
Chichester · New York · Brisbane · Toronto · Singapore

**Library of Congress Cataloging in Publication Data:**

Main entry under title:

Creep and shrinkage in concrete structures.
   (Wiley series in numerical methods in engineering)
   Revision of papers presented at the International
Symposium on Fundamental Research on Creep and
Shrinkage of Concrete, held in Sept. 1980, at the
Swiss Federal Institute of Technology in Lausanne, and
co-sponsored by Northwestern University and U.S. National
Science Foundation.
   'A Wiley-Interscience publication.'
   Includes index.
   1. Concrete—Creep—Addresses, essays, lectures.
   2. Concrete—Expansion and contraction—Addresses,
essays, lectures. I. Bažant, Z. P.    II. Wittmann,
F. H. (Folker H.)    III. International Symposium on
Fundamental Research on Creep and Shrinkage of Concrete
(1980: Lausanne, Switzerland)    IV. Series
   TA440.C73   1982    620.1'3633    82-4766
                             AACR2
ISBN 0 471 10409 4


**British Library Cataloguing in Publication Data:**

Creep and shrinkage in concrete structures.—
   (Wiley series in numerical methods in engineering)
   1. Concrete structures—Defects—Data processing
   I. Bažant, Z. P.    II. Wittmann, F. H.
   624.1'834    TA681.5

ISBN 0 471 10409 4


Filmset and printed in Northern Ireland at The Universities Press (Belfast) Ltd. and bound at the Pitman Press, Bath, Avon.

# Contributing Authors

C. A. ANDERSON | *Los Alamos National Laboratory, Los Alamos New Mexico 87454, USA*

Z. P. BAŽANT | *Center for Concrete and Geomaterials, The Technological Institute, Northwestern University, Evanston, Illinois 60201, USA*

E. ÇINLAR | *Department of Industrial Engineering, Northwestern University, Evanston, Illinois 60201, USA*

M. A. DAYE | *Acres America Incorporated, Columbia, Maryland, USA*

W. H. DILGER | *Department of Civil Engineering, University of Calgary, Calgary 44, Alberta, Canada T2N 1N4*

J. W. DOUGILL | *Department of Civil Engineering, Imperial College of Science and Technology, London, UK*

B. L. MEYERS | *Bechtel Power Corporation, Gaithersburg, Maryland, USA*

S. E. PIHLAJAVAARA | *Concrete Laboratory, Technical Research Centre of Finland, Betonimiehenkuja 5, 0 2150 Espoo 15 Otaniemi, Finland*

C. D. POMEROY | *Research and Development Division, Cement and Concrete Association, Wexham Springs, Slough SL3 6PL, UK*

H. G. RUSSELL | *Research and Development Division, Portland Cement Association, Skokie, Illinois 60077, USA*

F. H. WITTMANN | *Department of Materials Science, Swiss Federal Institute of Technology, Lausanne, Switzerland*

J. F. YOUNG | *Department of Civil Engineering, University of Illinois, Urbana, Illinois 61801, USA*

# Contents

# Preface

The widespread availability of powerful computers and numerous easy-to-handle structural analysis codes is having a profound impact on materials science and structural mechanics. More realistic and more sophisticated mathematical models for the mechanical behaviour are becoming tractable, and many material parameters or influencing factors can now be taken into consideration. This is particularly true of creep and shrinkage of concrete—a rather complicated phenomenon whose detailed understanding is now beginning to emerge. The potential of advanced structural analysis is, however, not realized in the current practice, which relies mostly on oversimplified material laws, as embodied in current national and international codes and recommendations. To obtain full benefit from the large finite element programs now in existence, realistic material models must be provided as the input.

The literature contains thousands of papers addressing various aspects of creep and shrinkage. Yet, there exists an obvious gap between the typical materials science approach on the one hand and the methods developed and used by structural engineers on the other. To a large extent, this is due to a lack of communication among specialists in various disciplines, hampered not only by diversity of their training but also by the specialized vocabularies and methods of their fields.

In an attempt to bridge this gap, an International Symposium on Fundamental Research on Creep and Shrinkage of Concrete was held in September 1980 at the Laboratory for Building Materials of the Swiss Federal Institute of Technology in Lausanne, under the co-sponsorship of Northwestern University and U.S. National Science Foundation.[†] To cover the diverse aspects of the subject, ten invited lectures on well-defined selected topics were presented at the Symposium by the authors of the chapters in this volume. Subsequently, these contributions were substantially expanded, up-dated and revised in response to the discussions at the Symposium as well as to anonymous reviews solicited by the editors. This finally led to the present volume which aims at providing a state-of-the-art exposition and thus a basis for further development.

† Grant ENG-7820989

ix

The first part of the volume, consisting of four chapters authored by J. F. Young, J. W. Dougill, E. Çinlar, and S. I. Pihlajavaara, deals with general material properties that serve as foundations for the treatment of creep and shrinkage. This includes descriptions of the microstructure and macrostructure of the material, the physical processes involved, the probabilistic aspects of the deformation, and the movement of water within the porous material.

The second part of the volume consists of three chapters which were written by C. D. Pomeroy, F. H. Wittmann, and Z. P. Bažant to treat the measurement and modelling of creep and shrinkage. After an exposition of the experimental techniques, the real and apparent physical mechanisms of creep and shrinkage are analysed. Subsequently, the basic mathematical models, i.e. the constitutive equations along with the related basic aspects of modelling the creep and moisture effects in structures, are described.

The third and last part of the volume consists of three chapters by C. A. Anderson, W. H. Dilger, and H. G. Russell, B. L. Meyers and M. A. Daye, dealing with structural analysis and the determination of structural behaviour. In particular, two of these chapters explain and illustrate general numerical structural analysis of creep and shrinkage effects using the finite element method, as well as the special practical methods for structures treated as beams and frames of composite cross sections. Finally, the theoretical predictions must be compared with experience, and so the last chapter outlines the observations of creep and shrinkage on full-scale structures.

It is the intention of the editors to point out with this volume the specific problems of different disciplines involved in the research on creep and shrinkage, and especially, to emphasize their interrelationship. It is now obvious that further progress in this field depends to a large extent on an intensified exchange of ideas among those attacking the problem from the bases of materials science, continuum mechanics and thermodynamics, experimental methods, structural mechanics, and numerical analysis. It is hoped that the present compilation of information coming from different directions will stimulate further advances and thus contribute to a deeper and more global understanding of this complex subject.

Z. P. Bažant and F. H. Wittmann

*Evanston and Lausanne*
*August 1981*

# PART I
# FOUNDATIONS AND GENERAL ASPECTS

Creep and Shrinkage in Concrete Structures
Edited by Z. P. Bažant and F. H. Wittmann
© 1982 John Wiley & Sons Ltd

*Chapter 1*

# The Microstructure of Hardened Portland Cement Paste

*J. F. Young*

## 1.1  INTRODUCTION

The detailed structure of hardened Portland cement paste is extremely complex. It is an intimate, yet heterogeneous, mixture of diverse elements with widely varying properties and characteristics. To approach the problem sensibly it is necessary to develop a useful, descriptive organization that allows the components to be treated, as a first approximation, independently of each other. Diamond[17] emphasized three basic levels of microstructure: (1) the *atomic level*, which is concerned with crystal and molecular structure; (2) the *particle level*, which is principally concerned with the morphological aspects of crystals and particles; and (3) the *micromorphological level*, which describes how all the components fit together. However, in considering creep and shrinkage phenomena another useful approach is to consider the microstructure in terms of its individual components as given in Table 1.1. In the following discussion these five components will be considered in turn, but before doing so it is appropriate to give a brief overall view of the paste structure at the micromorphological level.

## 1.2  MICROMORPHOLOGY OF HYDRATED CEMENT PASTES

A typical fracture surface of a cement paste is shown in Figure 1.1. Because this is only 1 day old it shows a large amount of capillary porosity, but clearly shows the C–S–H† surrounding the calcium silicate grains and ettringite ($C_6A\bar{S}_3H_{32}$) forming between the grains from the hydration of the calcium aluminates. The spiny appearance of the silicate grains is due to masses of tiny needles of C–S–H radiating out from the surface. These may well be artifacts of drying[28,29] and are probably not an important feature with regard to paste properties. The predominant form of C–S–H at later ages is rather featureless and provides a thick coating around the grain

---

† Chemical formulae are given in the standard notation used by cement scientists: $C = CaO$, $S = SiO_2$, $A = Al_2O_3$, $\bar{S} = SO_3$, $H = H_2O$. C–S–H designates the amorphous calcium silicate hydrate which forms during cement hydration and is of variable stoichiometry.

Table 1.1    Components of hardened Portland cement paste (approximate order of importance)

| Component | Remarks |
|---|---|
| C–S–H | Includes micropores (<26 Å diameter) |
| Capillary pores | Mesopores and macropores |
| Calcium hydroxide | Only crystalline material |
| Calcium aluminate hydrates | Ettringite and monosulphoaluminate |
| Anhydrous residues | Unhydrated remnants of cement grains |

(Figure 1.2). Calcium hydroxide (portlandite) forms as large crystals that grow within the pore system (Figure 1.3). A mature paste can be considered as a continuous matrix of C–S–H in which is embedded large crystals of calcium hydroxide, small clusters of calcium sulphoaluminate hydrates, and unhydrated residues of the original grains (see Figure 1.4).

## 1.3   THE C–S–H COMPONENT

As mentioned above, C–S–H can be considered to be the continuous solid matrix of the hydrated system and therefore is of particular importance. An understanding of its properties must be central to a successful physicochemical explanation of the fundamental origins of creep and shrinkage

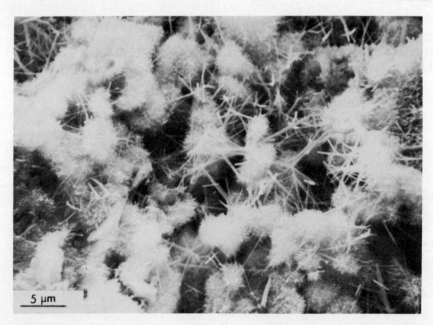

5 μm

**Figure 1.1**   Scanning electron micrograph of Portland cement paste hydrated for 1 day

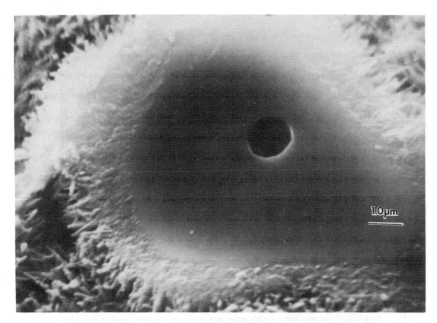

**Figure 1.2**  Scanning electron micrograph of partially hydrated tricalcium silicate grain showing the coating of C–S–H around the anhydrous core

**Figure 1.3**  Scanning electron micrograph of a tricalcium silicate paste hydrated for 10 days showing calcium hydroxide crystals (C) growing in the pore space. Anhydrous cores (A) are at the center of the hydrating grains

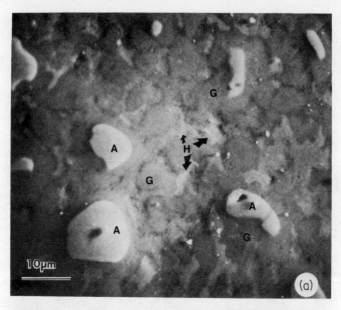

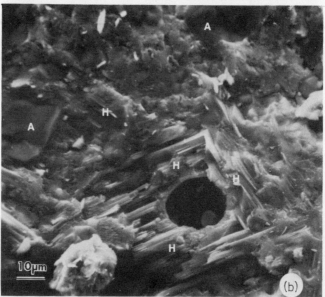

**Figure 1.4** (a) Cut and polished section of a mature tricalcium silicate paste showing residual anhydrous cores (A) embedded in hydrated matrix. Dark grey regions (G) are C–S–H, light grey areas (H) are calcium hydroxide. (Calcium sulphoaluminates are not present.) (b) Fractured surface of a mature tricalcium silicate paste showing anhydrous cores (A) and calcium hydroxide (H) embedded in a matrix of C–S–H. Note fracture passes indiscriminately through all these components

behaviour. Although considerable progress has been made in developing purely physical models,[54] it is my opinion that we cannot neglect the chemical structure of the material. Unfortunately C–S–H is essentially morphous material and it is therefore a difficult task to try to elucidate its basic structure.

### 1.3.1 Chemical characterizations

#### 1.3.1.1 Chemical stoichiometry

The chemical composition of C–S–H can be quite diverse. In the pure calcium silicate pastes the classical bulk chemical analyses of Brunauer and co-workers[12] suggested a composition of $C_3S_2H_3$. Recent microanalytical studies using electron optical methods[13,16,46,23,33] confirm that this C/S ratio is probably a good estimate of the average composition, but indicate considerable variation about the mean. A systematic difference in composition is claimed[46] between the acicular C–S–H and the underlying material and this could be the cause of time-dependent variations.[12] However, this difference is not substantiated by others and these studies emphasize the considerable experimental uncertainties that attend such analyses[18] such as lateral spread of the exciting electrons. Furthermore, in Portland cement paste compositional variations are much greater because of the ability of C–S–H to incorporate large quantities of impurity oxides: alumina and sulphate in particular. This was shown many years ago,[30,32,14] and recently the presence of both aluminum and sulphur has been confirmed by microanalysis.[33] In cement pastes the C/S ratio of C–S–H may vary as widely as 1.0–3.0 but with a mean value close to 1.5.

It would be of interest to know if there are correlations between chemical composition and structural features, such as morphology or degree of silicate polymerization (see below). Although there have been some efforts made to address these problems, there are many experimental difficulties involved in such studies and such correlations remain an open question.

#### 1.3.1.2 Silicate polymerization

The most significant development in recent times with regard to chemical composition has been the characterization of C–S–H as a silicate polymer. Two quite distinct methods have been developed for this determination. One involves the formation of trimethylsilyl derivatives[50] and subsequent analysis by chromatographic methods.[32,48] The other involves controlled hydrolysis and complexing of the silicic acids; the kinetics of complexation being analysed spectrophotometrically.[21,44] Both methods have shown that

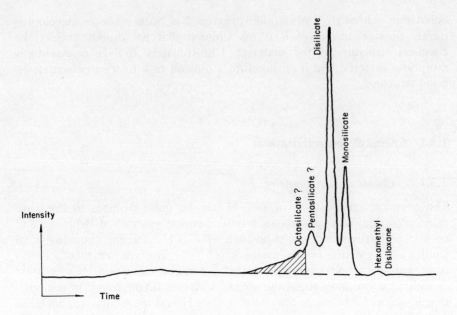

**Figure 1.5**   Gel permeation chromatogram of a derivatized alite paste (85% hydrated, ~40 days curing)[45]

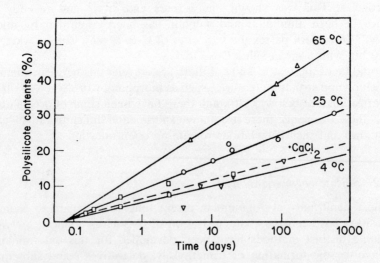

**Figure 1.6**   Effect of curing temperature and time of curing on the amount of polysilicate formed during hydration[56]

the dimeric silicate species ($Si_2O_7$) is a major component of C–S–H, as well as a 'polysilicate fraction'.† Gel permeation chromatographic analysis have shown that this fraction is polydisperse (see Figure 1.5). The major component of this fraction has been assigned to cyclic tetramer[32] or cyclic pentamer[48], but species containing up to 20 silicate units are also found. The question as to the structure of these high molecular weight species remains an open question but there is evidence[32] to suggest that these may not be linear chains but rather branched or cross-linked species of high connectivity. The degree of silicate polymerization, as measured by the total polysilicate content is known to be time and temperature dependent,[39,40,56] as shown in Figure 1.6.

## 1.3.2 Physical characterizations

### 1.3.2.1 Surface area

In the determination of the physical characteristics of C–S–H further difficulties become apparent. It has long been known that the measurement of surface area by physical adsorption is strongly dependent on the adsorbate used. For example, the surface area as measured by adsorption of water vapour is usually reported to be 5–10 times greater than that measured by the adsorption of nitrogen or organic molecules. However, it is clear from more recent work that the 'surface area' of C–S–H cannot be considered a unique property since its value depends on the choice of measurement, or conditions of drying (see Table 1.2). It can be noted from these values that

Table 1.2 Surface area of C–S–H materials measured by different techniques and under different conditions of drying

| Condition | Value ($m^2/g$) | Method of test | Reference |
|---|---|---|---|
| Saturated | ~1000 | Nuclear magnetic resonance | Barbić et al.[2] |
| Saturated | ~750 | Low angle X-ray scattering | Winslow and Diamond[53] |
| D-dried | 200–300 | Low angle X-ray scattering | Winslow and Diamond[53] |
| D-dried | 250–300 | Water vapour adsorption | |
| Solvent replacement | 150–200 | Nitrogen adsorption | Litvan[37] Lawrence et al[34] |
| Oven-dried | ~150 | Low angle X-ray scattering | Winslow and Diamond[53] |
| D-dried | 20–100 | Nitrogen adsorption | |

the very act of drying the material, in order to make conventional physical adsorption measurements, has a large impact on the structure of the

† Nomenclature is not standardized. Lentz originally used 'polysilicate' to refer to all material not analysed by gas chromatography; i.e. higher than four silicate units. Others have applied the term to the residue remaining after heating at 180 °C, and this probably includes only species having more than six or eight silicate tetrahedra linked together.

material. By suitable choice of drying procedures surface areas measured by nitrogen or water vapour adsorption can be very similar,[37] but even with a given drying method, variations in procedure can affect the results.[34] Direct observation of physical changes during drying have been observed by Jennings and Pratt[28,29] using a high voltage electron microscope.

### 1.3.2.2 Porosity

The same difficulties are faced when trying to measure the porosity of C–S–H. According to the definition in Table 1.1, the C–S–H component includes micropores (<26 Å diameter) and classical adsorption experiments with water vapour indicate that C–S–H has a high microporosity. But here again the same difficulties are encountered: different methods of measurements give different results and all involve pre-drying the material before the test. Recent results from our laboratories[43] involving drying and rewetting experiments and solvent replacement treatments lead to the conclusion that the porosity of a paste is significantly changed as water is removed. It appears that the pore structure of a saturated paste is more open than previously thought and that microporosity may only be created during drying. Similar conclusions have been reached from calorimetric studies of freezing,[49] and from adsorption studies. It is clear that water has a special role with regard to its interaction with C–S–H.

### 1.3.3 Structural models of C–S–H

Some idea of the chemical structure of C–S–H would be helpful in trying to deduce its role in creep and shrinkage phenomenon. Although purely physical models based on treating C–S–H as a xerogel† have proved to be useful,[54] there is increasing experimental evidence that the chemical nature of the material cannot be completely ignored and must be taken into account in more precise models. Even though the material is amorphous to X-rays there exists the possibility that short-range atomic order exists, and the nature and extent of this order still need to be resolved. Structural models must therefore be deduced indirectly from studies of the properties and behaviour of C–S–H and in the light of the foregoing discussion considerable divergence of views arise. It is quite common to consider the basic elements to be derived from a layer structure,[12,19] although the chief proponents differ radically concerning the structural features and properties of these layers. However, non-layer structures have been proposed[25,51,24] and it is perhaps prudent to consider this matter an open question even if a particular structural model is currently favoured.

---

† A xerogel is a colloidally disperse solid with a low water content, which is usually formed by drying (e.g. silica gel).

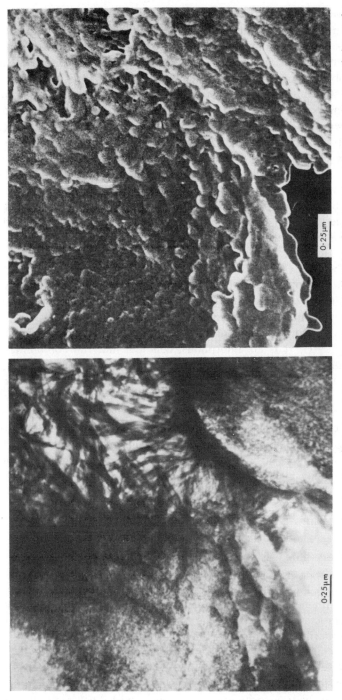

**Figure 1.7** Electron micrographs of an ion-thinned section of 11-month old tricalcium silicate paste. Left-hand side: scanning transmission mode; right-hand side: secondary electron mode. (Photographs courtesy of Dr B. J. Dalgleish and Prof. P. L. Pratt)

Perhaps the strongest evidence in favour of layer structures comes from electron microscopy studies of wet cement pastes, which permit the most direct observation. Jennings and Pratt[28,29] have shown that C–S–H exists as thin sheets or foils in the saturated state. Scanning electron microscopy (SEM) observations, which have hitherto been widely used, may show specious structures due to the strong drying conditions, especially in young pastes. For example, Type III C–S–H[17] which has been considered by some to represent non-layer structures may indeed be derived from thin foils that are collapsed on drying[15] (see Figure 1.7). High resolution electron micrographs of C–S–H, formed from hydrated calcium fluoroaluminate cement show, in addition to disturbed arrangements of lattice planes, a regular spacing over about 500 Å which strongly suggest a high degree of atomic ordering in localized areas.[52] If such areas are commonplace then it should be possible to obtain electron diffraction patterns using the micro-diffraction capabilities of modern instruments, and hence get a direct determination of structure. However, in ordinary cement paste such areas may be few and far between because of the large variability in chemical composition that occurs throughout the paste, and which tends to inhibit the formation of crystalline regions. Further application of the powerful capabilities of modern electron microscope should help to throw further light on the subject.

Based on the present state of our knowledge I would suggest the following view of C–S–H. It is an amorphous, colloidal material containing a large quantity of impurities (aluminium, iron, sulphur, magnesium, sodium, potassium, etc.) and which has considerable variability in local compositions and structures as determined by the local conditions pertaining during its formation. Large amounts of water are contained in an associated pore system. With time the material 'ages' to a lower state of free energy through polymerization of silicate units, with concomitant physical and chemical changes. Ageing is strongly affected by temperature and by removal of water, and may allow regions of short-range order to develop. However, such regions will be separated by completely amorphous material.

## 1.4 THE CAPILLARY PORE COMPONENT

The capillary pore system includes all pores >26 Å diameter, which are those in which menisci formation and associated capillary effects occur. Using the IUPAC classification[27] these can be divided into mesopores (26–500 Å diameter) and macropores (>500 Å diameter). The development of hydrostatic stress in a solid due to creation of menisci in the capillaries, will be primarily determined by the lower size range of the mesopores since its magnitude is inversely proportional to pore radius. Since physical adsorption of gases (water vapour or nitrogen) has been traditionally used to determine the distribution of mesoporosity, the same difficulties already discussed for microporosity and surface area exist. Such methods cannot be used to determine macroporosity. Mercury intrusion porosimetry, which also requires strong drying of the specimen and suffers the same disadvantages as adsorption measurements, is best suited for studying the macropore system and perhaps the larger mesopores, both of which can be expected to have little impact on drying shrinkage or creep. These various methods of determining capillary pore size distribution have been compared.[3] At the present time no direct method exists for measuring pore size distributions of a saturated paste. Probably the best approach is to try to deduce pore characteristics from studies of length and weight during drying and from solvent replacement experiments. Extrapolating measurements made on dried pastes to explain the behaviour of saturated pastes is a practice fraught with difficulties.

## 1.5  THE CALCIUM HYDROXIDE COMPONENT

In direct contrast to C–S–H, calcium hydroxide forms as stoichiometric, crystalline material. The crystals nucleate and grow within the water-filled

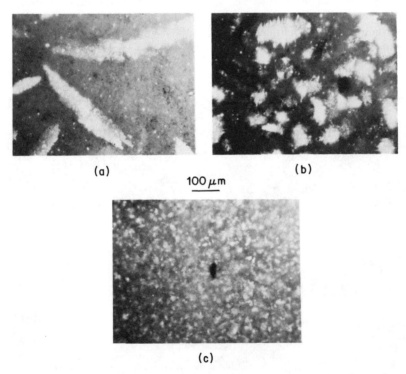

**Figure 1.8**  Optical micrographs of hydrated tricalcium silicate hydrated at various temperatures: (a) 5 °C; (b) 26 °C; (c) 5 °C initially then increased to 26 °C just before setting occurs. Light masses are CH crystals embedded in the grey-black ground mass of C–S–H. Intense black spots are voids (from Lawrence, *et al.*[35])

pores and are sufficiently large to be readily seen by optical microscopy. The size and morphology of the crystals can vary widely, however, depending on the temperature of hydration (Figure 1.8), the water:cement ratio, the presence of admixtures, or type of silicate.[7,8] However, the calcium hydroxide crystals are not completely pure; small quantities of silica and sulphate have been detected by microanalysis.[16,33,38] Also, the crystals tend to grow around and even completely occlude partially hydrated cement grains.

The crystal structure of $Ca(OH)_2$ is a layer structure such that there are atomic planes parallel to the 001 direction which have weaker bonds. The crystals therefore tend to cleave quite readily along these planes. Such cleavage commonly occurs during fracture of cement pastes (Figure 1.9) and is the cause of the familiar striated appearance of $Ca(OH)_2$ crystals in a fracture surface.

Reference can be found in the literature to amorphous calcium hydroxide, which is defined as the difference between the amount determined by quantitative X-ray diffraction analysis and other methods of determination. If appropriate corrections are made for preferred orientation[10,11] then estimation by X-ray diffraction agrees well with thermal methods of analysis.[36] The discrepancy is thus with chemical extraction methods. In my opinion amorphous calcium hydroxide is best considered to be a lime-rich fraction of C–S–H that is readily extracted during analysis.

**Figure 1.9**  Pronounced cleavage fracture of $Ca(OH)_2$ in a hydrated tricalcium silicate paste

## 1.6 THE CALCIUM ALUMINATE COMPONENT

This part of the paste includes the hydration products derived directly from tricalcium aluminate and the ferrite phase ($C_4AF$) which are present in Portland cement. The fraction of material that is involved is probably considerably less than that calculated from the known $C_3A$ and $C_4AF$ and $C\bar{S}H_2$ contents of a cement using standard hydration equations, because much of the aluminate and sulphate forms part of the C–S–H component. Although ettringite forms first during the $C_3A$–$C\bar{S}H_2$ reaction and is present in young cement pastes (see Figure 1.1), it is transient phase which is converted to monosulphoaluminate ($C_4A\bar{S}H_{12}$). This latter compound is one of a series of hexagonal hydrates of the general formula $C_4AXH_{12}$ where X is various anions, which can form either pure compounds or solid solutions with other anions. It has usually been assumed that the principal species are $SO_4^{2-}$ and $OH^-$, but $Al(OH)_4^-$, $CO_3^{2-}$ are probably also present; while anions such as $Cl^-$, $NO_3^-$, $HCO_2^-$, etc., which may be introduced into cement paste as chemical admixtures, can also participate in solid solutions. More recently silicate substitution (perhaps as $H_2SiO_4^{2-}$) has also been shown to occur.[47]

These hexagonal hydrates are all based on calcium aluminate layer structures (Figure 1.10) with the anions occupying into layer positions, along with water molecules, in order to balance the net positive changes of the layers. Well-crystallized samples develop as well-formed, thin, hexagonal plates with strong basal reflections in their X-ray diffraction patterns. In cement pastes the basal reflections are much broader and the morphologies are less distinct, indicating considerable disorder. Typically clusters ('rosettes') of small, platey crystals occur, without a definite hexagonal morphology (Figure 1.11). Water may be lost quite readily from between the layers during drying allowing collapse to a smaller basal spacing.

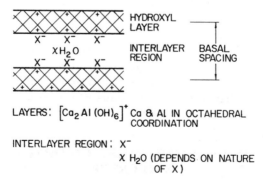

**Figure 1.10**  Schematic representation of the hexagonal calcium aluminate hydrate phases

**Figure 1.11** Hexagonal calcium aluminates formed during hydration of Portland cement clinker in a $CaSO_4$ solution

## 1.7  THE ANHYDROUS RESIDUE COMPONENT

Hydration slows down as the hydrating silicate grains become coated with layers of C–S–H which form diffusion barriers. The kinetics and local space requirements are such that the larger silicate grains never completely hydrate even when a paste is kept in water indefinitely. The minimum size of grain required to ensure that anhydrous grains persist in a paste will depend on the extent of moist curing (i.e. degree of hydration) and the water:cement ratio. Decreasing either of these parameters will increase the proportion of anhydrous material to hydrated material.

## 1.8  EFFECTS OF MICROSTRUCTURE ON DRYING SHRINKAGE AND CREEP

Mention will be made here of some selected experimental studies concerning the effects of microstructure on drying shrinkage and creep. A major contribution to microstructure–deformation relationships was made by Parrott[39–41] who showed that the effects of heat treatment reduce creep and shrinkage by increasing the polysilicate content of the paste. The polysilicate content was also used as a microstructural parameter in another study linking shrinkage with microstructural properties.[6] In these two studies different methods of silicate analysis were used.

### 1.8.1 Silicate polymerization

The use of silicate polymerization analysis to predict the potential creep of a paste after any predetermined history looks to be a promising approach. High degrees of silicate polymerization appear to result in a more stable, less deformable C–S–H better able to resist stresses imposed during drying.[6] A higher degree of silicate polymerization could be regarded as increasing the overall bonding of C–S–H and raising the energy needed to activate creep centres.[22] It has been suggested[42] that silicate polymerization of a cement paste, while under load, will cause creep to occur, and this concept can be used to explain transitional thermal creep (raising the temperature raises the rate of polymerization) and age effects (the rate of polymerization decreases with time).

### 1.8.2 Surface area and pore size distributions

Reference to Table 1.2 shows the large decrease in surface area, as measured by low angle X-ray scattering.[53] Decreases in surface area, determined by nitrogen adsorption have also been observed after drying.[26,4,5,43] It has been suggested[4] that changes in surface area reflect changes in pore size distributions associated with drying that may be the cause of irreversible shrinkage. However, on resaturation the surface area can be restored to its original value.[53,43] Therefore most of the changes in pore size distribution that occur as drying are not permanent and the original pore size distribution can be restored. Irrecoverable shrinkage can be attributed to a relatively small permanent pore collapse accompanying the rearrangement of C–S–H particles.[45]

The pore size distribution of a paste (as measured by nitrogen adsorption) is considered important in determining shrinkage behaviour.[9,6] For example, the addition of calcium chloride, which increases irreversible shrinkage on first drying[20,5] has a much finer mesopore structure[55] and a higher contribution to shrinkage attributable to the capillary component.[56,5,6] Conversely, curing at elevated temperature coarsens the pore structure and reduces capillary pore contributions to shrinkage.[56] However, in this case the increase in silicate polymerization, that also occurs when the curing temperature is increased, may be the more important factor in reducing shrinkage. It should also be noted that when calcium chloride is present the degree of silicate polymerization is reduced (see Figure 1.6) and this may contribute to the observed increase in shrinkage.

The major drawback in trying to correlate pore characteristics with shrinkage is that the paste must be dried very strongly in order to make the measurements. The drying conditions are more severe than those under study. The pore structure does not represent that of the saturated paste, although it is generally assumed that changes in that occur in the pore structure of the saturated paste will somehow be reflected in the pore structure of the dried paste.

### 1.8.2   Calcium hydroxide

It might be expected that since calcium hydroxide crystals do not lose moisture during drying, and hence do not shrink, they might act to restrain the deforming C–S–H. However, a study in which the size and morphology of the calcium hydroxide was varied over a wide range revealed no correlations with shrinkage.[6] Shrinkage can be correlated with parameters describing only the pore component and the C–S–H component. Studies are underway to determine if pastes from which calcium hydroxide is removed by leaching have higher shrinkage.

### 1.8.4   Calcium aluminate hydrates

There are reports that the $C_3A$ content of a cement affects its creep and drying shrinkage so that it is likely that the calcium aluminate component makes some contribution to creep and shrinkage. The structure of the hexagonal hydrates is consistent with such an assumption, but no systematic studies have been done to identify the extent or nature of this contribution. We do now know to what extent alumina and sulphate substitution in C–S–H may affect its contributions to shrinkage and creep. It is well known that an optimum sulphate content exists for minimum shrinkage and it occurs for creep also.[1] The origins of these phenomena most likely lie in the way the early formation of ettringite, and its conversion to the hexagonal hydrates affect the characteristics of the C–S–H component and the capillary pore component.

### 1.9   CONCLUSIONS

The complex and unstable character of C–S–H and the pore system, and the resulting difficulties in measuring their properties, makes the correlation of microstructure a difficult task. Although in the last several years considerable progress has been made, for example, the concept of the polysilicate content as a predictive parameter has emerged, the difficulties of characterization are too great to allow a series of microstructural parameters to completely predict deformation behaviour. However, the recognition of the unstable nature of C–S–H and the variability of the pore size distribution seems to be an important step in our quest for useful microstructural characterizations of hydrated cement pastes that can be used to explain and predict engineering properties.

## ACKNOWLEDGEMENTS

This paper was prepared as part of the project funded by the National Science Foundation Grant No. Eng 78-09529.

## REFERENCES

1. Alexander, K. M., Wardlaw, J., and Ivanusec, I. (1979), 'The influence of $SO_3$ content of Portland cement on the creep and other physical properties of concrete', *Cem. Concr. Res.*, **9**, 451–60.
2. Barbić, L., Kocuvan, I., Blinć, R., and Lahajnar, G. (1980), 'The determination of specific surface of cement and its hydrates by nuclear spin–lattice relaxation', *9th Tech. Meet. Yugoslav Cement and Asbestos Cement Producers, Ohrid, Yugoslavia*, 1980. Barbić, L., Kocuvan, I., Blinc, R., Lahajnar, G., Merljak, P., and Zupančič, I. (1982), 'The Determination of Surface Development in Cement Pastes by Nuclear Magnetic Resonance' *J. Am. Ceram. Soc.*, **65**, 25–31.
3. Bentur, A. (1980), 'The pore structure of hydrated cementitious compounds of different chemical composition', *J. Am. Ceram. Soc.*, **63**, 381–6.
4. Bentur, A., Berger, R. L., Lawrence, F. V., Milestone, N. B., Mindess, S., and Young, J. F. (1979a), 'Creep and drying shrinkage of calcium silicate pastes. III. A hypothesis of irreversible strains', *Cem. Concr. Res.*, **9**, 83–96.
5. Bentur, A., Milestone, N. B., Young, J. F., and Mindess, S. (1979b), 'Creep and Drying Shrinkage of Calcium Silicate Pastes. IV. Effects of Accelerated Curing', *Cem. Concr. Res.*, **9**, 161–70.
6. Bentur, A., Kung, J. H., Berger, R. L., Young, J. F., Milestone, N. B., Mindess, S., and Lawrence, F. V. (1980), 'Influence of microstructure on the creep and drying shrinkage of calcium silicate pastes', *Proc. 7th Int. Congr. Chem. Cem., Paris*, Vol. III, Editions Septima, pp. VI. 26–31.
7. Berger, R. L., and McGregor, J. D. (1972), 'Influence of admixtures on the morphology of calcium hydroxide formed during tricalcium silicate hydration', *Cem. Concr. Res.*, **2**, 43–57.
8. Berger, R. L., and McGregor, J. D. (1973), 'Growth of calcium hydroxide crystals formed during the hydration of tricalcium silicate. Effect of temperature and water:solids ratio', *J. Am. Ceram. Soc.*, **56**, 73–9.
9. Berger, R. L., Kung, J. H., and Young, J. F. (1976), 'Influence of calcium chloride on the drying shrinkage of alite paste', *J. Test. Eval.*, **4**, 85–93.
10. Bezjak, A., and Jelenić, I. (1965), 'Correction of preferential crystallite orientation in X-ray quantitative analysis. Quantitative X-ray determination of $Ca(OH)_2$', *Croatia Chem. Acta*, **37**, 255–64.
11. Bezjak, A., Gačeša, T., and Jelenić, I. (1967), 'Influence of particle size in

quantitative X-ray analysis of substances with a pronounced disposition for orientation of crystallites. Quantitative X-ray determination of calcium hydroxide', *Croatia Chem. Acta*, **39**, 109–18.

12. Brunauer, S., and Kantro, D. L. (1964), 'The hydration of tricalcium silicate and $\beta$-dicalcium silicate from 5 °C to 50 °C', in, H. F. W. Taylor, (Ed.) *The Chemistry of Cements*, Chap. 1, Academic Press, London, pp. 287–309.

13. Chatterji, S., and Thaulow, N. (1980), 'Estimation of the chemical composition of precipitated phase by the electron beam microanalytical technique', *X-Ray Spectrometry*, **9**, 5–7.

14. Copeland, L. E., Bodor, E., Chang, T. N., and Weise, C. H. (1967), 'Reactions of tobermorite gel with aluminates, ferrites, and sulfates', *J. PCA Res. Dev. Lab.*, **9**, 61–74.

15. Dalgleish, B. J., Pratt, P. L., and Moss, R. I. (1980), 'Preparation techniques and the microscopical examination of Portland cement paste and C₃S', *Cem. Concr. Res.*, **10**, 665–76.

16. Diamond, S. (1972), 'Identification of hydrated cement constituents using a scanning electron microscope-energy dispersive X-ray spectrometer combination', *Cem. Concr. Res.*, **2**, 617–32.

17. Diamond, S. (1976), 'Cement paste microstructure—an overview at several levels', in *Hydraulic Cement Pastes: Their Structure and Properties*, Cement and Concrete Assoc., Wexham Springs, Slough, UK, pp. 2–30.

18. Diamond, S., Young, J. F., and Lawrence, F. V. (1974), 'Scanning electron microscopy-energy dispersive X-ray analysis of cement constitutents—some cautions', *Cem. Concr. Res.*, **4**, 899–914.

19. Feldman, R. F., and Sereda, P. J. (1970), 'A new model for hydrated Portland cement and its practical implications', *Eng. J.*, **53**(8/9), 53–9.

20. Feldman, R. F., and Swenson, E. G. (1975), 'Volume changes on first drying of hydrated Portland cement with and without admixtures', *Cem. Concr. Res.*, **5**, 25–36.

21. Funk, H., and Drydrych, R. (1966), 'The degrees of anion condensation in silicic acids and silicates', in *Symposium on Structure of Portland Cement Paste and Concrete*, Highway Res. Bd. Spec. Rep. 90, Washington, DC, pp. 284–90.

22. Gamble, B. R., and Illston, J. M. (1976), 'Rate of deformation of cement paste and concrete during regimes of variable stress, moisture content and temperature', in *Hardened Cement Paste: Its Structure and Properties*, Cement and Concrete Assoc., Wexham Springs, Slough, UK, pp. 297–311.

23. Gard, J. A., Mohan, K., Taylor, H. F. W., and Cliff, G. (1980), 'Analytical electron microscopy of cement pastes. I. Tricalcium silicate pastes', *J. Am. Ceram. Soc.*, **63**, 336–7.

24. Gimblett, F. G. R., Singh, K. S. W., and Amin, Z. M. (1980), 'Influence of pretreatment on the microstructure of calcium silicate hydrate gels', *Proc. 7th Int. Symp. on Chem. Cem., Paris, 1980*, Vol. II, Editions Septima, pp. II.225–31.

25. Grudemo, Å. (1972), 'On the development of hydrate crystal morphology in silicate cement binders', *RILEM-d'INSA Symposium, Toulouse*, Rapp. 9:75, Cement och Betoninstitut, Stockholm, Sweden.

26. Hunt, C. M., Tomes, L. A., and Blaine, R. L. (1960), 'Some effects of ageing on the surface area of Portland cement paste', *J. Res. Natl. Bur. Stand.*, **64A**, 163–69.

27. IUPAC (1972), 'Manual of symbols and terminology, Appendix 2, Part 1, Colloid and surface chemistry', *Pure Appl. Chem.*, **31**, 578.

28. Jennings, H. M., and Pratt, P. L. (1979), 'On the hydration of Portland cement', *Proc. Br. Ceram. Soc.*, No. 28, 179–93.

29. Jennings, H. M., and Pratt, P. L. (1980), 'The use of a high-voltage electron microscope and gas reaction cell for the microstructural investigation of wet Portland cement', *J. Mater. Sci. Lett.*, **15**, 250–3.
30. Kalousek, G. L. (1957), 'Crystal chemistry of hydrous calcium silicates. I. Substition of aluminum in lattice of tobermorite', *J. Am. Ceram. Soc.*, **40**, 74–80.
31. Kalousek, G. L. (1965), 'Analyzing $SO_3$-bearing phases in hydrating cements', *Mater. Res.*, **5**, 292–304.
32. Lachowski, E. E. (1979), 'Trimethylsilylation as a tool for the study of cement pastes. II. Quantitative analysis of the silicate fraction of Portland cement pastes', *Cem. Concr. Res.*, **9**, 343–52.
33. Lachowski, E. E., Mohan, K., Taylor, H. F. W., and Moore, A. E. (1980), 'Analytical electron microscopy of cement pastes. II. Pastes of Portland cement and clinkers', *J. Am. Ceram. Soc.*, **63**, 447–52.
34. Lawrence, C. D., Gimblett, F. G. R., and Singh, K. S. W. (1980), 'Sorption of $N_2$ and $n$-$C_4H_{10}$ on hydrated cements', *Proc. 7th Int. Symp. Chem. Cem., Paris, 1980*, Vol. III, Editions Septima, pp. VI.141–6.
35. Lawrence, F. V., Young, J. F., and Berger, R. L. (1977), 'Hydration and properties of calcium silicate pastes', *Cem. Concr. Res.*, **7**, 369–78.
36. Lehmann, H., Locher, F. W., and Prussog, D. (1970), 'Quantitative determination of calcium hydroxide in hydrated cement', *Tonind. Ztg.*, **94**, 230–5.
37. Litvan, G. G. (1976), 'Variability of the nitrogen surface area of hydrated cement paste', *Cem. Concr. Res.*, **6**, 139–44.
38. Moore, A. E. (1980), 'Structure and composition of compounds in some fully hydrated cement pastes', *Proc. 7th Int. Congr. Chem. Cem., Paris*, Vol. III, Editions Septima, pp. VI.97–102.
39. Parrott, L. J. (1976), 'Effect of a heat cycle during moist curing upon the deformation of hardened cement paste', in *Hydraulic Cement Pastes: Their Structure and Properties*, Cement and Concrete Assoc., Wexham Springs, Slough, UK, pp. 189–203.
40. Parrott, L. J. (1977a), 'Basic creep, drying creep and shrinkage of a mature cement paste after a heat cycle', *Cem. Concr. Res.*, **7**, 597–604.
41. Parrott, L. J. (1977b), 'Recoverable and irrecoverable deformation of heat cured cement paste', *Mag. Concr. Res.*, **29**, 26–30.
42. Parrott, L. J. (1978), 'A study of basic creep in relation to phase changes in cement paste', *RILEM Int. Colloq. Adv. Theory, Mechanism and Phemenology of Creep.*
43. Parrott, L. J., Hansen, W., and Berger, R. L. (1980), 'Effect of drying and rewetting upon the pore structure of hydrated alite paste', *Cem. Concr. Res.*, **10**, 647–56.
44. Parrott, L. J., and Taylor, M. G. (1979), 'A development of the molybdate complexing method for the analysis of silicate mixtures', *Cem. Concr. Res.*, **9**, 483–8.
45. Parrott, L. J., and Young, J. F. (1980), 'Shrinkage and swelling of two hydrated alite pastes', *Int. Symp. on Fundamental Research on Creep and Shrinkage of Concrete, Lausanne, Sept. 1980.*
46. Rayment, D. L., and Majumdar, D. A. (1980), 'The composition of C–S–H phase(s) in hydrated $C_3S$ paste', *Proc. 7th Int. Conf. Chem. Cem., Paris, 1980*, Vol. III, Editions Septima, pp. II.64–70.
47. Regourd, M., Hornain, H., and Mortureux, B. (1976), 'Evidence of calcium silicoaluminates in hydrated mixtures of tricalcium silicate and tricalcium aluminate', *Cem. Concr. Res.*, **6**, 733–40.

48. Sarkar, A. K., and Roy, D. M. (1979), 'A new characterization technique for trimethylsilylated products of old cement pastes', *Cem. Concr. Res.*, **9**, 343–52.
49. Sellevold, E., and Bager, D. H. (1980), 'Some implications of calorimetric ice formation results for frost resistance testing of cement products', *Int. Colloq. Frost Resistance of Concr.*, Vienna. Tech. Rep. 86/80, 27 pp. Dept. Civil Eng., Tech. Univ. Denmark, Lyngby.
50. Tamás, F. D., Sarkar, A. K., and Roy, D. M. (1976), 'Effect of variables upon the silylation products of hydrated cements', in *Hydraulic Cement Pastes: Their Structure and Properties*, Cement and Concrete Assoc., Wexham Springs, Slough, UK, pp. 55–72.
51. Taylor, H. F. W. (1979), 'Cement hydration reactions: the silicate phases', in Ed. J. Skalny *Cement Production and Use*, The Engineering Foundation, New York, pp. 107–16.
52. Uchikawa, H., Uchida, S., and Mihara, Y. (1978), 'Characterization of hydrated ultra rapid hardening cement pastes', *Il Cemento*, **78**, 59–70.
53. Winslow, D. N., and Diamond, S. (1974), 'Specific surface of hardened Portland cement paste as determined by small-angle X-ray scattering', *J. Am. Ceram. Soc.*, **57**, 193–7.
54. Wittmann, F. H. (1976), 'The structure of hardened cement paste—a basis for a better understanding of the materials properties', in *Hydraulic Cement Pastes: Their Structure and Properties*, Cement and Concrete Assoc., Wexham Springs, Slough, UK, pp. 96–117.
55. Young, J. F. (1974), 'Capillary porosity in hydrated tricalcium silicate pastes', *J. Powder Technol.*, **9**, 173–9.
56. Young, J. F., Berger, R. L., and Bentur, A. (1978), 'Shrinkage of tricalcium silicate pastes: superposition of several mechanisms', *Il Cemento*, **75**, 391–8.

Creep and Shrinkage in Concrete Structures
Edited by Z. P. Bažant and F. H. Wittmann
© 1982 John Wiley & Sons Ltd

*Chapter 2*

# Mechanics of Concrete Systems: Current Approaches to Assessing Material Behaviour and Some Possible Extensions

*J. W. Dougill*

## 2.1  INTRODUCTION

In this chapter, the object is to review procedures used for estimating the behaviour of concrete systems starting from a knowledge of the behaviour of the constituents. A structural mechanics view is adopted so that behaviour is described in the terms commonly used in the analysis of concrete structures. Almost invariably, analysis is conceived in continuum terms and assuming that the structure of the material can be ignored. Thus, analysis is undertaken in terms of conventional stress and strain and the material properties used in analysis are chosen to link these quantities with the other relevant variables such as time or temperature.

In exploring a structural material, experiments are done to see how it behaves. Theories of behaviour are adopted or devised, more or less to fit the observed phenomena. Having done this, quantities are identified to characterize the material. These are the material properties. With incomplete descriptions of complex behaviour, it is unlikely that there will be only one way available to describe the material. Thus, as suggested in Figure 2.1, the material properties are conditioned by the idealization used as well as by the material itself.

With composite materials, such as concrete, the complex link between phenomena, theory and descriptive parameters encourages a more detailed study of the material based on the properties of the component materials and how these interact. Theoretical studies of this sort tend to concentrate on the particular mechanisms or physical processes which are thought to have most influence on the type of overall behaviour being considered. Thus, different models are developed to describe a range of phenomena exhibited by a single material but in different circumstances. A continuum approach can be adopted in devising these models. Alternatively, a block

23

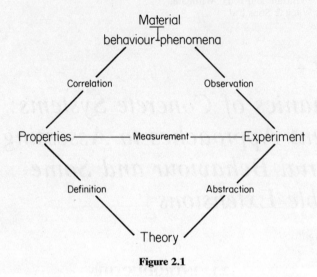

**Figure 2.1**

building micromechanics approach can be used. So far, it is this last method that has been employed most extensively for concrete. We start by looking at this and then move on to consider other forms of model that might be useful in describing concrete behaviour.

## 2.2  SOLID MODELS FOR CONCRETE SYSTEMS

A variety of structural models have been used to represent different aspects of concrete behaviour and some of the more important of these are shown in Figure 2.2. From this it will be seen that the various models can be separated into two main groups depending on whether attention is focused on non-linear behaviour, including breakdown and failure, or on linear behaviour encompassing elasticity, shrinkage, creep and thermal effects.

Although this chapter is necessarily set in the context of creep and shrinkage, it is still useful to look at the models that are primarily concerned with failure. Here the form of model is determined by the view taken of whether tensile breakdown and cracking or shear bond failure and slip is the most important mechanism contributing to non-linearity and failure. Thus, concentration on tensile failure leads to the lattice models of Reinius[1,2] and Baker[3] whereas the shear bond view provides the basis for models by Brandtzaeg,[4] Shah and Winter,[5] Taylor[6] and others.

### 2.2.1  Lattice models

Both Reinius and Baker proposed models for concrete, shown in Figure 2.3, based on stiff particles connected by bar linkages. In Reinius' model, the aggregate is represented by rigid single sized spherical particles in a body centred cubic array. Two kinds of link are used to connect the particles. The long link along the edges of the cube take only tension or compression whilst short links made of several inclined bars are used to transmit both shear and normal force between the centre and outer particles. With a suitable choice of member properties and allowing bar failure in both tension and compression, overall behaviour similar to that of concrete in compression was obtained.

The main disadvantage of the model is that the role of the aggregate is not well defined and that it is difficult to express the stiffness and strength of the linkages in terms of known properties of the cement paste. In spite of this, it is surprising that, with the increase in computational capacity since the 1950's, the model has not received more attention in studying behaviour under combined stresses. In contrast to this, there has been significant recent activity in geotechnics in direct modelling of granular media,[7,8] although here the emphasis has been on frictional behaviour at the contact surfaces and without consideration of cohesive bond. In Baker's model[3] attention is focused on the effect of the aggregate in developing thrust rings which are stabilized by adjacent rings of aggregate and the tension developed in pockets of mortar enclosed by the rings. The form of the lattice is determined by the direction of the forces transmitted between pieces of aggregate and through the mortar. The way Baker envisaged this is shown in Figure 2.3. We note that in its original form, the lattice geometry depended on the state of stress. With purely volumetric (or all round stress) the thrust ring is

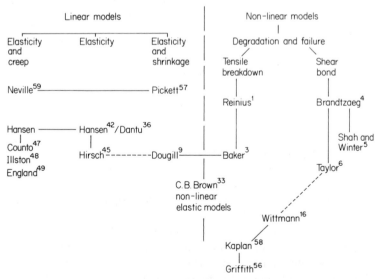

**Figure 2.2** Classification of structural models for concrete behaviour

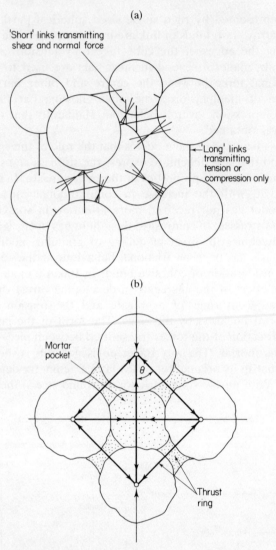

**Figure 2.3**  (a) Reinius' model for concrete (after
Reinius[2]); (b) Baker's lattice element (after Baker[3])

symmetric, carries most of the compression and so relieves the material in
the mortar pocket of most of its stress. With uniaxial compression, the lines
of thrust are more nearly in the direction of the applied load and the angle $\theta$
in thus somewhat less than 45°. In addition, Baker suggested that the thrust
ring angle $\theta$ would change with time, due to the effect of creep and
redistribution of load between the mortar and the stiff aggregate structure.

The variable geometry property is inconvenient in analysis of general loading systems. Accordingly, later applications concerned with thermal incompatibility[9] and three-dimensional states of stress[10,11] were based on a straightforward lattice with a constant angle of 45°. However, as with Reinius' model, there is a fundamental difficulty in assigning appropriate values for the properties of the lattice members. Because of this, it is doubtful whether Baker's lattice can be developed into a satisfactory quantitative tool. If this limitation is accepted, the model still has considerable heuristic value in developing an understanding of the influence of the aggregate on the load capacity of concrete systems.

A more general look at the use of lattice structures in analysis leads to a comparison between Bakers' and the Hrennikoff/McHenry analogies[12,13] designed for use in two-dimensional stress analysis in the way finite elements are today. In these models, a unit cell of bar members is used to replace an element of an elastic continuum. With simple lattice geometries there are not enough variables available to allow a complete match with the elastic properties to be obtained. For instance, with Baker's model and $\theta = 45°$, Poisson's ratio is restricted to a value of 1/3 for an isotropic material or 1/4 if a three-dimensional version is used. The more complicated geometry of Reinius' model avoids this difficulty and allows greater freedom in specifying the initial elastic behaviour. Besides this, Reinius' model includes another feature that could be of interest. A value must be given for the ratio of the particle size to the length of the unit cell. Thus, results from an analysis using assemblies of Reinius' type cells should include effects due to aggregate size and stress gradient; aspects on which there is still speculation and uncertainty.

### 2.2.2 Shear bond and crack blunting

Baker's approach to the effect of aggregate inclusions is essentially based on statics and on providing a solution for the internal forces that are in equilibrium with the applied loads. For composites, comprising inclusions in a perfectly plastic matrix, this approach could be developed to provide lower bound estimates for strength under combined stresses in the way proposed by Drucker and Chen.[14] In plasticity it is noteworthy that regular arrays of particles are not necessarily as effective a reinforcement as randomly distributed inclusions.[15] This situation also obtains with brittle or strain softening matrices in which a regular array of particles may allow unrestricted crack extension in a particular direction as opposed to likely crack blunting with random placement of particles. This effect is recognized in Wittmann's[16] simulation of crack propagation in concrete which considers growth in the direction of an applied compressive stress from inclined cracks at the

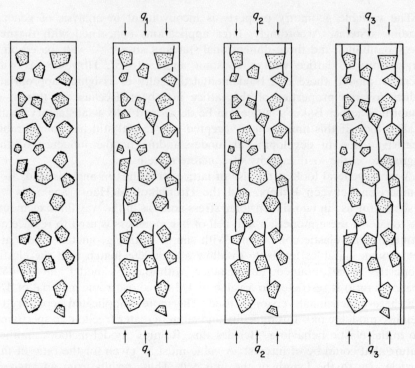

**Figure 2.4** Modelling of crack development in normal strength concrete (after Witt-mann[16])

aggregate matrix interface. This model depends on expressions for equilibrium crack length developed in terms of the long range compressive stress field and local stress intensity factors. There is an approximation here involving interaction effects between neighbouring particles of aggregate. However, the model is interesting and the results informative on the development and effect of cracking (Figure 2.4).

Wittmann's analysis shows how difficult it is to isolate phenomena and concentrate, for instance, on either tensile failure or shear bond as the dominant mechanism. This is done however in Brandtzaeg's model for concrete in triaxial compression. This was one of the earliest idealizations used for concrete, but it fits well into the context of modern developments.

### 2.2.3 Brandtzaeg and shear bond models

Brandtzaeg[4] considered a material made up of linear elastic elements containing planes of weakness at which yield occurred while the stresses in the element satisfied the Mohr–Coulomb criterion. A sample of material was taken to contain a large number of well distributed elements so that all possible directions for the planes of weakness were represented equally. On this basis, samples were analysed for surface loadings giving axially symmetric triaxial compression with average principal stresses $\sigma_1$, $\sigma_2 = \sigma_3$. Before yield occurred in any element, behaviour was linear elastic. Following yield, it was supposed that the strain throughout the sample was determined by the deformation of the elastic region and so was taken to be uniform throughout. With increasing load, more elements yielded causing an overall loss in stiffness and a non-linear stress–strain response.

Because of the need to satisfy the boundary conditions, the local stresses in the elastic region differ from those in elements where yield has occurred. For instance, in uniaxial compression ($\sigma_1$ prescribed, $\sigma_2 = \sigma_3 = 0$) tensile stresses are induced in the elastic region in a direction at right angles to the applied compression. Brandtzaeg supposed that this could cause cracking and so set a strength limit to this local tensile stress. In his view, cracking would render the uniform strain assumption untenable and limit the range of application of his theory. Accordingly, he defined the overall axial stress, $\sigma_1$, when cracking occurs to be the 'critical stress' and regarded the material as being 'disorganized' for behaviour beyond this point.

Below the critical stress, Brandtzaeg's analysis gave results which compared very well with contemporary experimental data.[17] In uniaxial compression the critical stress occurred when the sample volume was near its minimum and the applied stress was around 80% of the peak value. The analysis also showed that triaxial compression caused an increase in critical stress and that the critical stress in equal biaxial compression was slightly greater than in simple compression. Clearly, there are similarities between Brandtzaeg's 'critical stress' and Newman's original view of 'discontinuity'[18] although the latter now refers, in Brandtzaeg's terms, to the first appearance of slip and non-linearity rather than to the onset of disorganization.[19]

Brandtzaeg's model is very attractive. It fits well with ideas on shear bond failure[20,21] if the planes of weakness are taken to represent the aggregate–cement paste interface. The thrust ring effect from the aggregate reinforcement is not considered. However, from a mechanics point of view, the model is the most complete of those discussed so far. It links stress and strain through comparatively few material properties (angle of friction, cohesion, local tensile strength) using assumptions that can be easily generalized for other situations. Thus despite its seniority, Brandtzaeg's model continues to attract attention with Shah and Winter[5] using a simplified version in their model of compression failure and Taylor[6,22] extending the original approach to consider three independent principal stresses.

### 2.2.4 Brandtzaeg and modern analysis

It is of interest to set Brandtzaeg's assumptions into a more up-to-date context. First, the idea of separate elements constrained to the same deformation field is central to the overlay device sometimes used in conjunction with the finite element method of analysis.[23] In this, two or more elements, with different properties, share the same nodes and so occupy the same space within a body. The combined stiffness of these elements provides the composite behaviour of the material in the region defined by the nodes.

For example, consider a piece of a two-dimensional sheet-like material which conforms to Brandtzaeg's assumptions. It would be convenient to treat this as a pile of elastic elements of the same size, each containing a line of weakness in a different direction, and with each element sharing the same nodes and deformation field in the finite element specification.

Because overall behaviour is non-linear, the analysis would need to be undertaken incrementally. The stiffness matrix required is that connecting increments of stress and strain. In Brandtzaeg's formulation, the deformation is determined by the stresses in the elastic region so that the incremental stiffness matrix for the yielding elements would be zero. A distinct improvement on this would be to use the theory of plasticity and adopt a flow rule for plastic deformation in a layer corresponding to, or associated with, the Mohr–Coulomb yield criterion. An analysis of this sort would then be embedded in current techniques used in incremental plasticity. Here, the use of independent linear loading functions is already established[24,25] and a similar approach dealing with degradation rather than slip has been developed.[26,27] This takes us into the realm of the continuum models which will be discussed later. However, it should be noted that by adopting the overlay method and using viscoelastic elements coupled with plastic slip, Brandtzaeg's original view of material behaviour could be extended to include time dependence under moderately high stress conditions. This is again an area of study needing further investigation.

### 2.2.5  Heterogeneity

Before discussing models concerned with linear behaviour, it is relevant to compare the effects of initial heterogeneity in brittle materials on behaviour at low stresses and when breakdown becomes significant. A view of this is provided by Burt and Dougill's study of failure in a model heterogeneous medium made up of a random network of linear elastic brittle members.[28,29]

The network comprised a plane pin-jointed frame with either a random geometry or a repetitive lattice structure of the sort used by McHenry and Baker. The values of stiffness and strength for each member were allocated to give a normal distribution of member properties throughout the sample with specified mean and standard deviation. The resulting structure was analysed for a sequence of increments of boundary displacement; at each stage reducing the stiffness of 'failed' members to zero and using an iterative

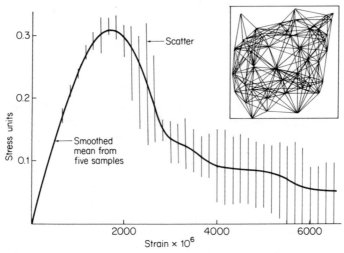

**Figure 2.5** Stress–strain curve for samples of a brittle heterogeneous material

scheme to restore equilibrium. A typical random network and its response under uniaxial loading is shown in Figure 2.5.

Because of the way the model is specified, it is possible to examine different samples of material each having the same internal geometry and distribution of member properties, but in which the properties of individual members vary from sample to sample. The difference in behaviour observed in testing five such macroscopically identical samples is shown by the scatter about the load displacement curve in Figure 2.5.

A remarkable feature of this result is the lack of scatter in the results at low stress and before there is any significant effect of member breakage. It seems that, with samples having the same integral descriptions of material properties, any local variation in the material has only a minor influence on initial stiffness. No similar study has been attempted to look into the effects due to differences in local geometry. However, it seems reasonable to assume that, in devising models to describe nominally linear behaviour, it should be possible to use a less detailed description of structure and material composition than if degradation and failure were to be described. It is therefore not altogether surprising that the relatively simple models used for elasticity, creep and shrinkage provide far better quantitative predictions of behaviour than do the relatively complex models used for failure.

## 2.3 GENERAL RELATIONS BETWEEN ELASTIC MODULI

In dealing with elastic moduli the most common model adopted is that of a matrix of one material which contains dispersed inclusions of a second material. The composite is taken to be homogeneous in the sense that the inclusion concentration can be taken to be constant within any volume, however small. In these terms, concrete is viewed as being made of aggregate particles in a mortar matrix.

Models having more than two phases have also been developed which would allow separate consideration of different kinds of aggregate within the matrix.[30] However, we restrict our attention, here, to only two phases and adopt a notation in which suffices 1 and 2 refer quantities to the first and second phase respectively with the composite quantity being left without a suffix. We use g for the volumetric concentration, so that $g_1 + g_2 = 1$ everywhere.

The problem of determining values for the moduli of a composite from the properties and proportions of its constituents has an extensive literature.[31,33] We shall not however attempt to cover the whole of this field but rather to concentrate on particular aspects relevant to work on concrete.

First, it is of interest that a number of general results can be obtained for isotropic materials knowing the volumetric composition but without specifying further the detailed geometry of the material. Thus, use of the theorem of minimum potential energy and the assumption of a uniform strain field throughout a sample leads to upper bound estimates $(K^V, G^V)$ for the bulk and shear moduli $(K, G)$ of the composite i.e.

$$K^V = g_1 K_1 + g_2 K_2 \geqslant K$$
$$G^V = g_1 G_1 + g_2 G_2 \geqslant G \tag{2.1}$$

These approximate values for the composite moduli have been attributed to Voigt. Alternative estimates $(K^R, G^R)$ originally from Reuss, can be derived using complementary energy and a uniform stress field. These are lower bounds to the composite moduli so that,

$$1/K^R = g_1/K_1 + g_2/K_2 \quad \text{with } K^R \leqslant K$$
and $\tag{2.2}$
$$1/G^R = g_1/G_1 + g_2/G_2 \quad \text{with } G^R \leqslant G$$

In general the bounds are wide and not informative for a material like concrete in which the phase moduli differ substantially. Because of this, there has been effort devoted to improving the bounds and deriving better estimates for the moduli $K$ and $G$.

One approach to this is from an exact result, derived by Hill[32], for the bulk modulus of a composite in which the shear moduli are the same in each phase. For this case:

$$G_1 = G_2 = G$$
and $\tag{2.3}$
$$\frac{K}{K^R} = \left(1 + \frac{4GK^V}{3K_1 K_2}\right)\left(1 + \frac{4GK^R}{3K_1 K_2}\right)^{-1}$$

where $K^V$ and $K^R$ are given in equations (2.1) and (2.2). The argument follows that if the shear moduli of the phases are different with $G_1 > G_2$, the bulk modulus of the composite must lie between the values of $K$ found from equation (2.3) for $G = G_1$ and $G = G_2$. This proposition was proved by Hill and so leads to the bounds

$$\frac{4G_2 K^V + 3K_1 K_2}{4G_2 K^R + 3K_1 K_2} \leqslant \frac{K}{K^R} \leqslant \frac{4G_1 K^V + 3K_1 K_2}{4G_1 K^R + 3K_1 K_2} \tag{2.4}$$

Bounds equivalent to (2.4) were derived about the same time as Hill by Hashin and Shrikman[34] using a different approach. In addition, they established bounds on the shear modulus but only when $G_1 - G_2$ and $K_1 - K_2$ are of the same sign. This limitation is not particularly restrictive when dealing with real materials and when differences between the phase moduli are substantial.

The Hashin–Shrikman/Hill bounds are the best available without further appeal to the geometry of the material. However, further work has been undertaken to derive estimates for the moduli under slightly less general conditions.

### 2.3.1 Spherical coated particle models

The first approach is due to Hashin.[35] Imagine a material composed of spherical particles each of which is surrounded by a shell of matrix material. The ratio of the shell thickness to the particle diameter is constant and determines the local volumetric composition. If the right grading is used, together with vanishingly small particles, it is possible to fill any given space with coated spheres leaving no residual porosity. If the composite so formed is subject to hydrostatic stress, every composite element of the material is deformed in a similar way. Thus, for this material, an exact value of the bulk modulus can be obtained using the known solution for a thick-walled sphere. Hashin recognized this and obtained two different results, identical to the bounds (2.4), depending on which of the two phases was identified as the inclusion. In particular, if the particle is made of the stiffer material (with properties $K_1$, $G_1$) the spherical model material has a stiffness identical to the lower bound given in (2.4). Thus, although not giving any additional information on stiffness, the spherical coated particle approach does lead to physical models corresponding to the bounding solutions.

### 2.3.2 Self-consistent models

The other main approach can be termed the 'self-consistent' method and depends on an assumption concerning the average stress or strain in the inclusions. It is supposed that this average stress is the same as would occur

Table 2.1    Data for self-consistent mechanics using spherical inclusion

| | Surrounding medium 's' | Inclusion 'i' | 'Concrete' inclusion $\nu = \nu_i = \nu_s = 0.2$ |
|---|---|---|---|
| Deviatoric stress $\sigma'$ | $\sigma'_s$ | $\sigma'_i = \dfrac{\sigma'_s}{\beta + (1-\beta)G_s/G_i}$ | $\sigma'_i = 2\gamma\sigma'_s$ |
| Volumetric stress $\sigma^*$ | $\sigma^*_s$ | $\sigma^*_i = \dfrac{\sigma^*_s}{\alpha + (1-\alpha)K_s/K_i}$ | $\sigma^*_i = 2\gamma\sigma^*_s$ |
| Free thermal strain $\theta_i$ in inclusion | zero | $e'_i = 0$ $e^*_i = \dfrac{3K_i\theta_i}{3K_i + 4G_s}$ | $e^*_i = \gamma\theta_i$ |
| Free thermal strain $\theta_s$ in surrounding medium | $e' = 0$ $e^* = \theta_s$ | $e_i = 0$ $e^*_i = \dfrac{4G_s\theta_s}{3K_i + 4G_s}$ | $e^*_i = (1-\gamma)\theta_s$ |

$$\alpha = \frac{1}{3}\frac{1+\nu}{1-\nu} = \frac{3K}{3K+4G} \qquad \beta = \frac{2}{15}\frac{4-5\nu}{1-\nu} = \frac{6}{5}\frac{K+2G}{3K+4G} \qquad \gamma = \frac{E_i}{E_i + E_s}$$

The practical range for values of $\alpha$ and $\beta$ is shown in Fig. 2.6

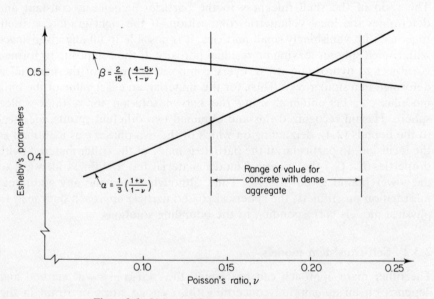

**Figure 2.6**   Values of Eshelby's parameters $\alpha$ and $\beta$

within a single particle surrounded by an infinite matrix, which has the same properties as the composite material and is subject to the same average stress as the composite. Provided solutions for the stresses in the inclusion are known the procedure is fairly straightforward, at least for linear materials. The average stresses and strains are related by the law of mixtures[36,32] and the respective elastic constants. The self-consistent approximation then provides the additional information necessary to solve these equations and obtain expressions for the moduli. The procedure has been used in studying the properties of aggregates of crystals,[37] elastic composites,[30,38] in plasticity[39] and for metallic creep.[40] The fundamental solutions for the stresses in the inclusion are usually taken from Eshelby's paper dealing with the ellipsoidal inclusion.[41] A feature of these results is that a long range stress produces a uniform stress in an isolated inclusion. The particular results are simplified when the particle is spherical and both the matrix and the inclusion are isotropic. Results of this sort are given in Table 2.1 in a form useful in applications. Some examples follow in a later section.

## 2.4  SOLID MODELS FOR CONCRETE

In considering concrete, both Dantu[36] and Hansen[42] suggested using stratified series and parallel models comprising layers of aggregate and paste or mortar to represent the stiffness of concrete. This amounted to putting Poisson's ratio equal to zero for the concrete, and each of the constituent phases, and adopting the Voigt assumption of uniform strain in the parallel model and Reuss' uniform stress assumption for the series model. In this way, values for Young's modulus were obtained having the same form as the estimates for the moduli given in equations (2.1) and (2.2). The wide difference between these values led Hansen and others to develop more detailed models and better estimates for Young's modulus. These have been reviewed and compared by Manns[43] whilst Hansen[44] has recommended a number of practical formulae for different situations.

The simplest improvement on the original Dantu–Hansen values followed work by Hirsch[45,46] based on an assumed periodic variation of direct stress through a concrete sample containing a regular arrangement of aggregate. The resulting expression for the reciprocal of Young's modulus could be written as the weighted average of the Dantu–Hansen values. In addition, it turned out that equal weighing fitted experimental results quite well, so leading to the expression

$$\frac{1}{E} = \frac{1}{2}\left(\frac{1}{E^{V}} + \frac{1}{E^{R}}\right) \tag{2.5}$$

Here, the moduli $E^{V}$ and $E^{R}$ correspond to the parallel and series results, respectively given by equations (2.1) and (2.2), with $E$ substituted for $K$ or

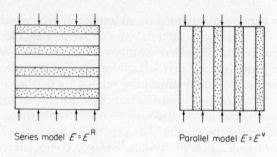

Series model $E = E^R$       Parallel model $E = E^V$

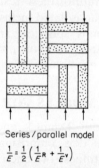

Series / parallel model

$$\frac{1}{E} = \frac{1}{2}\left(\frac{1}{E^R} + \frac{1}{E^V}\right)$$

**Figure 2.7** Layer models for composite elastic behaviour

G. The solid model corresponding to this result consists of a series–parallel
arrangement combining the Hansen models in an isotropic two-dimensional
array as shown in Figure 2.7.

With the exception of Hirsch's work, most other contributions have been
based on inclusion models rather than layered systems. These can give
acceptable results for a range of inclusion stiffness from fully rigid to the
zero stiffness appropriate to a porous material. Typical of these are the
models used by Counto,[47] Illston[48] and England.[49] Although these inclusion
models are very much in the spirit of Hashin's coated spherical particle, the
analysis is usually simplified by ignoring Poisson's ratio and considering only
the strains in the direction of an applied compressive load. Although this
one-dimensional treatment renders the analysis approximate, it has the great
advantage that other constitutive laws, besides linear elasticity, can be
introduced easily. Thus, the inclusion models have been extended to provide
information on creep and shrinkage. In doing this, there is an opportunity to
fit physical processes into a simple structural context and so extend the
range of phenomena being modelled. For instance, Illston[48] arranged for
pore space and redistribution of material in his model in order to represent

the effect of hydration in increasing the amount and stiffness of the paste phase.

This direct link with the behaviour of the constituent materials is not so evident in time-dependent extensions of the layer models. For instance, in considering creep it might seem a convenient simplification to regard one of the two phases as elastic and the other as viscous. If this is done, the Dantu–Hansen models become the Maxwell and Kelvin bodies of viscoelasticity whilst the combined model is transformed into a Burger's body. Of course, the behaviour of hardened cement paste cannot be represented satisfactorily by a single dashpot: it is not a fluid and the paste exhibits an instantaneous elastic strain on loading. Thus, although the Burger's body is recognized as being a convenient and flexible model for concrete,[50] it is difficult to link the viscoelastic properties of the elements in the model with the behaviour of the individual constituents. Hansen has attempted this, but it would be more usual for the viscoelastic parameters to be assigned on the basis of tests on the concrete as a whole.

## 2.5 DANTU AND THE PARTIAL STRESS CONCEPT

The use of solid models allows separate considerations to be given to the behaviour and properties of each phase in a composite whilst at the same time making provision for structural interaction between the phases. Dantu[36] attempted to do much the same thing, without the use of a structural model, by separating the average stress and strain at a point in a composite into components attributed to the separate phases. In doing this, Dantu effectively adopted, for concrete, an analytical model consisting of an isotropic continuum formed of two different materials that coexist with constant volumetric composition at every point within the composite. On this basis, if $\sigma_{ij}$, $\varepsilon_{ij}$ are the conventional stress and strain components at a point in the combined material the 'average' stress and strain $(s_{ij}, e_{ij})$ can be defined, relative to some volume $V$ by

$$s_{ij} = \frac{1}{V}\int_V \sigma_{ij}\, dV \quad \text{and} \quad e_{ij} = \frac{1}{V}\int_V \varepsilon_{ij}\, dV \tag{2.6}$$

The expression for the average stress in the composite continuum can be formally expressed in terms of integrals throughout the separate phases.[51] When taken with the definition (2.6) this leads to a law of mixtures relationship between the macroscopic average and the partial stresses in each phase. A similar result holds for the strains so that

$$s_{ij} = g_1(s_{ij})_1 + g_2(s_{ij})_2$$

and

$$e_{ij} = g_1(e_{ij})_1 + g_2(e_{ij})_2$$

$$\tag{2.7}$$

where $g_1$ and $g_2$ are again the volume fractions of the separate phases with $g_1 + g_2 = 1$.

In what follows, we shall consider a statistically homogeneous material and uniform states of macroscopic stress and strain. In this situation, there is no need to rely on Dantu's continuum model. Equation (2.7) applies equally well to the averages throughout a finite region of a composite where, for instance, the partial stress can be interpreted as being the average stress in each phase.[32] It follows, at least for linear materials, that the average stress and strain in any phase can be linked by the macroscopic constitutive law for the phase material.

If we wish to find the partial stresses $(s_{ij})_1$ and $(s_{ij})_2$, constitutive relations are needed for each phase and also for the composite material. Dantu's model does not give the combined behaviour. Accordingly the approach can be used only to interpret behaviour of concrete in situations where full details of the properties of the concrete and its constituents are known.[51] This is a severe limitation. However, results of such studies are sufficiently interesting to encourage attempts to extend Dantu's work so that predictions can be made without having to anticipate overall behaviour. One way of doing this is by adopting the self-consistent model.

## 2.6  SELF-CONSISTENT AVERAGE STRESS THEORY

In this section we take concrete and its constituents to be linear, elastic or viscoelastic, materials and use the results already given in Table 2.1 to provide a self-consistent approximation to the average stress in one of the phases.

All the materials are taken to be isotropic. Accordingly, it is convenient to work in terms of the volumetric and deviatoric components of the various tensors of stress and strain. As a matter of notation, we write

$$s^* = s_{kk}/3 \quad \text{and} \quad s'_{ij} = s_{ij} - \delta_{ij}s^* \tag{2.8}$$

to define the volumetric component, $\delta_{ij}s^*$, and the deviatoric component, $s'_{ij}$, of the tensor $s_{ij}$. With the deviatoric component, it will be usual to omit the suffices defining the particular element of the tensor unless an expression needs to be developed in suffix notation. Thus, $s'_1$ refers to a typical deviatoric component of the average stress in phase 1. In applications to concrete, phase 1 will be taken to be the aggregate and phase 2 the mortar or paste.

### 2.6.1  Elasticity

For isotropic linear elastic behaviour of the concrete

$$s^* = 3Ke^* \quad \text{and} \quad s' = 2Ge' \tag{2.9}$$

while similar expressions obtain for the two component materials. These may be combined with equation (2.7) to obtain expressions for the average stresses in one or other phase, i.e.

$$s_1^* = \frac{1}{g_1} \frac{K_1}{K} \left( \frac{K_2 - K}{K_2 - K_1} \right) s^* \qquad s_1' = \frac{1}{g_1} \frac{G_1}{G} \left( \frac{G_2 - G}{G_2 - G_1} \right) s' \qquad (2.10)$$

Equation (2.10) are similar to those derived by Dantu. To take the theory further, the self-consistent approximation is used to provide an expression for the average stress in the inclusions making up phase 1. From Table 2.1,

$$s_1^* = \frac{s^*}{\alpha + (1-\alpha)K/K_1} \quad \text{and} \quad s_1' = \frac{s}{\beta + (1-\beta)G/G_1} \qquad (2.11)$$

where $\alpha$ and $\beta$ are Eshelby's functions of Poisson's ratio shown in Figure 2.6. On substituting these results into equations (2.10) and rearranging, the following equations connecting the various moduli are obtained:[38]

$$\frac{g_1}{K - K_2} + \frac{g_2}{K - K_1} = \frac{\alpha}{K}$$

$$\frac{g_1}{G - G_2} + \frac{g_2}{G - G_1} = \frac{\beta}{G} \qquad (2.12)$$

In order to find $K$ and $G$, the above equations must be taken together with the definitions of $\alpha$ and $\beta$ given in Table 2.1. This is not straightforward. However the limiting cases corresponding to rigid inclusions and to open pores are of interest. With rigid inclusions $K_1 = G_1 = \infty$ so that

$$K = \frac{\alpha}{\alpha - g_1} K_2 \quad \text{and} \quad G = \frac{\beta}{\beta - g_1} G_2 \qquad (2.13)$$

This suggests that the material as a whole becomes rigid, and so cannot be deformed, when the inclusion concentration $g_1 = \alpha = \beta$. From Figure 2.6 it is clear that this can only occur when $\alpha = \beta = 0.5$ and when the volume concentration of inclusion is 50%.

Other than the restriction to spherical particles, no geometrical arguments have been invoked in the analysis. However, on physical grounds, it is clear that the material must lock into a rigid state at an inclusion concentration below 100% because of the interference between particles. With different packings of single sized particles, the possible range for the aggregate volume fraction varies from 53–74%. It is of interest, but maybe fortuitous, that the self-consistent model suggests that the material will lock at an inclusion volume concentration approximately the same as that for the loosest packing of uniform spherees.

Following this, it would appear that a similar volume concentration constraint should apply with porous material ($K_1 = G_1 = 0$) when the overall stiffness is zero. For the porous material,

$$K = \frac{1-g_1-\alpha}{1-\alpha}K_2 \quad \text{and} \quad G = \frac{1-g_1-\beta}{1-\beta}G_2 \tag{2.14}$$

which would suggest zero stiffness at 50% porosity.

### 2.6.2  The $\nu = 0.2$ approximation

For normal weight dense concretes, Poisson's ratio lies in a rather narrow range from around 0.14 to 0.23.[52] Because of this, it is of interest that equations (2.12) are uncoupled for $\nu_1 = \nu_2 = \nu = 0.2$ with $\alpha = \beta = 0.5$. It is therefore a convenient simplification to adopt these values for concrete irrespective of composition. This approximation reduces equations (2.12) to

$$\frac{g_1}{E-E_2} + \frac{g_2}{E-E_1} = \frac{1}{2E} \tag{2.15}$$

A more convenient form for calculation follows after defining non-dimensional moduli:

$$M = \frac{E}{(E_1 E_2)^{1/2}} \qquad M^R = \frac{E^R}{(E_1 E_2)^{1/2}} \qquad M^V = \frac{E^V}{(E_1 E_2)^{1/2}}$$

so that

$$M^2 - M(M^V - 1/M^R) - 1 = 0 \tag{2.16}$$

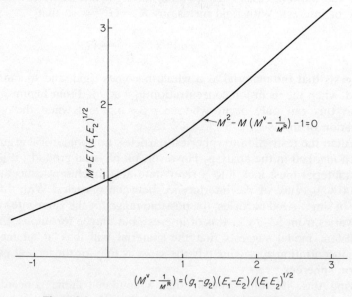

**Figure 2.8**   Young's modulus from the self-consistent model

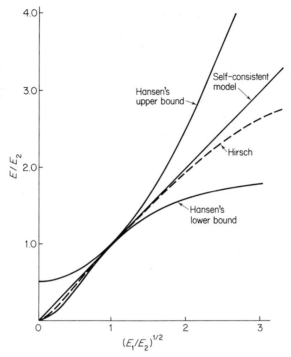

**Figure 2.9**  Estimates of Young's modulus for concrete with 50% aggregate concentration

with

$$M^V - \frac{1}{M^R} = (g_1 - g_2)(E_1 - E_2)(E_1 E_2)^{-1/2}$$

The solution of equation (2.16) can be read directly from the graph in Figure 2.8. For a 50% aggregate concentration, $g_1 = g_2$, and the equations lead to

$$E = (E_1 E_2)^{1/2} \qquad (2.17)$$

As can be seen from Figure 2.9, this result compares well with other approximations used for estimating Young's modulus for concrete.

The uncoupling of equations (2.12), through putting the various Poisson's ratios equal to 0.2, means that the ratio $K/G$ is the same for each phase and that identical equations are obtained for each modulus $K$, $G$ or $E$. In some ways, this helps to explain the success of the solid models used for concrete: success which has mostly been achieved in a simplified analysis and in spite of ignoring the effects of Poisson's ratio.

In what follows, we adopt the simplification implied by putting $\alpha = \beta = 0.5$ so that the results are approximate and most relevant to dense concrete.

### 2.6.3  Shrinkage and thermal expansion

It is of interest to use the self-consistent approach to determine the thermal expansion or shrinkage of concrete in terms of the unrestrained deformation of the constituents. Let $\theta^*$ be the average strain which occurs in the concrete as a whole, due to shrinkage or thermal expansion in the absence of stress. Similarly, let $\theta_1^*$ and $\theta_2^*$ be the free strains that would occur in the two phases in the absence of restraint due to structural interaction. The problem is concerned only with volumetric stresses and strains. Accordingly, the mixture relations (2.7) become

$$0 = g_1 s_1^* + g_2 s_2^* \quad \text{and} \quad \theta^* = g_1 e_1^* + g_2 e_2^* \tag{2.18}$$

which are to be solved using the stress–strain relations

$$e_1^* = s_1^*/3K_1 + \theta_1^* \quad \text{and} \quad e_2^* = s_2^*/3K_2 + \theta_2^* \tag{2.19}$$

Equations (2.18) and (2.19) can be solved to give expressions for the average stresses in the two phases, i.e. in phase 1:

$$s_1^* = \frac{3}{g_1} \frac{K_1 K_2}{K_2 - K_1} (\theta^* - g_1 \theta_1^* - g_2 \theta_2^*) \tag{2.20}$$

The term in brackets, in this expression, can be taken as a measure of the strain incompatibility that is causing stresses in the aggregate and paste. It will be noted that, if the overall strain (thermal expansion or shrinkage) is linked to the free strain in the constituents by the law of mixtures, the average stresses are all zero. Another peculiarity occurs if a homogeneous body $(K_1 = K_2)$ is considered. Apparently, unless the stresses are to be infinite for arbitrary temperature rise,

$$\theta^* = g_1 \theta_1^* + g_2 \theta_2^*; \tag{2.21}$$

but of course, this must be so for the conditions of uniform macroscopic stress and strain being considered.

Now, returning to equation (2.20) and using the stress–strain relations (2.19), we find the average strain in the inclusion,

$$e_1^* = \theta_1^* + \frac{1}{g_1} \frac{K_2}{K_2 - K_1} (\theta^* - g_1 \theta_1^* - \theta_2 g_2^*) \tag{2.22}$$

Next, an alternative expression for $e_1^*$ is obtained by adopting the self-consistent approximation and using the results in Table 2.1 to give

$$e_1^* - \theta_1^* = \frac{4G}{3K_1 + 4G} (\theta^* - \theta_1^*) \tag{2.23}$$

After substituting for $e_1^*$ and some rearrangement, the following expression connecting the unrestrained deformation is obtained:

$$\frac{\theta^* - \theta_1^*}{\theta^* - \theta_2^*} = -\frac{g_2 K_2}{g_1 K_1}\left(\frac{3K_1 + 4G}{3K_2 + 4G}\right) \qquad (2.24)$$

Following this, the result can be simplified when dealing with concrete by putting $\nu_1 = \nu_2 = \nu = 0.2$ to obtain

$$\frac{\theta^* - \theta_1^*}{\theta^* - \theta_2^*} = -\frac{g_2 E_2}{g_1 E_1}\left(\frac{E_1 + E}{E_2 + E}\right) \qquad (2.25)$$

On putting $\theta_1^*$ equal zero, the equation relates the shrinkage of the concrete $\theta^*$ to that of the paste $\theta_2^*$. So, after using (2.16) and rearranging

$$\theta^* = \frac{E_2(E_1 - E)}{E(E_1 - E_2)}\theta_2^* \qquad (2.26)$$

or

$$\theta^* = g_2 A_2 \theta_2^* \qquad (2.27)$$

where $A_2$ is a coefficient giving the overall deformation in terms of the 'lack of fit' in phase 2. In determining $A_2$, the overall Young's modulus of the concrete, $E$, is required. This can be found from equation (2.16) or directly from Figure 2.8. Some typical results are shown in Figure 2.10. When the aggregate is taken to be rigid, with $\alpha = 0.5$,

$$\theta^* = (1 - 2g_1)\theta_2^* \qquad (2.28)$$

again suggesting that the material locks up at an aggregate volume concentration of 50%.

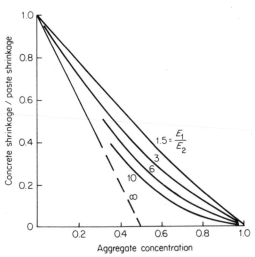

**Figure 2.10** Shrinkage prediction from the self-consistent model with $\nu = 0.2$

## 2.6.4   Creep and stress redistribution

Consider a loaded concrete sample under an equal triaxial compressive stress $s^* = p$. This load is maintained so that the concrete creeps with consequent redistribution of stress between the aggregate and the cement paste. We now use the self-consistent theory to investigate the extent of this redistribution.

In calculating creep under variable stress, a simplified view based on the 'rate of creep' method of analysis will be used. In addition, no account will be taken of changes in the elastic properties of the materials during the period under load. In these terms, the constitutive relations become

$$\dot{e}^* = \dot{s}^*/3K + s^*\dot{c}^*$$
$$\dot{e}_1^* = \dot{s}_1^*/3K_1 \qquad\qquad (2.29)$$
$$\dot{e}_2^* = \dot{s}_2^*/3K_2 + s_2^*\dot{c}_2^*$$

where $c^*$ is the specific creep in equal triaxial compression and a superior dot is used to indicate differentiation with respect to time. Using these with equation (2.7) leads to an expression for the average stress in the cement paste

$$\dot{s}_2^* = \frac{1}{g_2}\frac{K_2}{K}\frac{K_1-K}{K_1-K_2}\dot{s}^* + \frac{3}{g_2}\frac{K_1 K_2}{K_1-K_2}(s^*\dot{c}^* - g_2 s_2^*\dot{c}_2^*)$$

with
$$\qquad\qquad (2.30)$$

$$s_2^* = \frac{1}{g_2}\frac{K_2}{K}\frac{K_1-K}{K_1-K_2}p \quad \text{at } t=0$$

During the creep test, the load is maintained constant so that:

$$\dot{s}_2^* = \frac{3}{g_2}\frac{K_1 K_2}{K_1-K_2}(p\dot{c}^* - g_2 s_2^*\dot{c}_2^*) \qquad\qquad (2.31)$$

Although it would be within the scope of the self-consistent approach, we will not attempt to find an expression for the concrete specific creep $c^*$ in terms of $c_2^*$. However, suppose that, after some time, the rate of creep in the concrete is proportional to that in the paste and

$$\dot{c}^* = g_2 B_2 \dot{c}_2^* \qquad\qquad (2.32)$$

with $B_2$ independent of time. Equation (2.31) then becomes

$$\frac{ds_2^*}{dc_2^*} = \frac{3K_1 K_2}{K_1-K_2}(pB_2 - s_2^*) \qquad\qquad (2.33)$$

which suggests that, if redistribution occurs together with the condition (2.32), the stress in the paste changes monotonically with $c_2^*$ and tends to a steady state value

$$s_2^* = B_2 p \qquad\qquad (2.34)$$

It is of interest that if $A_2 = B_2$ there is no redistribution of internal stress due to creep under hydrostatic loading. The difference in these quantities is probably dependent on Poisson's ratio and the ratio of the creep function for shear and bulk compression. Because of this, it would appear that the $\nu = 0.2$ assumption might not be appropriate in a more detailed analysis.

## 2.7 CONCLUDING DISCUSSION

A number of different approaches to the micromechanics of concrete systems have been reviewed. The aim has been to explore how these approaches can be used to feed knowledge of behaviour of individual constituents into models for overall material behaviour. Accordingly, no consideration has been given to purely statistical methods or to continuum models devised to account for behaviour at a phenomenological level without regard to structure or composition.

Again no consideration has been given to studies of composite behaviour involving detailed numerical analysis and the finite element method. Most approaches of this kind have been two dimensional and based on a simplified view of cracking and shear bond. Although adding to a general understanding of what occurs under load, the approach is not directly helpful in developing constitutive relations other than as a way of obtaining results of material behaviour in a form similar to those from physical experiments. Interpretation of results obtained for local conditions is obviously dependent on the local geometry adopted whilst the overall results are very much affected by any simplification reducing the scale of the problem from three to two dimensions. In a sense, results from these finite element studies are almost too detailed due to the attempt to mirror complex behaviour in a complex way. Because of this, there is some attraction in looking at the simpler older models to see how these ideas and mechanisms can contribute to descriptions of materials that are likely to be useful in analysis. The problems here are not solely to do with creep and shrinkage. It is rather the character of the model adopted which is of prime importance in providing the necessary flexibility to enable different modes of behaviour to be included.

Of the non-linear models, Brandtzaeg's shear bond theory is still the most complete and the approach can be fitted into the context of modern analysis of continua using plasticity theory and finite elements.

In dealing with elasticity and shrinkage, most success has been achieved using inclusion models. However, it seems that the continuum viewpoint can again be adopted to provide essentially similar results through the use of self-consistent mechanics. There are attractions in doing this, provided that the separate elements of the material can still be identified within the continuum.

Looking to the future, it would be informative to extend the use of the average stress concept and self-consistent theory to include interfacial effects, growth of material within the phases and possible the effects of load bearing water and local moisture migration. To do this, it will probably be necessary to abandon the view of Dantu's statistical stress as being the average stress in each phase and to take up the continuum mixture model in which each phase is continuous and pervades the entire region occupied by the material. Theories of this sort already exist (see for instance Bedford and Stern[53]) but have not so far been applied to concrete.

One difficulty with the continuum view that is highlighted, but not solved, by the solid models is the need to resolve whether aggregate size is of importance in descriptions of local structural behaviour. In the continuum studies, the problem has been avoided by considering macroscopically uniform states of stress and strain and taking averages over a large volume. In Dantu's analysis, the difficulty emerges when leaving $V$ undefined in the definition of statistical stress in equations (2.6). Attempts have been made to resolve this problem in continuum analysis by introducing couple stresses and microrotations as additional variables. In its simplest form, in applications in isotropic elasticity, this requires the use of two additional elastic constants.[54,55] One of these is a modulus of local curvature and the other is a characteristic length, possibly related to grain size of the material. However, although the theory has been extended to plasticity and viscoelasticity, there is still a lack of experimental evidence to confirm that the approach is relevant to the behaviour of real materials. On the other hand, if a view of material behaviour at the structural level is required, possibly using mixture theory, it would seem that these effects should be included. This is a further area where structural models could be devised to lead the way in exploratory investigation. It is considered that the combined effects of creep and bond breakdown could also be investigated in this way using an updated version of Brandtzaeg's original model for concrete.

## REFERENCES

1. Reinius, E. (1955), 'En teori om betongens deformation och brott', *Betong,* No. 1, 15–43.
2. Reinius, E. (1956), 'A theory of the deformation and the failure of concrete', *Mag. Concr. Res.*, **8,** No. 24, 157–60.
3. Baker, A. L. L. (1959), 'An analysis of deformation and failure characteristics of concrete', *Mag. Concr. Res.*, **11,** No. 33, 119–28.
4. Brandtzaeg, A. (1927), 'Failure of a material composed of non-isotropic elements', *Nor. Vidensk. Selsk. Skr.*, No. 2.
5. Shah, S. P., and Winter, G. (1966), 'Inelastic behavior and fracture of concrete', *Causes, Mechanisms and Control of Cracking in Concrete*, ACI Publ. SP-20, Am. Concr. Inst., Detroit, pp. 5–28.

6. Taylor, M. A. (1969), 'A physical–mathematical model for the deformation and failure of brittle materials under general states of stress', Univ. Calif. Berkeley, Dept. Civil Eng., Div. SESM Report under contract W-7405-eng-26, Nov. 1969.
7. Cundall, P. A. (1976), 'Explicit finite difference methods in geomechanics', *Proc. 2nd Int. Conf. on Numerical Methods in Geomechanics, Blacksburg*, pp. 132–50.
8. Cundall, P. A., and Strack, O. D. L. (1979), 'A discrete numerical model for granular assemblies', *Geotechnique*, **29**, No. 1, 47–65.
9. Dougill, J. W. (1961), 'The effects of thermal incompatibility and shrinkage on the strength of concrete', *Mag. Concr. Res.*, **13**, No. 39, 119–24.
10. Anson, M. (1964), 'An investigation into a hypothetical deformation and failure mechanism for concrete', *Mag. Concr. Res.*, **16**, No. 47, pp. 73–82.
11. Baker, A. L. L. (1970), 'A criterion of concrete failure', *Proc. Inst. Civ. Eng.*, **47**, 285–95.
12. Hrennikoff, A. (1941), 'Solution of problems in elasticity by the framework method', *J. Appl. Mech.*, **8**, No. 4, Dec.
13. McHenry, D. (1943), 'A lattice analogy for the solution of stress problems', *J. Inst. Civ. Eng.*, No. 2, 59–82.
14. Drucker, D. C. and Chen, W. F. (1968), 'On the use of simple discontinuous fields to bound limit loads', *Proc. Conf. on Engineering Plasticity*, Cambridge University Press, Cambridge, pp. 129–45.
15. Drucker, D. C. (1959), 'On minimum weight design and strength of non-homogeneous plastic bodies', in W. Olszak (Ed.), *Proc. IUTAM Symp. on Non-Homogeneity in Elasticity and Plasticity, Warsaw 1958*, Pergamon, Oxford, pp. 139–46.
16. Wittmann, F. H., (1979), 'Micromechanics of achieving high strength and other superior properties, session I report: high strength concrete', in S. P. Shah (Ed.), *Proc. Workshop at Univ. Illinois at Chicago Circle, Dec. 1979*, pp. 8–30.
17. Richart, F., Brandtzaeg, A., and Brown, R. L. (1928), 'A study of the failure of concrete under combined compressive stresses', *Bull. No. 185, Univ. Illinois, Eng. Exp. Stn, Urbana Ill.*
18. Newman, K. (1965), 'The structure and engineering properties of concrete', in J. R. Rydzewski (Ed.), *Theory of Arch Dams*, Pergamon, Oxford, pp. 683–712.
19. Newman, K., and Newman, J. B. (1971), 'Failure theories and design criteria for plain concrete', *Proc. Conf. on Structure Solid Mechanics Eng. Design, Southampton 1969*, Wiley, Chichester Part II, pp. 963–95.
20. Taylor, M. A., and Broms, B. B. (1964), 'Shear bond strength between coarse aggregate and cement paste or mortar', *Proc. Am. Concr. Inst.*, **61**, 939–56.
21. Alexander, K. M., Wardlaw, J., and Gilbert, D. J. (1968), 'Aggregate cement bond, cement paste strength and the strength of concrete', *Proc. Conf. on The Structure of Concrete, London 1956*, Cem. and Concr. Assoc., London, pp. 59–81.
22. Taylor, M. A. (1971), 'A theory for the deformation and failure of cement pastes, mortars and concretes under general states of stress', *Proc. ACI*, **68**, 756.
23. Zienkiewicz, O. C., Nayak, G. C., and Owen, D. R. J. (1972), 'Composite and overlay models in numerical analysis of elasto-plastic continua', in A. Sawczuk (Ed.), *Proc. Int. Symp. on the Foundations of Plasticity*, Noordhoff, pp. 107–23.
24. Sanders, J. L. Jr (1954), 'Plastic stress strain relations based on linear loading functions', *Proc. 2nd US Conf. on Appl. Mech.*, pp. 455–60.
25. Batdorf, S. B., and Budiansky, B. (1949), 'A mathematical theory of plasticity based on the concept of slip', *NACA Tech. Note No. 1871, Washington*, p. 33.

26. Dougill, J. W. (1976), 'On stable progressively fracturing solids', *Z. Angew. Math. Phys.*, **27**, No. 4, 423–37.
27. Dougill, J. W., and Rida, M. A. (1980), 'Further consideration of progressively fracturing solids', *J. Eng. Mech. Div. ASCE*, Oct.
28. Dougill, J. W., and Burt, N. J. (1973), 'Progressive failure of random network structures', Paper VI-422, *Proc. 3rd. Int. Conf. on Fracture, München, April 1973.*
29. Burt, N. J., and Dougill, J. W. (1977), 'Progressive failure in a model heterogeneous medium', *J. Eng. Mech. Div. ASCE*, **103**, No. EM3 365–76.
30. Budiansky, B. (1965), 'On the elastic moduli of some heterogeneous materials', *J. Mech. Phys. Solids*, **13**, 223–7.
31. Hashin, Z. (1964), 'Theory of mechanical behaviour of heterogeneous media', *Appl. Mech. Rev.*, **17**, No. 1, 1–9.
32. Hill, R. (1963), 'Elastic properties of reinforced solids: some theoretical principles', *J. Mech. Phys. Solids*, **11**, 357–72.
33. Brown, C. B. (1968), 'Models for concrete stiffness with full and zero contiguity', *Proc. Conf. on The Structure of Concrete, London 1965*, Cem. and Concr. Assoc., London, pp. 3–15.
34. Hashin, Z., and Shtrikman, S. (1963), 'A variational approach to the theory of the elastic behaviour of multiphase materials', *J. Mech. Phys. Solids*, **11**, 127–40.
35. Hashin, Z. (1962), 'The elastic moduli of heterogeneous materials', *J. Appl. Mech.*, **29**, No. 1, 143–50.
36. Dantu, P. (1958), 'Etude des contraintes dans les milieux hétérogènes: application au beton', *Ann. Inst. Batim. Trav. Publics*, **11**, No. 121, 54–98.
37. Hershey, A. V. (1954), 'The elasticity of an isotropic aggregate of anisotropic cubic crystals', *J. Appl. Mech.*, **21**, 236–40.
38. Hill, R. (1965), 'A self consistent mechanics of composite materials', *J. Mech. Phys. Solids*, **13**, 213–22.
39. Budiansky, B., and Wu, T. T. (1962), 'Theoretical prediction of plastic strains of polycrystals', *Proc. 4th US Natl Congr. on Applied Mechanics*, ASME, pp. 1175–85.
40. Brown, G. M. (1970), 'A self-consistent polycrystalline model for creep under combined stress states', *J. Mech. Phys. Solids*, **18**, 367–81.
41. Eshelby, J. D. (1957), 'The determination of the elastic field of an ellipsoidal inclusion and related problems', *Proc. R. Soc.* A, **241**, 376–96.
42. Hansen, T. C. (1958), 'Creep of concrete—a discussion of some fundamental problems', *Bull. (Medd.) Swed. Cem. Concr. Res. Inst.*, Stockholm, No. 33, pp. 1–48.
43. Manns, W. (1971), 'Influence of elasticity of cement paste and aggregate on the elastic behaviour of concrete and mortar', in M. Teeni (Ed.), *Proc. Conf. on Eng. Design, Southampton 1969*, Wiley, Chichester, pp. 667–80.
44. Hansen, T. C. (1968), 'Theories of multi-phase materials applied to concrete, cement mortar and cement paste', *Proc. Conf. on The Structure of Concrete, London 1965*, Cem. and Concr. Assoc. London, pp. 16–23.
45. Hirsch, T. J. (1962), 'Modulus of elasticity of concrete affected by elastic moduli of cement paste, matrix and aggregate', *Proc. ACI*, **59**, No. 3, 427–52.
46. Dougill, J. W. (1962), Discussion of a paper by T. J. Hirsch (Ref. 45)', *Proc. ACI*, **59**, No. 9, 1363–5.
47. Counto, U. J. (1964), 'The effect of elastic modulus of the aggregate on the elastic modulus, creep and creep recovery of concrete', *Mag. Concr. Res.*, **16**, No. 48, 129–38.

48. Illston, J. M. (1968), 'The delayed elastic deformation of concrete as a composite material', *Proc. Conf. on The Structure of Concrete and its Behaviour under Load, London 1965.*, Cem. and Concr. Assoc., London, pp. 24–36.
49. England, G. L. (1965), 'Method of estimating creep and shrinkage strains in concrete from the properties of the constituent materials', *Proc. ACI*, **62,** No. 11, 1411–20.
50. Hansen, T. C. (1960), 'Creep and stress—relaxation of concrete. A theoretical and experimental investigation', *Proc. (Handl.) Swed. Cem. Concr. Res. Inst., Stockholm,* No. 31, 1–112.
51. Dougill, J. W. (1970), 'Some results for the average stresses induced in the principal components of concrete', *Mag. Concr. Res.*, **22,** No. 72, 133–42.
52. Anson, M., and Newman, K. (1966), 'The effect of mix proportions and methods of testing on Poisson's ratio for mortars and concretes', *Mag. Concr. Res.*, **18,** No. 56, 115–30.
53. Bedford, A., and Stern, M. (1972), 'A multi-continuum theory for composite elastic materials', *Acta Mech.*, **14,** 85–102.
54. Mindlin, R. D. (1963), 'Influence of couple-stresses on stress concentrations', *Exp. Mech.*, **3,** No. 1, 1–7.
55. Koiter, W. T. (1964), 'Couple-stresses in the theory of elasticity', *Proc. R. Ned. Acad. Sci., Ser. B,* **67,** 17–44.
56. Griffith, A. A. (1920), 'The phenomena of rupture and flow in solids', *Phil. Trans. Royal Soc. London A,* **221,** 163–198.
57. Pickett, G. (1956), 'Effect of aggregate on shrinkage of concrete and a hypothesis concerning shrinkage', *Proc. Am. Concr. Inst.*, **52,** 581–590.
58. Kaplan, M. F. (1961), 'Crack propagation and the fracture of concrete', *Proc. Am. Concr. Inst.*, **58,** 591–610.
59. Neville, A. M. (1964), 'Creep of concrete as a function of its cement paste content', *Mag. Concr. Res.*, **16,** No. 46, 21–30.

Creep and Shrinkage in Concrete Structures
Edited by Z. P. Bažant and F. H. Wittmann
© 1982 John Wiley & Sons Ltd

*Chapter 3*

# Probabilistic Approach to Deformations of Concrete†

*E. Çinlar*

## 3.1  INTRODUCTION

The deformations of apparently identical concrete members under identical environmental conditions and subjected to identical stress histories show deviations amounting up to 20 or 30 per cent of the average values. Our purpose is to discuss the probabilistic approach to modelling such variations with special emphasis on time dependence.

Although the basic model here seems to be in agreement with the experimental data (see Çinlar *et al.*[5] for this), there can be other models that fit the same data as well. In fact, since our approach is mechanical-probabilistic, and since the creep mechanism itself is not yet understood completely, there will certainly be better models built in future. Our hope is that this chapter proves useful as a guide to such future endeavours. For these reasons, the thinking process leading to the probabilistic model is illustrated in depth by giving the details on the use of mathematical theory in narrowing down the choices, the use of physical theory to formulate the mathematical model, and the necessity of distinguishing the mathematical model from the physical reality that it approximates.

### 3.1.1  Deformations considered

When concrete is subjected to sustained loads, in addition to the nearly instantaneous elastic deformation at the time of stress application, it shows further deformations continuously in time. Moreover, concrete undergoes stress-independent deformations such as shrinkage or swelling. Although it is common practice to distinguish between these different types of deformations by calling them instantaneous deformation, creep deformation, shrinkage, etc., such distinctions are impossible to make in the presence of

† Research supported by the Air Force Office of Scientific Research through their Grant No. AFOSR-80-0252 and by the National Science Foundation Grants Nos ENG77-02529 and ENG77-06767.

randomness. Thus, we will be concerned with the total deformations directly, and we will distinguish between creep and shrinkage, etc. only as they affect the probability law governing the deformations.

### 3.1.2 Simplifying assumptions

Throughout this chapter we visualize the concrete member under study as a cylindrical column (even though the shape of its cross-sections has no importance). We consider only *uniaxial compressive stresses*, which we assume to be the same for all cross-sections; this neglects the lateral stresses and strains that are invariably present, and also the weight of the column itself. We assume that all *stresses* involved are *within the linear range*, that is, less than about 0.4 of the compression strength. We take the concrete to be *macroscopically homogeneous*, that is, the material properties do not vary by position within the column. Also, since we are taking the concrete member under study as already cast, the randomness introduced by the process of concrete making is being suppressed. Finally, we assume that the *height* of the column is *large* compared with the average size of the aggregates, say, by a factor of 20 at least.

We do not assume anything *a priori* about temperature, humidity, and stress conditions. We aim at a general model that would allow such factors to vary randomly over time.

Throughout, the *origin of the time axis* will be the time of setting of the concrete.

### 3.1.3 Point of view

For the purposes of this subsection, we suppose that the environmental and stress conditions are all deterministic, that is, they are known (with certainty) in advance for all time to come. Suppose the height of the column is unity at time 0, and suppose that the only randomness we are interested in lies with the total deformation process.

We say that the deformation process follows the trajectory $w$ if $w$ is some function on $[0, \infty)$ and if the height of the column at time $t$ is exactly $1 - w(t)$, $t \geq 0$. Of course, at time 0, we do not know the trajectory that the deformation process will follow. Moreover, for any specific function $w$, our estimate (made at time 0) of the chances that the deformation process will follow $w$ would be zero.

Accordingly, we are led to consider the set $W$ of all possible trajectories $w$ and to speak of the chances that the deformation process will follow a trajectory belonging to some collection $F$ of functions. Thus, for various $F$ contained in $W$, we speak of the probability of $F$, denoted by $P(F)$.

Constructing a probabilistic model of the deformation process is primarily the specification of these numbers $P(F)$ for sufficiently many interesting $F$.

It can be shown that this task reduces to specifying $P(F)$ for every $F$ having the form

$$F = \{w \in W: w(t_1) \leq x_1, \ldots, w(t_n) \leq x_n\}$$

for some time values $t_1, \ldots, t_n$ and deformation values $x_1, \ldots, x_n$. This latter task is rarely done explicitly. Instead, it is done by combining certain qualitative notions (such as independence) of probability theory with reasoning based on the physics of the real phenomena involved. While doing so, one considers more primary random phenomena such as movements of solid particles within the micropores under stress and formations of bonds, etc.

Therefore, we will take the point of view that there is a very large probability space $(\Omega, \mathcal{H}, \mathbb{P})$ in the background that allows all such related random phenomena to be discussed together. In particular, then, there must be some functions $Z_t$ from $\Omega$ into the real line such that, for every $w$ in $W$, there is $\omega$ in $\Omega$ with the property $Z_t(\omega) = w(t)$ for all $t \geq 0$. Now, the earlier task of specifying $P(F)$ (for the basic $F$ mentioned in the preceding paragraph) becomes that of specifying $\mathbb{P}(H)$ for $H$ having the form

$$H = \{\omega \in \Omega: Z_{t_1}(\omega) \leq x_1, \ldots, Z_{t_n}(\omega) \leq x_n\}.$$

With this set-up, it is clear that the deformation process is precisely the stochastic process $(Z_t)_{t \geq 0}$, and constructing a probabilistic model of the deformation process becomes that of specifying $\mathbb{P}(H)$ for $H$ having the basic form above.

### 3.1.4 Prerequisites

We will use the language of probability theory and stochastic processes quite extensively. The appendix to this chapter contains a rapid review of most of the more important notions that we need and fixes the terminology and notations that are to be used in this chapter. Ideally, the reader should have some background in stochastic processes. If not, he can still profit from this chapter provided that he is willing to accept many things on faith. To make the chapter accessible to such readers, many details are left out and some technical points are carefully camouflaged. Otherwise, we have not sacrificed anything from the precision and rigor of the mathematical statements.

### 3.1.5 Outline

In Section 3.2 we discuss the variation in deformation with height and argue that, as a function of height, what we have is a process with stationary independent increments. This implies that for fixed times $t$, the distributions of the strains must be infinitely divisible, thus suggesting that the increments over time might be independent. In Section 3.3 we hypothesize that this is indeed true and give reasons to justify it. We view the deformation value at a time as the sum of a deterministic term and an infinite sum of random jumps of molecular magnitudes. The crucial hypothesis is that the point process formed by the times and magnitudes of jumps can be approximated by a Poisson random measure. In Section 3.3 we further argue that the mean measure of this Poisson random measure has a certain form, which argument leads to locally gamma processes (which generally have continuous distributions).

In Section 3.4 the special case of basic creep is examined; the objective here is to clarify the formulae already provided in Section 3.3 and to gain better insight into the physical meanings of some of the parameter functions. With the same objectives, we discuss the case of shrinkage and dilations in the absence of applied loads in Section 3.5. Finally, Section 3.6 is devoted to some observations on the functional forms of the parameters identifying the probability law of the deformation process, and Section 3.7 is devoted to the general method for handling randomness in environmental and stress conditions.

In the appendix we give a very quick introduction to stochastic processes of the type that the present work makes use of.

## 3.2  HEIGHT DEPENDENCE

In this section we aim to characterize the variation of the deformation process with the position along the axis of the cylindrical column. Recall the simplifying assumptions we introduced in Section 3.1.2. Since we are considering only uniaxial stresses, which are further assumed to be the same at all points of all cross-sections, the deformations are also along the axis.

### 3.2.1  Conditions

Further, we assume throughout this section that the environmental conditions such as temperature and humidity are deterministic, that is, they are known in advance with certainty, even though they may be varying with time. Similarly, we assume the stress history to be deterministic.

### 3.2.2  Notations

Let $(\Omega, \mathcal{H}, \mathbb{P})$ be an appropriate probability space. Recall that time 0 corresponds to the time of setting of the concrete. Considering the cross-section perpendicular to the axis at point $p$ initially, let $Z_t^p(\omega)$ denote the total deformation at time $t$ associated with that cross-section assuming that the random outcome turns out to be $\omega$. (Then, the position of the same cross-section at time $t$ is $p - Z_t^p(\omega)$.) We make the conventions that $Z_0^p(\omega) = 0$ for all $p$ and all $\omega$, and that $t \to Z_t^p(\omega)$ and $p \to Z_t^p(\omega)$ are both right-continuous and left-hand limited.

It is implicit in this notation that each outcome $\omega$ describes completely where each point $p$ is at all times $t$ corresponding to that outcome $\omega$. We further assume that $\mathcal{H}$ is chosen appropriately large, so that $\omega \to Z_t^p(\omega)$ is a random variable for every $p$ and every $t$.

### 3.2.3 Stationarity and independence of increments in $p$

Let time $t > 0$ be fixed, and put $Y_p = Z_t^p$. We are interested in the stochastic process $(Y_p)_{p \geq 0}$. For $0 < p_1 < p_2 < \cdots < p_n$, where $p_n$ is equal to the height of the column, we may think of the concrete cylinder as $n$ cylinders of initial heights $p_1, p_2 - p_1, \ldots, p_n - p_{n-1}$ put on top of each other. The deformation values, at time $t$, for these imaginary cylinders are $Y_{p_1}, Y_{p_2} - Y_{p_1}, \ldots, Y_{p_n} - Y_{p_{n-1}}$.

Although there may be slight interactions across the planes separating these $n$ cylinders, the effect of such interactions on the total deformations should be negligible as long as the cylinders are not too thin compared with the aggregate size. Thus, if $p_1, p_2 - p_1, \ldots, p_n - p_{n-1}$ are not too small, then the random variables $Y_{p_1}, Y_{p_2} - Y_{p_1}, \ldots, Y_{p_n} - Y_{p_{n-1}}$ should be nearly independent. Moreover, since we have assumed macroscopic homogeneity in material properties, and since the stresses across cross-sections are all the same, the distribution of $Y_{p+q} - Y_p$ should not depend on $p$. Hence, the following hypothesis concerning the probability law of $(Y_p)$ is well justified.

### 3.2.4 Hypothesis

For fixed time $t$, the stochastic process $p \to Z_t^p$ has stationary and independent increments.

### 3.2.5 Probability law of $Z_t^p$

It follows from the preceding hypothesis that, in particular, the distribution of $Z_t^p$ is infinitely divisible. The general form of all such distributions is known. We now argue that, within that class, the appropriate distribution is characterized by the further property that $p \to Z_t^p$ is of bounded variation (that is, it can be written as the difference of two increasing processes—one of which expresses the stress effects and the other the swelling).

Suppose that the concrete is kept under water so that there is no shrinkage. Then, $Z_t^p$ must be positive and increasing with $p$. Conversely, if there is no stress, depending on humidity level, there is either shrinkage or swelling only. In general, both effects are present, and $p \to Z_t^p$ might fluctuate due to creep and shrinkage and swelling effects that may be alternating in time and have stochastic variability along the axis. But, the probability law of $Z_t^p$ must be such that the special cases of increasing with $p$ or decreasing with $p$ can be explained through simple changes in parameters, not necessitating changes in the fundamental form of the probability law. Hence, $p \to Z_t^p$ must have the probability law of a process with bounded variation. From the known results of probability theory we obtain the following.

### 3.2.6 Fourier transform of $Z_t^p$

For any fixed $t$,

$$\mathbb{E}\left[\exp\left(i\lambda Z_t^p\right)\right] = \exp\left(ik(t)p + p\int_{-\infty}^{\infty}\mu_t(dz)(e^{i\lambda z}-1)\right)$$

for all $\lambda$ real and $p$, where $k(t)$ is some constant and $\mu_t$ is some measure on $(0, \infty)$ such that $\int \mu_t(dz)|z| < \infty$, where the last integrability condition reflects the finiteness of the means (see Section 3.2.7).

Note that $Z_t^p$ is not equal to $pZ_t^1$, and $Z_t^p$ does not even have the same distribution as $pZ_t^1$, except in the degenerate case where $\mu_t$ is identically zero. Thus, the deterministic rule associating the deformation of a column of height $p$ to that of a column of unit height is *no longer true*. It is replaced by the rule expressed in Hypothesis 3.2.5, which in particular states that the deformation associated with a column of height $p = 5$, for instance, is the sum of five independent random variables each one of which has the same distribution as the deformation associated with a column of unit height. The deterministic rule is true, however, for the expected values and variances. $\mathbb{E}[Z_t^p] = p\mathbb{E}[Z_t^1]$ and $\text{Var}\,[Z_t^p] = p\,\text{Var}\,[Z_t^1]$.

### 3.2.7 Means and variances

For any $t \geq 0$ and $p > 0$,

$$\mathbb{E}[Z_t^p] = p\left(k(t) + \int_{-\infty}^{\infty}\mu_t(dz)z\right)$$

$$\text{Var}\,[Z_t^p] = p\left(\int_{-\infty}^{\infty}\mu_t(dz)z^2\right)$$

### 3.2.8 More on the law of $Z_t^p$

Hypothesis 3.2.5 together with Section 3.2.6 specifies the law of $(Z_t^p)$ for fixed $t$ up to a choice of $k(t)$ and $\mu_t$. We will be able to say more about the latter two when we consider the time evolution of deformations in the next section.

Note that the probability law of $Z_t^p$ is completely specified through Section 3.2.6 once that of $Z_t^1$ is specified. Hence, from now on we will concentrate on $Z_t = Z_t^1$ only.

### 3.2.9 Deformations at different times

Before we consider the time dependence of the strain process $(Z_t)_{t \geq 0}$ in the next section, we list the following observations. First, for any fixed $t$, the distribution of $Z_t$ is *infinitely divisible* (this follows from Hypothesis 3.2.4 and is made more explicit in Section 3.2.6). Second, and this is a generalization of the first, all the arguments leading to Hypothesis 3.2.4 go through, *mutadis mutandis*, for the vector valued process $p \to \mathbf{Y}_p = (Z_{t_1}^p, Z_{t_2}^p - Z_{t_1}^p, \ldots, Z_{t_m}^p - Z_{t_{m-1}}^p)$, where $0 \leq t_1 < t_2 < \cdots < t_m$ are fixed times. It follows that the process $p \to \mathbf{Y}_p$ has stationary and independent increments, and therefore, the distribution of $\mathbf{Y}_p$ is infinitely divisible for any $p$. To reiterate it for $p = 1$, we see that $(Z_{t_1}, Z_{t_2} - Z_{t_1}, \ldots, Z_{t_m} - Z_{t_{m-1}})$ has an infinitely divisible distribution. This *suggests* that the process $t \to Z_t$ has independent increments. We shall argue in the next section that this is indeed so.

### 3.3 TIME DEPENDENCE

Our aim is to develop a model of the time dependence of deformations of concrete through considerations based on its microstructure. Much of what we have to say on the microstructure is based on Bažant[1]; we refer to Bažant[1] and references listed there for further details.

Throughout this section, the simplifying assumptions of Section 3.1.2 and the environmental and stress conditions of Section 3.2.1 are in force. Our notations are as in Section 3.2.2, except that we will work only with $Z_t = Z_t^1$ (which is sufficient in view of the results of Section 3.2).

### 3.3.1 Macropores and micropores

Concrete consists of aggregate and sand embedded in a matrix of hardened cement paste. This matrix is highly porous, and its main solid component is the cement gel. The largest pores are called *macropores*; they are of round shape, contain capillary water, and are connected by a system of thinner pores. The thinnest pores, called *micropores*, are the gaps between the 'sheets' of solid matter that make up the cement gel. These sheets are of colloidal dimensions with average thickness of about 30 Å; the gaps between them average about 15 Å.

### 3.3.2  Within the micropores

The micropores are essentially laminar in shape, and some may even be tubular. They contain adsorbed layers of water molecules that are held strongly by the solid walls. They also contain solid particles that bridge the gaps between the opposing walls and form bonds. These particles can exert a significant amount of disjoining pressure on the pore walls, and consequently, can carry significant amounts of compressive transverse stresses.

### 3.3.3  Migrations

When a load is applied on concrete, most of the resulting compression across the laminar micropores is carried by the solid particles. At a given location of a micropore, the effect of high enough transverse pressures is to push some of the bonded particles above their activation energy barriers, thus causing them to migrate to locations of lower stress. This eventually reduces the transverse pressure and stops the migrations at that location for a while. At the same time, there is a loss of mass and thickness in the adsorbed layers, and the loss in thickness contributes something to the total deformation. In the case of migrations into a location, these processes are reversed.

Thus, we use the term 'migrations' to signify two related events whose effects on deformations are opposites: first, the breaking up of bonds and the movement of particles that become de-bonded; and second, the formation of new bonds by particles which inhibit further movements and retard local deformations.

### 3.3.4  Migration size

Moreover, we think of a migration as involving a number of particles that form a bond or bonds and where movements occur during a small enough interval of time as to be considered simultaneous. We are forced to think along these lines because of the hindered nature of movements within micropores, which causes dependence between the movements (of individual particles) taking place during a period of peak transverse pressure.

### 3.3.5  Role of water

The preceding argument is further enhanced when we consider the role of water within the micropores. The presence of water molecules is essential for migrations to be possible; without them there is almost no creep. The higher the water content, the higher the number of particles migrating, and therefore the higher is the rate of creep. When the water content changes (in either direction), large numbers of water molecules move through the micropores, thus giving solid particles greater mobility and increasing the creep rate. Even if there is no load, this mechanism will cause some creep.

### 3.3.6 Transverse pressures

We view the transverse pressure at a location as randomly fluctuating over time, even though these transverse pressures averaged over all locations may be only decreasing or only increasing. Moreover, at a given location, periods of peak transverse pressure (during which most of the de-bonding occurs) are likely to be of short duration compared with the times between the peak periods (during which most of the particles come together to eventually form bonds that carry little stresses at first).

### 3.3.7 Role of temperature

Since high temperature means generally higher levels of energy, and thus more chances for particles to jump over their activation energy barriers, higher temperatures give both water and solid particles higher mobility. Thus, as far as the effect at a given location is concerned, higher temperatures mean a greater rate of migrations, and therefore, a greater creep rate at the macroscopic scale. However, at the macroscopic level, higher temperatures cause faster aging, which in turn decreases the rate of creep.

### 3.3.8 Modelling the migrations at a location

Recall that our use of the term migration is in a technical sense and is supposed to suggest movements of groups of particles. Since the effect of a migration upon the total deformation is bound to depend on the size of the migration as well as the geometry etc. involved at the location of that migration, different migrations will affect the deformation process differently. Thus, in accordance with this and with the arguments of Sections 3.3.3–6, we characterize the $i$th migration occurring at a location $x$ through its time $T_i^x(\omega)$ and its effect $Q_i^x(\omega)$ upon the total deformation, both depending on the outcome $\omega$ of random phenomena involved. We make the convention that $Q_i^x(\omega)$ is positive or negative depending on whether the deformation effect is to increase or to decrease the deformation. We define

$$N^x(\omega, B) = \sum_{i=1}^{\infty} 1_B(T_i^x(\omega), Q_i^x(\omega))$$

for every Borel set $B \subset [0, \infty) \times (-\infty, \infty)$; that is, $N^x(\omega, B)$ is the number of 'migrations' occurring at location $x$ whose time and deformation effects belong to the set $B$.

### 3.3.9   Superposition over all locations

Consider the collection of points $(T_i^x(\omega), Q_i^x(\omega))_{i \geqslant 1}$ as a subset of $[0, \infty) \times (-\infty, \infty)$, and take the union of those subsets for all $x$. The result is the superposition of all the migration times and effects without regard to location. The number of points belonging to $B$ in the superposition is

$$N(\omega, B) = \sum_x N^x(\omega, B)$$

We shall argue shortly that $N$ is approximately a Poisson random measure (see Section 3.3.11 ff.).

### 3.3.10   Relationship to deformations

Recall that $Z_t(\omega)$ is the total deformation amount at time $t$ corresponding to the outcome $\omega$ (for the column of unit height). We then have

$$Z_t(\omega) = k(t) + \sum_x \sum_i Q_i^x(\omega) 1_{[0,t]}(T_i^x(\omega))$$

$$= k(t) + \int_{[0,t] \times \mathbb{R}} N(\omega; ds, dz) z$$

for all times $t$ and outcomes $\omega$. Here, $k(t)$ is as in Section 3.2.6, and stands for the elastic deformation, having nothing to do with the migrations. This value $k(t)$ can be computed by the usual deterministic methods.

### 3.3.11   Random measure N

Concerning the random measure $N$ defined in Section 3.3.9 we note two important features.

First, the contribution $N^x$ of the location $x$ to the sum $N$ must be negligible. This is because the number of locations contributing to $N$ is very large, and because the same location $x$ cannot experience too many migrations in short periods of time. In particular, this argument is satisfied if we think of locations $x$ as 'creep centres' in which case there is at most one 'creep activity' at $x$ over all time (this seems to be the point of view of the Munich group).

Second, the interactions between two locations $x$ and $y$ must be almost non-existent if $x$ and $y$ are sufficiently apart. For this purpose, since 1 mm of distance cuts across nearly 100,000 micropores, a distance of 1 mm between $x$ and $y$ is sufficient to give us stochastic independence between $N^x$ and $N^y$.

It follows from these two features and a result due to Sze[6] that the random variable $N(B)$ has approximately the Poisson distribution and therefore, that $N$ is a Poisson random measure. (The results in Sze[6] were partly motivated by the present study, and are stronger versions of the classical theorems on the convergence of superpositions of point processes to Poisson processes; the latter were reviewed in Çinlar[2].) Thus, the following hypothesis seems well justified:

The random measure $N$ is Poisson

### 3.3.12 Recapitulation

In justifying the preceding hypothesis, the following played significant roles. First, the environmental factors and macroscopic stress conditions were assumed to be deterministic; otherwise, if any of these is random, $N^x$ and $N^y$ become stochastically dependent no matter how far apart $x$ and $y$ are due to their joint dependence on such conditions. Second, the amount of stress carried by the micropores altogether is assumed to be deterministic. This is not strictly true since the exact amount of hydration is not known. Thus, this amounts to assuming a deterministic hydration process. Third, within the argument of negligibility for $N^x$, there is the assumption that the same location does not experience too many migrations over time. Thus, our arguments are likely to fail if there are rapid fluctuations in temperature, humidity or stress.

### 3.3.13 Independence of the increments of $Z_t$

This follows from the hypothesis in Section 3.3.11, the formula for $Z_t$ given in Section 3.3.10, and the well known properties of Poisson random measures. Thus, the probability law of the process $t \to Z_t$ is completely specified once the distributions of the increments $Z_u - Z_t$ are specified for all $t \leqslant u$. In terms of Fourier transforms, we have

$$\mathbb{E}\{\exp[i\lambda(Z_u - Z_t)]\} = \exp\left(i\lambda[k(u) - k(t)] + \int_{(t,u]\times\mathbb{R}} \nu(ds, dz)(e^{i\lambda z} - 1)\right)$$

where $k$ is the same function as in Section 3.3.10 and where $\nu$ is the mean measure of the Poisson random measure; that is, for any Borel subset $B$ of $(0, \infty)\times(-\infty, \infty)$,

$$\nu(B) = \mathbb{E}[N(B)]$$

Comparing the Fourier transform above with that of Section 3.2.6, we see that, in Section 3.2.6, $\mu_t(A) = \nu((0, t]\times A)$ for any Borel subset $A$ of $(-\infty, \infty)$.

### 3.3.14 The mean measure

Among infinitely divisible distributions, the distribution of $Z_t - Z_u$ is specified by the exact choice of $\nu$. The Fourier transform in Section 3.3.13, or more specifically, the formula in Section 3.3.10 imply that $\nu$ must satisfy certain integrability conditions. For example, since $Z_t$ is bounded by 1, the expected value of $|Z_t|$ must be bounded by 1. Thus, whatever $\nu$ we pick, $\nu$ must satisfy $\int \nu(ds, dz)|z| \leq 1$. But the real insight into $\nu$ must come from its definition: $\nu(dt, dz)$ is the expected number of migrations occurring during $(t, t+dt)$ whose contributions to the deformation process have been within $(z, z+dz)$ in each case.

This interpretation shows that, in the variable $z$, $\nu(dt, dz)$ must be absolutely continuous with respect to the Lebesgue measure. Thus, we can write

$$\nu(dt, dz) = n(dt)f(t, z)\,dz + g(t, z)\,dt\,dz$$

for some positive function $f$ and $g$ and some positive measure $n$. It is possible that $n$ is singular, that is, the increasing function $t \to n((0, t])$ might not be differentiable. This is certainly the case if the stress function $t \to \sigma(t)$ has jumps. Indeed, the only sudden changes that can cause sudden deformations seem to be related to stress. Thus, we may suppose that $n(dt) = |\sigma(dt)|$ with the further convention that if $\sigma(dt)$ is positive then $f(t, z) = 0$ for all $z \leq 0$, and conversely, if $\sigma(dt)$ is negative then $f(t, z) = 0$ for all $z \geq 0$. This yields

$$\nu(dt, dz) = |\sigma(dt)|\,f(t, z)\,dz + g(t, z)\,dt\,dz$$

where the first term on the right accounts for the instantaneous effect of stress changes, whereas the second term accounts for the long term effects of stress—these will become clearer.

### 3.3.15  Dependence on $z$

From the expected value interpretation given for $\nu$, it follows that the dependence of $f(t, z)$ and $g(t, z)$ on $z$ should have the same functional form. The effect $z$ is largely dependent on the number of particles involved in a migration. This number is likely to be quite small in most cases, and the frequency of migrations of large size should decrease exponentially fast with size. This is consistent with the known results about micropore size distribution. It follows that $f(t, z)$ and $g(t, z)$ should decrease exponentially with $|z|$. On the other hand, for $z$ near 0, the frequencies $f(t, z)$ and $g(t, z)$ should be very large; otherwise, if $f(t, z)$ and $g(t, z)$ are integrable over a neighborhood of $z = 0$, then the process $(Z_t)$ would have only finitely many jumps during finite intervals of time, which is against our view of deformations as being the sum of a very large number of very small jumps. Thus, we will take the behaviour of $f(t, z)$ and $g(t, z)$ near $z = 0$ to be of the form $|z|^{-1}$. (Incidentally, assuming a form $|z|^{-\alpha}$, we must have $\alpha = 1$: for $\alpha < 1$ the integral over $(0, \varepsilon)$ is finite and this means not enough jumps, and for $\alpha > 1$, we have the integral of $z \cdot |z|^{-\alpha}$ over $(0, \varepsilon)$ infinite and this means infinite expectations.) We put the consequences of this reasoning in Section 3.3.17 below as an hypothesis after introducing some notations.

### 3.3.16 Notations

We let $h_t$, $T_t$, and $V_t$ denote the pore humidity, temperature, and cement gel volume at time $t$. Then, $\dot{h}_t$ will denote the derivative of $h$ at $t$. We let $a_t$ denote the aging factor at $t$; this depends (deterministically) on the histories of temperature, humidity, stress, etc. during $(0, t]$. We denote by $\sigma(t)$ the compressive stress being applied to the concrete at time $t$ (negative values, then, mean tensile stress), and let $\sigma_t$ denote that portion of the stress being carried by the micropores. The function $t \rightarrow \sigma(t)$ can be written as

$$\sigma(t) = \sigma^+(t) - \sigma^-(t) \quad t \geq 0$$

uniquely, so that both $\sigma^+(t)$ and $\sigma^-(t)$ are increasing in $t$ with the times of increase being disjoint for the two. This assumes that $t \rightarrow \sigma(t)$ is of bounded variation, but this should hold in almost all real cases.

### 3.3.17 Hypothesis

For all $t \geq 0$ and $-\infty < z < \infty$, $z \neq 0$, the mean measure $\nu$ of the Poisson random measure $N$ satisfies

$$\nu(dt, dz) = \alpha_z(dt) \frac{1}{|z|} \exp\left(\frac{-|z|}{c_t}\right) dz$$

where

$$\alpha_z(dt) = \begin{cases} \sigma^+(dt)f_t + g_t^+ \, dt & \text{if } z > 0 \\ \sigma^-(dt)f_t + g_t^- \, dt & \text{if } z < 0 \end{cases}$$

and where

$$c_t = c(T_t, h_t)$$
$$f_t = f(a_t, V_t, T_t, h_t)$$
$$g_t^+ = g^+(\sigma_t, a_t, V_t, T_t, h_t, \dot{h}_t)$$
$$g_t^- = g^-(\sigma_t, a_t, V_t, T_t, h_t, \dot{h}_t)$$

for some appropriate positive Borel functions $c$, $f$, $g^+$, $g^-$.

### 3.3.18 Remarks on the hypothesis

Other than the already mentioned factors determining the form of $\nu(dt, dz)$, the preceding hypothesis implies that $c_t$, $f_t$, $g_t^+$, $g_t^-$ depend on time $t$ only through factors such as $T_t$, $h_t$, etc., and not all of them either. In the next sections we will give more details concerning the functions $f$, $g^+$, $g^-$, and $c$, which will give more meaning to them.

### 3.3.19  Decomposition of Z

Hypotheses 3.3.11 and 3.3.17 together with the formula for $Z_t$ given in Section 3.3.10 suggest that we can decompose $Z$ as

$$Z_t = k(t) + X_t - Y_t$$

where the processes $(X_t)_{t \geqslant 0}$ and $(Y_t)_{t \geqslant 0}$ are both increasing and are stochastically independent. Clearly, $X_t$ is the shrinkage plus compression, and $Y_t$ is the swelling plus extension. For any $\omega$, the times of increase of $t \to X_t(\omega)$ and $t \to Y_t(\omega)$ are disjoint. In some cases, for example, in the case of basic creep, $Y_t$ may be negligible. But, if the stresses are compressive and humidity is increasing, for example, it is quite likely that both $X$ and $Y$ are significant.

### 3.3.20  Process X

It follows from Section 3.3.11, 3.3.17, and 3.3.10 that the process $X$ is an increasing pure jump process with independent increments. It has the representation

$$X_t(\omega) = \int_{(0,t] \times (0,\infty)} L(\omega; ds, dz) z$$

where $L$ is the restriction of $N$ onto $(0, \infty) \times (0, \infty)$. Thus, $L$ is a Poisson random measure on $(0, \infty) \times (0, \infty)$ whose mean measure is

$$\lambda(ds, dz) = (\sigma^+(ds)f_s + g_s^+ \, ds) \frac{1}{z} \exp\left(\frac{-z}{c_s}\right) dz.$$

Hence, the process $X$ is a local gamma process with shape function $p$ and scale factor $(c_s)$, where

$$p(t) = \int_0^t \sigma^+(ds)f_s + \int_0^t g_t^+ \, ds \quad t \geqslant 0$$

See Section A.3.8 of the Appendix for the definition of local gamma processes, and Çinlar[4] for more details.

### 3.3.21  Process Y

Similarly, the process $(Y_t)$ is an increasing pure jump process with independent increments, and can be represented as

$$Y_t(\omega) = \int_{(0,t] \times (0,\infty)} M(\omega; ds, dz) z$$

where $M$ is the reflection of the restriction of $N$ onto $(0, \infty) \times (-\infty, 0)$, that is, $M(B) = N(-B)$ for $B \subset (0, \infty) \times (0, \infty)$. Then, $M$ is a Poisson random measure on $(0, \infty) \times (0, \infty)$ whose mean measure is

$$\mu(ds, dz) = (\sigma^-(ds)f_s + g_s^- \, ds) \frac{1}{z} \exp\left(\frac{-z}{c_s}\right) dz$$

and it follows that $(Y_t)$ is a local gamma process with shape function

$$q(t) = \int_0^t \sigma^-(ds)f_s + \int_0^t g_s^- \, ds \quad t \geq 0$$

and scale factor $(c_s)$.

### 3.3.22 Distributions

In general, there is no explicit expression for the distributions of $X_t$, $Y_t$, $Z_t$. In the special case where $c_t = c_0$ for all $t \geq 0$ (for example, if the temperature and humidity are constant), then $X_t$ and $Y_t$ are both gamma distributed with the scale factor $c_0$ and shape parameters $p(t)$ and $q(t)$ given in Sections 3.3.20, 3.3.21, and hence their Fourier transforms are

$$\mathbb{E}[\exp(i\lambda X_t)] = (1 - ic_0\lambda)^{-p(t)}$$
$$\mathbb{E}[\exp(i\lambda Y_t)] = (1 - ic_0\lambda)^{-q(t)}$$

Then, since $X_t$ and $Y_t$ are independent,

$$\mathbb{E}[\exp(i\lambda Z_t)] = \mathbb{E}[e^{i\lambda k(t)} \exp(i\lambda X_t) \exp(-i\lambda Y_t)]$$
$$= e^{i\lambda k(t)}(1 - ic_0\lambda)^{-p(t)}(1 + ic_0\lambda)^{-q(t)}$$

The densities corresponding to $X_t$ and $Y_t$ may be found in Section A.1.6 of the Appendix; the density corresponding to $Z_t$ is basically the convolution of the two.

### 3.3.23 Means and variances

It follows from Section 3.3.20, 3.3.21, and the results on Poisson random measures (see Section A.3 of the Appendix) that

$$\mathbb{E}[X_t] = \int_0^t p(ds)c_s \qquad \mathbb{E}[Y_t] = \int_0^t q(ds)c_s$$

$$\text{Var}[X_t] = \int_0^t p(ds)c_s^2 \qquad \text{Var}[Y_t] = \int_0^t q(ds)c_s^2$$

Since $Z_t = k(t) + X_t - Y_t$,

$$\mathbb{E}[Z_t] = k(t) + \mathbb{E}[X_t] - \mathbb{E}[Y_t] = k(t) + \int_0^t r(ds)c_s$$

where, with $g_s = g_s^+ - g_s^-$, and recalling that $\sigma = \sigma^+ - \sigma^-$,

$$r(t) = p(t) - q(t) = \int_0^t \sigma(\mathrm{d}s) f_s + \int_0^t g_s \, \mathrm{d}s$$

Since $X_t$ and $Y_t$ are independent,

$$\mathrm{Var}\,[Z_t] = \mathrm{Var}\,[X_t] + \mathrm{Var}\,[Y_t] = \int_0^t |r(\mathrm{d}s)| c_s^2$$

where

$$|r(\mathrm{d}s)| = p(\mathrm{d}s) + q(\mathrm{d}s)$$

### 3.3.24  Discussion of the model

A stochastic model is an approximation of reality, and the real test of its validity lies with how well its predictions fit the real data. Unfortunately, the data available so far are inadequate for a statistical test of the two main hypotheses.

We made an effort to justify our hypotheses by reasonings based on our understanding of the physics of the phenomena involved. Since our knowledge of the microstructure contains serious gaps, since there seem to be serious differences of opinion among experts regarding it, and since we are unlikely to ever understand it as fully as required, it is important to note the following concerning our basic model.

First, just as the same mathematical model may fit more than one real phenomena, our model might be true even if our justifications are faulty. Put in purely mathematical terms, our model merely states that

$$Z_t = k(t) + X_t - Z_t \quad t \geq 0$$

where $k$ is a deterministic function, $(X_t)$ and $(Y_t)$ are local gamma processes, and $(X_t)$ and $(Y_t)$ are independent. Physically, this amounts to assuming that:

(a)  creep activity taking place in different locations at different times are stochastically independent; and
(b)  the frequency of creep activities contributing $z$ amount of deformation has the form constant $\times \exp(-|z|/\text{constant})/|z|$, where the constants depend on time.

Local gamma processes give gamma related distributions. Among infinitely divisible distributions they seem to be the most appropriate. Gaussian distributions are unacceptable because the paths $t \to Z_t(\omega)$ are of bounded variation (which is not true of processes with independent increments related to Gaussian distributions). Similarly, one can exclude all stable distributions with index greater than or equal to one. Stable distributions with index less than one are excluded, since they yield infinite expectations. Compound Poisson (and Poisson) distributions are excluded because the paths $t \to Z_t(\omega)$ should not have sizable jumps. Furthermore, some experimental data seem to collaborate this choice. Since the gamma density with shape parameter $p < 1$ is strongly non-symmetric with a lot of weight near 0, actual value of a random variable with such a density is smaller than the expected value with high probability. This may explain the observed tendency for large deviations from the mean for times close to the time of loading. For large $t$ this effect disappears since the gamma density with a large shape parameter is nearly symmetric like the Gaussian density.

Thus, we believe that among infinitely divisible distributions, gamma related ones are the most appropriate for our purposes. Although our arguments for infinite divisibility seem beyond reproach, there is a simple minded theoretical objection that is worth answering. We have claimed twice, in Section 3.2 and again in Section 3.3, that $Z_t$ ought to have an infinitely divisible distribution $\varphi$. On the other hand:

(a) obviously $|Z_t| \le 1$, and $\varphi(1) - \varphi(-1) = 1$;
(b) there is no infinitely divisible distribution $\varphi$ with bounded support, that is, $\varphi(1) - \varphi(-1) = 1$ is impossible.

This incompatibility is present in almost all applications of probability theory to physical situations; for example, for the Gaussian distribution $\gamma$ we always have $\gamma(1) - \gamma(-1) > 0$ no matter what the mean and variance are, but this does not stop us from using it as the distribution of random variables which are bounded by 1. The reasonability of our claim of infinite divisibility and the non-serious nature of the objection above to it can be seen at once if we consider the magnitudes involved. Creep deformations etc. are usually of the order $10^{-6}$, whereas the bound we are seeking is 1, therefore $\varphi(1) - \varphi(-1) = 1 - \varepsilon$ where $\varepsilon$ is quite likely to be of the order $10^{-20}$. Such a discrepancy between mathematical and physical situations cannot be considered serious.

### 3.4 BASIC CREEP CASE

Our objective is to provide some insight into the natures of the factors $k$, $f$, $g^+$, $g^-$, $c$ in Hypothesis 3.3.17 by limiting our attentions to the case of basic creep. Throughout this section, the notations, conditions, and set-up of Sections 3.2 and 3.3 are in force.

#### 3.4.1 Further conditions

In this section we assume, further, that the temperature is kept constant at some reference temperature $T_0$, that the humidity is $h_t = 1$ for all $t$ (and there is no moisture exchange), and assume a simple form for the stress function:

$$\sigma(t) = \begin{cases} 0 & \text{if } t < t_0 \\ \sigma & \text{if } t \geq t_0 \end{cases}$$

where $t_0 > 0$ is some fixed time and $\sigma$ is some fixed number. Under these conditions, in Section 3.3.17, we have

$$c_t = c_0 \qquad \sigma^-(t) = 0 \qquad \sigma^+(t) = \sigma(t)$$

$f_t$ depends only on $a_t$ and $V_t$, and $g_t^+$ and $g_t^-$ depend only on $\sigma_t$, $a_t$, and $V_t$.

### 3.4.2 Negligibility of $g_t^-$

If $g_t^-$ is ever strictly positive, in the decomposition of $(Z_t)$ given in Section 3.3.19, the process $(Y_t)$ will not vanish. Consequently, the deformation trajectories $t \to Z_t(\omega)$ will have downward jumps. Even though each such jump is of molecular dimensions, over time they are likely to add to something noticeable at the macroscopic scale. Such is known not to be the case with basic creep. Hence, $g_t^-$ must be negligibly small for all $t$, and we may exclude the term $g_t^-$ from further considerations. Then, the basic decomposition of the deformation process $(Z_t)$ becomes

$$Z_t = k(t) + X_t$$

where $X$ is as described in Section 3.3.20.

### 3.4.3 Deterministic term $k(t)$

The total deformation is the simple sum of the deformations taking place in the space occupied by micropores and those taking place in the space occupied by aggregates and the sheets of solid matter making up the cement gel. Since the deformations in the micropore space are accounted for in the term $X_t$, the term $k(t)$ is due to the deformations associated in the aggregates plus the sheets.

The load carried by aggregates is $\sigma(t) - \sigma_t$ at time $t$, and the volume of the aggregates is some constant $V_a$ (which is also the average area since the height is unity). The load carried by the cement gel is $\sigma_t$, and its volume is $V_t$ at time $t$. (Note that total volume is $V_a + V_t + V_m$ where $V_m$ is the volume of space occupied by macropores.) Thus,

$$k(t) = \frac{\sigma(t) - \sigma_t}{E_a V_a} + \frac{\sigma_t}{E_c V_t}$$

where $E_a$ and $E_c$ are the modulus of elasticities for the aggregates and *cement sheets* respectively. The terms $E_a$ and $E_c$ themselves depend on the concrete mix and the concrete making process itself; therefore, their constancy in this model (and thus the deterministic character of $k(t)$) is because we have suppressed the randomness introduced by the concrete technology. See Section 3.7 for more on this point.

This still leaves $\sigma_t$ to be determined. It is clear that $\sigma_t = 0$ for $t < t_0$, and $\sigma_{t_0}$ should be almost equal to $\sigma(t_0) = \sigma$ since almost all of the applied load is carried by the cement gel at the time $t_0$ of application. For $t > t_0$, $\sigma_t$ must be decreasing in $t$. Some further thoughts on this will be put in Section 3.5.

### 3.4.4 Effect of $f$

Since we are taking $\sigma(dt) = \sigma^+(dt)$ to be 0 for all $t$ except $t = t_0$, the contribution of the term $f$ to the total deformation is some random amount $X_{t_0}^f$ at time $t_0$ (which remains the same for all $t > t_0$). It follows from Section 3.3.20 (see Section A.3.8 in the Appendix and p. 471 in Çinlar[4]) that $X_{t_0}^f$ has the gamma distribution with shape parameter $p(dt_0) = \sigma^+(dt_0)f_{t_0} = \sigma f_{t_0}$ and scale factor $c_{t_0} = c_0$; thus

$$\mathbb{E}[X_{t_0}^f] = \sigma f_{t_0} c_0 \qquad \mathrm{Var}\,[X_{t_0}^f] = \sigma f_{t_0}(c_0)^2$$
$$\mathbb{E}[\exp(i\lambda X_{t_0}^f)] = (1 - ic_0\lambda)^{-\sigma f_{t_0}}$$

We set

$$X_t^f = \begin{cases} 0 & \text{if } t < t_0 \\ X_{t_0}^f & \text{if } t \geq t_0 \end{cases}$$

The process $(X_t^f)$ represents the effect of the forcing function $f$ upon the process $(X_t)$.

### 3.4.5 Effect of $g^+$

Since the effect of $g^-$ is only on the process $(Y_t)$ even when $g^-$ is non-negligible, we may denote the effect of $g^+$ on $(X_t)$ by $(X_t^g)$ without causing ambiguities. Clearly, we have

$$X_t = X_t^f + X_t^g$$

and the stochastic processes $(X_t^f)$ and $(X_t^g)$ are independent.

It follows from Section 3.3.20 that $(X_t^g)$ is a local gamma process with shape function $t \rightarrow \int_0^t g_s^+ \, ds$ and scale factor $c_t = c_0$. Thus,

$$\mathbb{E}[X_t^g] = \int_0^t g_s^+ c_0 \, ds \qquad \mathrm{Var}\,[X_t^g] = \int_0^t g_s^+(c_0)^2 \, ds$$

and

$$\mathbb{E}[\exp(i\lambda X_t^g)] = \exp\left(\int_0^t g_s^+ \log(ic_s\lambda - 1)\right) = (1 - ic_0\lambda)^{-\gamma^+(t)}$$

where $\gamma^+(t) = \int_0^t g_s^+ \, ds$. Because $c_s = c_0$ for all $s$, the distribution of $X_t^g$ is the gamma distribution with parameters $\gamma^+(t)$ and $c_0$ for shape and scale factor.

### 3.4.6 Summary

In the case of basic creep with a stress function which has only one jump, we see that we can decompose the deformation process $Z$ as

$$Z_t = k(t) + X_t^f + X_t^g$$

Here, $k$ is *deterministic elastic* response, $(X_t^f)$ is the random *instantaneous* response to loading, and $(X_t^g)$ is the random *creep* response to loading.

## 3.5 SHRINKAGE AND THERMAL DILATATIONS

Our objective is to provide some further insight into the model outlined in Section 3.3 by limiting ourselves to a case that is totally opposite to that of basic creep. Throughout this section, the notations, conditions and set-up of Sections 3.2 and 3.3 are still in force.

### 3.5.1 Further conditions

In this section we assume that there is no load on the concrete, that is, $\sigma(t) = 0$ for all $t$. Then, $\sigma_t = 0$ for all $t$ also.

It follows that the contribution of $f$ vanishes, and we are left with

$$Z_t = k(t) + X_t^g - Y_t^g \quad t \geq 0$$

where $k(t)$ is the deterministic 'dilatation' response to temperature and humidity, $(X_t^g)$ is the random *shrinkage* response, and $(Y_t^g)$ is the random *dilatation* response to temperature and humidity variations over time.

### 3.5.2 Independence of $X^g$ and $Y^g$

The process $(X_t^g)$ and $(Y_t^g)$ are stochastically independent since $(X_t)$ and $(Y_t)$ are so (see Section 3.3.19), as long as the humidity and temperature are deterministic, that is, are known with certainty, as we have been assuming. The distinction between stochastic and deterministic independence is worth keeping in mind here, since the latter is of course not true.

It is worth noting that

$$\mathbb{E}[Z_t] = k(t) + \mathbb{E}[X_t^g] - \mathbb{E}[Y_t^g]$$

may be continuously increasing (so that, in deterministic terms, there is only shrinkage) even though the deformation trajectory $t \to Z_t(\omega)$ fluctuates wildly for any outcome $\omega$.

### 3.5.3  Process $X^g$

This is a local gamma process with shape function $\int g_t^+$ and scale factor $c_t$. Of course, in the present case, $g_t^+$ is a function of $a_t$, $V_t$, $T_t$, $h_t$, $\dot{h}_t$ only. Thus,

$$\mathbb{E}[X_t^g] = \int_0^t g_s^+ c_s \, ds \qquad \text{Var}\,[X_t^g] = \int_0^t g_s^+ c_s^2 \, ds$$

and

$$\mathbb{E}\,[\exp{(i\lambda X_t^g)}] = \exp{\left( \int_0^t g_s^+ \log{(ic_s\lambda - 1)}\, ds \right)}$$

Since $c_t = c(T_t, h_t)$ depends on $t$ now, the distribution of $X_t^g$ cannot be stated explicitly.

### 3.5.4  Process $Y^g$

This is again a local gamma process; its shape function is $\int g_s^-$ and scale factor $c_t$. Expressions for $\mathbb{E}[Y_t^g]$, $\text{Var}\,[Y_t^g]$, and $\mathbb{E}[\exp{(i\lambda Y_t^g)}]$ can be obtained from the respective formulae above for $X_t^g$ upon replacing $g^+$ by $g^-$ throughout.

## 3.6  REMARKS ON PARAMETERS

Our aim is to discuss the functional forms of parameter functions $f_t$, $g_t^+$, $g_t^-$, and $c_t$ in the hope that their meanings become clearer. This should also help other workers in correcting our mistakes.

### 3.6.1  General comments

The formulae for expected values and variances given in Sections 3.3, 3.4, and 3.5 are always in terms of the parameter functions, and there is no randomness in them. Therefore, such formulae should be helpful in determining the forms of the parameter functions through deterministic mechanical considerations and statistical techniques. In particular, when $c_t = c_0$ is constant, such considerations are exactly as in deterministic studies. Note that our stochastic model does not eliminate the need for macroscopic deterministic studies. Also, note that the stochastic model here is not any more difficult to use than the deterministic ones: if one has a good understanding of the *expected behaviour*, all one needs is the extra term $c_t$ (which can be solved from the variance formulae) in order to completely specify the distributions of the random variables involved.

### 3.6.2   Scale factor $c_t$

Going back to the discussions of Sections 3.3.15–17, we note that $1/c_t$ is the *rate of decrease* of the *frequency* of migrations of size effect y for large y. Thus, it is related to the *variability* of the number of particles migrating as a group. This in turn is related to the number of *particles* that are likely to be at a micropore.

Of course, the number of particles available for migrations at a micropore is random; the distribution of that number depends on humidity. Different particles at a micropore have different random energies; the distribution of the energy of a particle is well known to be Maxwellian with the mean equal to temperature. It follows that $c_t$, which is an average over all micropores, should depend on the humidity $h_t$ and temperature $T_t$.

The reasoning above might suggest that $c_t$ also depends on $\sigma_t$, but it is not so. Of course, the transverse pressure at time $t$ is some random amount whose expected value is $\sigma_t$, and the transverse pressure is the force behind the migrations. But, the way we visualize it, the number of particles at a location (available for migration) is determined previous to the time at which the transverse pressure reaches a peak to force some of those particles out. Further, when the pressure reaches a peak, it forces a *proportion* of the particles to migrate. This proportion does depend on the pressure, but the variability by size cannot depend on it.

In Çinlar[4] and Çinlar *et al.*[5], it was expressed that the term $c_t$ would depend on $t$ even in the case of basic creep, even though the statistical work in Çinlar *et al.*[5] indicated that it can be taken to be constant. We are changing our view on this matter.

### 3.6.3   Factor $f$

The factor $f_t$ is a measure of the instantaneous response to a load of unit magnitude applied at $t$ (more precisely, $f_t c_t$ is the expected value of that random instantaneous response). Therefore, $f_t$ is a function of $a_t$, $V_t$, $T_t$, $h_t$ but not of $\sigma_t$ and $\dot{h}_t$. Hence, $f$ should be picked so that

$$f(a, V, T, h) = g^+(\sigma, a, V, T, h, \dot{h}) \quad \text{when } \sigma = 1, \dot{h} = 0$$

In using the same factor $f_t$ whether the load is compressive or opposite, we are implicitly assuming that responses in both cases are the same in magnitude (but, of course, not in direction). This may be faulty. If the instantaneous response to tensile stresses is markedly different, we should have

$$\alpha_y(dt) = \begin{cases} \sigma^+(dt)f_t^+ + g_t^+\, dt & \text{if } y > 0 \\ \sigma^-(dt)f_t^- + g_t^-\, dt & \text{if } y < 0 \end{cases}$$

in Section 3.3.17. The appropriate changes elsewhere are easy to do.

### 3.6.4  Factors $g^+$, $g^-$

We do not have much to say on this except that the dependence of $g^+$ on $\sigma_t$, $T_t$, $h_t$, $\dot{h}_t$ is probably very difficult to obtain. For, $g_t^+$ is a measure of the rate of *migrations causing de-bonding*, whereas $g_t^-$ is a measure of the rate of *migrations causing bonds*. When the conditions are ripe, within the same concrete, there will be *local* compressions and dilatations simultaneously. Such local variations are likely to be at the *macro*scale, say, within regions of mm$^3$ size. But how such things can be observed eludes me. Entirely similar remarks hold for $g^-$.

### 3.6.5  Stress $\sigma_t$

This is a troublesome quantity. It is the average of the transverse pressure field within the cement gel, and by that token, should be a deterministic quantity as we have been assuming all along. On the other hand, it is somehow related to creep, and could therefore be random. To settle the matter, suppose the same load is applied to a large number of identical specimens kept under identical environmental factors. By averaging the creep values obtained for different specimens we obtain average transverse pressures through some computations. Now consider the 100 specimens whose creep values are greatest, and the 100 specimens whose creep values are the lowest. Is $\sigma_t$ the same for both groups? Above, we have been assuming that the answer is yes.

### 3.6.6  Volume $V_t$

By this we mean the total volume less the volume of the space occupied by by aggregates and macropores. Thus, $V_t$ *includes* the volume occupied by *micropores*.

### 3.6.7  Aging factor $a_t$

This is best thought of as an abstract concept that simplifies computations. Surely, no amount of microscopic examination of a given cement gel (even if it were physically possible) will be able to determine some number $a_t$ for that cement gel. But one can give it an almost physical meaning. For each molecule $i$ making up the sheets of cement gel, let $\tau_i$ be the random time of its formation. Then, at time $t$, the 'average age' is

$$\frac{\sum_i \tau_i 1_{(0,t]}(\tau_i)}{\sum_i 1_{(0,t]}(\tau_i)}$$

Presumably this should be done first assuming basic creep conditions and no load. The average value so obtained will be the *reference age*, denoted by $a(t)$. Then, one could look for time transformations that relate the aging under differing conditions of stress, temperature, etc. to that reference age.

For example, taking $h_t$ and $T_t$ variable, but no stresses, the age at time $t$ in our formulation would be

$$a_t = a(s_t)$$

where 'equivalent time' $s_t$ is

$$s_t = \int_0^t \beta_T \beta_h \, dt$$

for some functions $\beta_T$ and $\beta_h$ that are examined before elsewhere; see Bažant,[1] p. 26.

### 3.6.8 Principle of superposition

This refers to the additivity of deformations due to additive stresses in deterministic terms. In stochastic terms, it can be rephrased as follows.

Let $\sigma_1(t)$ and $\sigma_2(t)$ be two stress functions, and let $\Phi_1$ and $\Phi_2$ be the joint distribution of $(Z_{t_1}, \ldots, Z_{t_n})$ under the stress conditions $\sigma_1(t)$ and $\sigma_2(t)$ respectively. Then, the joint distribution of $(Z_{t_1}, \ldots, Z_{t_n})$ under the stress condition $\sigma(t) = \sigma_1(t) + \sigma_2(t)$ is

$$\Phi = \Phi_1 * \Phi_2$$

Here the asterisk denotes convolution:

$$\Phi(B) = \int_{\mathbb{R}^n} \Phi_1(dx)\Phi_2(B - x) \quad B \subset \mathbb{R}^n, \text{ Borel}$$

This can be restated in intuitive terms as follows: the deformation process under the condition $\sigma(t) = \sigma_1(t) + \sigma_2(t)$ has the same probability law as the *sum* of two independent deformation processes, one of which is subjected to $\sigma_1(t)$ and the other to $\sigma_2(t)$.

In order for this principle to hold, in our model the following must hold:

$$\sigma_t = \sigma(t)\gamma_t \qquad g_t^+ = \sigma(t)\gamma_t^+ \qquad g_t^- = \sigma(t)\gamma_t^-$$

where neither $\gamma_t$ nor $\gamma_t^+$ nor $\gamma_t^-$ depend on stresses (not even through aging or volume). These conditions are not fulfilled, and the principle of superposition does not hold in general. However, the principle might prove useful as an approximation of reality.

## 3.7 RANDOMIZATION OF CONDITIONS

Our objective is to mention briefly the general problem involved when the environmental conditions and/or stress conditions are random.

### 3.7.1 Problem

When any one of the temperature, humidity, or stress is random, *none* of the *independence* hypotheses or results is true. Moreover, *none* of the formulae for expected values, or variances, or Fourier transforms is true.

### 3.7.2 Conditioning

To explain the principle involved, assume that only temperature is random. So, we assume $(T_t)$ is a stochastic process. Suppose we have computed, for example, the expected value of the deformation $Z_t$ under the conditions of Sections 3.2 and 3.3 for every possible trajectory $w(t)$, $t \geq 0$, that the *temperature process* might follow. For each such trajectory $w$, we will then have some number that we denote by

$$\eta_t(w) = \mathbb{E}[Z_t \mid T = w]$$

read the *conditional expectation* of $Z_t$ given that the temperature $T_u$ is equal to $w(u)$ for all $u \geq 0$. This is a functional of $w$ (a function of a function), which depends on $w$ only through its values $w(u)$ for $u \leq t$.

    If the outcome $\omega$ turns up, the trajectory for temperature is $T(\omega): u \to T_u(\omega)$, and therefore, for someone who knows the temperatures but has no knowledge of the deformations, the expectation of the deformation at time $t$ is

$$\hat{Z}_t(\omega) = \eta_t(T(\omega))$$

For someone who does not know the temperature history, the expected value of $Z_t$ would be the weighted average of $\hat{Z}_t(\omega)$ where the mass at $\omega$ is $\mathbb{P}(\mathrm{d}\omega)$. So,

$$\mathbb{E}[Z_t] = \int_\Omega \mathbb{P}(\mathrm{d}\omega)\hat{Z}_t(\omega) = \mathbb{E}[\hat{Z}_t]$$

We simplify all this by writing

$$\hat{Z}_t = \mathbb{E}[Z_t \mid \mathscr{T}]$$

where $\mathscr{T}$ stands for 'temperature history'. Then, we can write

$$\mathbb{E}[Z_t] = \mathbb{E}[\mathbb{E}[Z_t \mid \mathscr{T}]]$$

### 3.7.3 Sample computation

Suppose all the conditions of the basic creep case discussed in Section 3.4 except that the load applied, $\sigma$, is now considered random. Let us denote it by $S$. Suppose that its distribution

$$\Psi(\sigma) = \mathbb{P}\{S \leq \sigma\} \quad \sigma \geq 0$$

is known. For simplicity, let us further assume that the principle of super-position holds, so that $\sigma_t = \sigma(t)\gamma(t)$, $g_t^+ = \sigma(t)\hat{g}_t$ where $\hat{g}_t$ does not depend on stresses. Then, for the deformation at $t \geq t_0$ we have

$$\mathbb{E}[Z_t \mid S] = \frac{S - S\gamma(t)}{E_a V_a} + \frac{S\gamma(t)}{E_c V_t} + Sf_{t_0} c_0 + S \int_{t_0}^{t} \hat{g}_s c_0$$

where the three terms on the right are obtained from the formulae of Sections 3.4,3, 3.4.4, and 3.4.5 for the expectations by merely putting in $S$ for $\sigma$. Now, this has the form

$$\mathbb{E}[Z_t \mid S] = S \cdot m_t \quad t \geq t_0$$

where $m_t$ is a constant. Thus,

$$\mathbb{E}[Z_t] = \mathbb{E}[m_t S] = m_t \mathbb{E}[S] = m_t \int_0^{\infty} \Psi(d\sigma)\sigma \quad t \geq t_0$$

To give another example, we now compute the Fourier transform of $Z_t$. First, for the deterministic stress case, we note that

$$k(t) = \sigma k_t \qquad \gamma^+(t) = \sigma \gamma_t^+ = \sigma \int_{t_0}^{t} \hat{g}_s$$

where $k_t$ is chosen appropriately. So, for $t \geq t_0$,

$$\mathbb{E}[\exp(i\lambda Z_t) \mid S] = [\exp(i\lambda S k_t)](1 - ic_0 \lambda)^{-f_{t_0} S}(1 - ic_0 \lambda)^{-\gamma + S}$$

since in the deterministic case this is what we have with $\sigma$ replacing $S$. Note that this has the form

$$\mathbb{E}[\exp(i\lambda Z_t) \mid S] = [\exp(-i\lambda k_t)(1 - ic_0 \lambda)^{+f_{t_0} + \gamma_t^+}]^{-S} = (b_t(\lambda))^{-S}$$

so, if

$$\hat{\Phi}(\mu) = \int_0^{\infty} \Phi(d\sigma)e^{-\mu\sigma}$$

$$\mathbb{E}[\exp(i\lambda Z_t)] = \mathbb{E}[\mathbb{E}[\exp(i\lambda Z_t) \mid S]]$$

$$= \int_0^{\infty} \Phi(d\sigma) \exp[-S \log b_t(\lambda)] = \hat{\Phi}(\log b_t(\lambda))$$

### 3.7.4  General solution

In general, whatever factors are taken to be random, the procedure illustrated above can be used. First, we compute the quantity we are interested in under the condition that everything is deterministic. In the resulting expressions, replace every deterministic term by its random equivalent. Then, compute the expected value of the random variable so obtained.

Above, we have mentioned randomizations of temperature, humidity, stress. In practice, one will need further randomizations to take into account the randomness of the actual concrete mix (as opposed to the specified), and the actual process of concrete making.

## APPENDIX

### A.1  Probability spaces

Our aim is to review some elementary notions and fix some terminology. Most of what follows is standard.

#### A.1.1  Basic notions

The basic notion is that of a random experiment, an experiment whose outcome cannot be told in advance. It is modelled by a *probability space:* a triple $(\Omega, \mathcal{H}, \mathbb{P})$ where $\Omega$ is the set of all possible outcomes, $\mathcal{H}$ is a $\sigma$-algebra of subsets of $\Omega$, and $\mathbb{P}$ is a probability measure on $\mathcal{H}$. The elements of $\Omega$ are called outcomes; the generic element is denoted by $\omega$. The elements of $\mathcal{H}$ are called events; events are denoted by letters such as $H$ (capital eta). Every event is a subset of $\Omega$, but not every subset is an event. The probability measure $\mathbb{P}$ assigns a number $\mathbb{P}(H)$, called the probability of $H$, to every event $H$.

Intuitively, $\mathbb{P}(H)$ is a measure of the chance that the outcome of the random experiment belongs to $H$. In keeping with this interpretation, $\mathbb{P}(H)$ is a number in $[0, 1]$ for any event $H$, $\mathbb{P}(\Omega) = 1$, and $\mathbb{P}(H) = \sum_i \mathbb{P}(H_i)$ whenever the event $H$ is the union of disjoint events $H_1, H_2, \ldots$. As a consequence, the collection $\mathcal{H}$ of all events must have some closure properties: $\Omega$ must be an event; if $H$ is an event then so must its complement $\Omega \backslash H$; if $H_1, H_2, \ldots$ are events then so must their union $H = H_1 \cup \cdots$. By the phrase '$\mathcal{H}$ is a $\sigma$-algebra on $\Omega$' is meant precisely that $\mathcal{H}$ has these properties.

#### A.1.2  Random variables and stochastic processes

A real random variable is a function $X$ from $\Omega$ into $\mathbb{R} = (-\infty, \infty)$ such that $\{\omega \in \Omega : X(\omega) \leq x\}$ is an event for every $x$ in $(-\infty, \infty)$. A vector valued random variable is a mapping from $\Omega$ into $\mathbb{R}^d$ such that each component is a real random variable. A real valued stochastic process is a function $t \to X_t$ where each $X_t$ is a real random variable. The process $t \to X_t$ is said to be *positive* if $X_t(\omega) \geq 0$ for all $\omega$, *increasing* if $X_t(\omega) \leq X_u(\omega)$ for all $t < u$ and all $\omega$.

## A.1.3  Distributions

For a real random variable $X$, the numbers

$$\Phi(x) = \mathbb{P}\{X \leq x\} \quad -\infty < x < \infty$$

(where the right-hand side is short for $\mathbb{P}(\{\omega \in \Omega : X(\omega) \leq x\})$) are well defined, and the resulting function $\Phi$ on $(-\infty, \infty)$ is called the *distribution* function of $X$. If $\Phi$ is differentiable, then its derivative $\phi$ is called the *density* function of $X$.

## A.1.4  Expectations

If $X$ is a real random variable and if $f$ is a bounded Borel function on $(-\infty, \infty)$, then the *expected value* of $f(X)$ is

$$\mathbb{E}[f(X)] = \int_{\Omega} \mathbb{P}(d\omega) f(X(\omega)) = \int_{-\infty}^{\infty} \Phi(dx) f(x)$$

where the integrals are Lebesgue–Stieltjes type—that the two are equal can be shown. If $\Phi$ has a density, say $\phi$, then the last integral reduces to the ordinary integral $\int \phi(x) f(x)\, dx$. If $f$ is complex valued, say $f = f_1 + if_2$, then $\mathbb{E}[f(X)] = \mathbb{E}[f_1(X)] + i\mathbb{E}[f_2(X)]$.

If $f(x) = x$ we get the expected value, or mean, of $X$; if $f(x) = x^n$ we get the $n$th moment of $X$; if $f(x) = (x - \mu)^2$ where $\mu$ is the mean of $X$ we get the variance of $X$; if $f(x) = e^{i\lambda x}$ we get the Fourier transform of $X$. By the uniqueness theorems for Fourier transforms, since

$$\mathbb{E}[e^{i\lambda X}] = \int_{-\infty}^{\infty} \Phi(dx) e^{i\lambda x}$$

the Fourier transform of $X$ determines the distribution function of $X$ uniquely. If $X$ is positive, it is more convenient to use *Laplace transforms*: $\mathbb{E}[e^{-\lambda X}]$. The following are some examples.

## A.1.5  Poisson distribution

A random variable $X$ is said to have the Poisson distribution with mean $a$ if $X$ takes only the values $0, 1, 2, \ldots$ and

$$\mathbb{P}\{X = k\} = \frac{e^{-a} a^k}{k!} \quad k = 0, 1, \ldots.$$

Then,

$$\mathbb{E}[X] = a \quad \text{Var}[X] = a \quad \mathbb{E}[e^{i\lambda X}] = \exp[a(e^{i\lambda} - 1)]$$

### A.1.6   Gamma distribution

A random variable $X$ is said to have the gamma distribution with *shape parameter p* and *scale factor c* if $X$ takes values in $[0, \infty)$ and its density is

$$\phi(x) = \frac{e^{-x/c}(x/c)^{p-1}}{c\Gamma(p)} \qquad 0 \le x < \infty$$

where $\Gamma$ is the gamma function. Here $p > 0$ and $c > 0$ are fixed. Then, it is easy to show that

$$\mathbb{E}[X^n] = \frac{\Gamma(p+n)}{\Gamma(p)} c^n \qquad \mathbb{E}[e^{i\lambda X}] = (1 - ic\lambda)^{-p}$$

and in particular, since $\Gamma(x) = x\Gamma(x-1)$, we have

$$\mathbb{E}[X] = pc \qquad \mathrm{Var}[X] = pc^2$$

It is easy to show that $X = cY$, where $Y$ has the gamma distribution with shape parameter $p$ and scale factor 1. This explains why $c$ should be called the scale factor.

The shape of the density function $\phi$ is determined by $p$: if $p < 1$, then $\phi$ is strictly decreasing with $\phi(0+) = +\infty$ and $\phi(+\infty) = 0$; if $p = 1$, then $\phi$ is strictly decreasing from $\phi(0) = 1/c$ to $\phi(+\infty) = 0$; if $p > 1$, then $\phi(0) = 0$, increasing at first, and then decreasing to $\phi(\infty) = 0$. For $p$ large, $\phi$ looks like the normal density.

## A.2   Infinite divisibility

This concept seems to be intrinsically related to random deformations of materials. The following is a quick review. Throughout, $(\Omega, \mathcal{H}, \mathbb{P})$ is an appropriately large probability space.

### A.2.1   Independence

Two real random variables $X$ and $Y$ are said to be *independent* if

$$\mathbb{P}\{X \le x, \ Y \le y\} = \mathbb{P}\{X \le x\}\mathbb{P}\{Y \le y\}$$

for all $x$ and $y$. Intuitively, this means that knowing the value of $X$ does not help in guessing the quantities related to $Y$. The concept is very useful because, usually, one can ascertain the independence of $X$ and $Y$ without computations by looking into the random phenomena giving rise to $X$ and $Y$. Then, the formula above reduces the task of specifying the joint distribution (which is the left-hand side) to that of specifying the individual distribution functions. Also, it follows from the independence of $X$ and $Y$ that

$$\mathbb{E}[f(X)g(Y)] = \mathbb{E}[f(X)]\mathbb{E}[g(Y)]$$

for any bounded Borel functions $f$ and $g$. Taking $f(x) = g(x) = e^{i\lambda x}$, we see that

$$\mathbb{E}[e^{i\lambda(X+Y)}] = \mathbb{E}[e^{i\lambda X}]\mathbb{E}[e^{i\lambda Y}]$$

for any independent $X$ and $Y$, thus giving us a valuable formula for finding the Fourier transform (and through it, the distribution function) of $X + Y$.

The above notions and ideas can be extended to an arbitrary number of random variables.

### A.2.2  Infinite divisibility

A random variable $X$ is said to be infinitely divisible if for every integer $n$ there exist independent random variables $Y_1, \ldots, Y_n$ having the same distribution such that $X = Y_1 + \cdots + Y_n$. Then, if $\Phi_\lambda$ is the Fourier transform of $X$, for every $n$, $(\Phi_\lambda)^{1/n}$ must also be the Fourier transform of some distribution. This restricts the form that $\Phi_\lambda$ can have. All such $\Phi_\lambda$ are characterized by the Lévy–Khinchine formula. In particular, if $X$ is positive and infinitely divisible, then its Fourier transform has the form

$$\mathbb{E}[e^{i\lambda X}] = \exp\left(i\lambda k + \int_0^\infty \mu(dx)(e^{i\lambda x} - 1)\right)$$

where $k \geq 0$ is a constant and $\mu$ is a measure on $(0, \infty)$ satisfying $\int \mu(dx) \min(x, 1) < \infty$.

If $X$ has the Poisson distribution with mean $c$, then it is infinitely divisible, and in the formula above we have $k = 0$ and $\mu(dx) = c\delta_1(dx)$ where $\delta_1$ is the Dirac measure putting all its mass at 1 (so that $\int \delta_1(dx)f(x) = f(1)$ for any $f$).

If $X$ has the gamma distribution with shape parameter $p$ and scale factor $c$, then $X$ is infinitely divisible and we have $k = 0$ and

$$\mu(dx) = p\frac{e^{-x/c}}{x}\,dx \quad 0 < x < \infty$$

Another well known infinitely divisible distribution is the Gaussian one. A random variable $X$ with that distribution takes values in $(-\infty, \infty)$, and the above representation does not apply. If such an $X$ has mean 0 and variance 1, then its Fourier transform is $\Phi_\lambda = \exp(-\lambda^2/2)$. If mean is $k$ and variance $v^2$ then $\Phi_\lambda = \exp(i\lambda k - \lambda^2/2v^2)$.

### A.2.3  Processes with independent increments

A stochastic process $(X_t)_{t\geq 0}$ is said to have independent increments if $X_{t_1} - X_{t_0}, X_{t_2} - X_{t_1}, \ldots, X_{t_n} - X_{t_{n-1}}$ are independent random variables for any $n$ and any times $0 \leq t_0 < t_1 < t_2 < \cdots < t_n$. The probability law of such a process is then easy to specify once the distribution of $X_u - X_t$ is known for all $t < u$.

If, further, the distribution of $X_u - X_t$ is the same as that of $X_{u-t} - X_0$ for all $0 \le t < u$, then $(X_t)$ is said to have stationary and independent increments. If $(X_t)$ has stationary and independent increments, then every $X_t$ and every increment $X_u - X_t$ has an infinitely divisible distribution. This is obvious: if points $t = t_0 < t_1 < \cdots < t_n = u$ are picked equidistant, then $X_u - X_t = (X_{t_1} - X_{t_0}) + (X_{t_2} - X_{t_1}) + \cdots + (X_{t_n} - X_{t_{n-1}})$ and the terms on the right are independent and identically distributed. In the converse direction, given an infinitely divisible distribution $\Phi$, there is always a process $(X_t)$ with stationary and independent increments such that $X_1$ has the distribution $\Phi$.

If $(X_t)$ has independent (not necessarily stationary) increments, each increment $X_u - X_t$ has an infinitely divisible distribution under the condition that $P\{X_t \ne X_{t-}\} = 0$ for every $t$ (that is, the probability of a discontinuity at a pre-fixed time is zero). If $P\{X_{t-} \ne X_t\} > 0$ for some $t$, the same holds provided that the distribution of $X_t - X_{t-}$ is infinitely divisible for every such $t$. We will construct some examples of such processes in the next section.

## A.3 Poisson random measures

These play an important role as building blocks for other processes, and directly, as counting measures for random phenomena involving populations. As usual, $(\Omega, \mathcal{H}, \mathbb{P})$ is some fixed, appropriate, probability space.

### A.3.1 Random points

Let $D$ be a Borel subset of some $d$-dimensional Euclidean space. In our applications, $D = (0, \infty)$, $D = (0, \infty) \times (0, \infty)$ come up often. Let $Y_1, Y_2, \ldots$ be random variables taking values in $D$. For each $\omega \in \Omega$ and Borel subset $B$ of $D$, then

$$M(\omega, B) = \sum_n 1_B(Y_n(\omega))$$

is the number of points $Y_n(\omega)$ belonging to $B$. Then, for $\omega$ held fixed, $B \to M(\omega, B)$ is a measure on $D$ taking values like $0, 1, 2, \ldots$ only. Hence, the collection $M$ is called a *random counting* measure.

### A.3.2 Poisson random measures

Let $m$ be a positive $\sigma$-finite measure on $D$. The random counting measure $M$ is said to be a Poisson random measure with mean measure $m$ if

$$\mathbb{P}\{M(B) = n\} = \frac{e^{-m(B)} m(B)^n}{n!} \quad n = 0, 1, \ldots$$

for every Borel $B \subset D$. Thus, for every Borel subset $B$, the number of points belonging to $B$ has the Poisson distribution with mean $m(B)$. See Section A.1.5 of this Appendix.

In order for a random counting measure $M$ to be Poisson it is necessary and sufficient that:

(a)   there be no multiplicities, that is, $P\{Y_i = Y_j\} = 0$ for all $i, j$;
(b)   $M(B_1), \ldots, M(B_n)$ be independent whenever the Borel sets $B_1, \ldots, B_n$ are disjoint.

### A.3.3   Poisson processes

If $D = (0, \infty)$ and if $M$ is a Poisson random measure on $D$ with mean measure $m$, then $N_t = M((0, t])$ is a random variable with mean $a_t = m((0, t])$. Then, the stochastic process $t \to N_t$ is called a non-stationary Poisson process with mean function $(a_t)$. The process $(N_t)$ has independent increments (by virtue of independence of $M(B_1), \ldots, M(B_n)$ for $B_1 = (0, t_1]$, $B_2 = (t_1, t_2], \ldots, B_n = (t_{n-1}, t_n]$). If $a_t = at$ for some constant $a$, then $(N_t)$ is further stationary.

### A.3.4   Integrals

Let $M$ be a Poisson random measure on $D$ with mean measure $m$. Then, for any Borel function $f: D \to [0, \infty)$ and Borel set $B \subset D$,

$$X(\omega) = \int_B M(\omega, dx) f(x) = \sum_i 1_B(Y_i(\omega)) f(Y_i(\omega))$$

defines a random variable $X$.

If $f$ takes only finitely many values, say $v_1, \ldots, v_n$ on $B_1, \ldots, B_n$ respectively, then $X = v_1 M(B \cap B_1) + \cdots + v_n M(B \cap B_n)$, where $X_1 = M(B \cap B_1), \ldots, X_n = M(B \cap B_n)$ are independent Poisson distributed random variables with means $a_1 = m(B \cap B_1), \ldots, a_n = m(B \cap B_n)$. Thus, $\mathbb{E}[X] = v_1 a_1 + \cdots + v_n a_n$ which we recognize as the integral of $f$ with respect to the measure $m$ over the set $B$. Similar computations using the results listed in Section 3.1.5 yield

$$\mathbb{E}[X] = \int_B m(dx) f(x) \qquad \text{Var}[X] = \int_B m(dx) f(x)^2$$

and

$$\mathbb{E}[e^{i\lambda X}] = \exp\left( \int_B m(dx)(e^{i\lambda f(x)} - 1) \right)$$

Approximating arbitrary positive Borel functions by such simple functions and performing the limits, we obtain that the formulae above are *true for arbitrary* positive *f*.

### A.3.5 Finiteness of integrals

In the set-up of Section 3.3.4, for *f* positive, we get the expression for $\mathbb{E}[e^{-\lambda X}]$ upon replacing $i\lambda$ by $-\lambda$ in the Fourier transform. Now, $X < \infty$ with probability one if and only if $\mathbb{E}[e^{-\lambda X}]$ increases to 1 as $\lambda$ decreases to 0. This yields, since $e^{-f(x)} - 1$ can be bounded from above or below by $c \min (f(x), 1)$ by choosing the constant $c$ appropriately,

$$\mathbb{P}\{X < \infty\} = 1 \Leftrightarrow \int_B m(\mathrm{d}x) \min (f(x), 1) < \infty$$

This is useful in extending the integrals $\int M(\mathrm{d}x)f(x)$ to non-positive *f*. For arbitrary Borel function *f*, writing $f = f^+ - f^-$ where both $f^+$ and $f^-$ are positive, we have $\int M(\mathrm{d}x)f(x)$ is well defined and finite if both $\int M(\mathrm{d}x)f^+(x)$ and $\int M(\mathrm{d}x)f^-(x)$ are finite. Moreover, in that case, the formulae of Section 3.3.4 for expectations and Fourier transforms still hold as they stand.

### A.3.6 Increasing processes with independent increments

Let *M* be a Poisson random measure on $D = (0, \infty) \times (0, \infty)$ with some mean measure *m* satisfying

$$\int_{(0,t] \times (0,\infty)} m(\mathrm{d}s, \mathrm{d}y) \min (y, 1) < \infty, \quad t > 0.$$

Then, taking $f(s, y) = y$ for all $s$ and setting $B = (0, t] \times (0, \infty)$ within the set-up of Section 3.3.4 above, it follows from Section 3.3.5 that

$$X_t(\omega) = \int_{(0,t] \times (0,\infty)} M(\omega; \mathrm{d}s, \mathrm{d}y)y$$

defines a finite valued random variable $X_t$ for each *t*. Clearly, $t \to X_t(\omega)$ is increasing right continuous and every increase is due to a jump (note that $X_t(\omega) - X_{t-}(\omega) = y$ if $(t, y)$ is a point $Y_i(\omega)$ for some $i$ in the set-up of Section 3.3.1).

Let $0 < t_1 < t_2 < \cdots < t_n$, and set $B_1 = (0, t_1] \times (0, \infty)$, $B_2 = (t_1, t_2] \times (0, \infty), \ldots, B_n = (t_{n-1}, t_n] \times (0, \infty)$. Then, the integral of *f* (recall that $f(s, y) = y$) over $B_1, B_2, \ldots, B_n$ respectively yields $X_{t_1}, X_{t_2} - X_{t_1}, \ldots, X_{t_n} - X_{t_{n-1}}$. Since $B_1, \ldots, B_n$ are disjoint, the restrictions of *M* onto $B_1, \ldots, B_n$ yield independent random measures, which in turn implies that the increments $X_{t_1}, \ldots, X_{t_n} - X_{t_{n-1}}$ are independent. Thus, $(X_t)$ has *independent increments*.

The expressions for the expected value, variance, and Fourier transform of $X_u - X_t$ can be obtained from the appropriate formulae in Section 3.3.4 by taking $B = (t, u] \times (0, \infty)$ and $f(x) = y$ if $x = (s, y)$.

Let $k$ be an increasing deterministic function on $(0, \infty)$, let $W_1, W_2, \ldots$ be positive independent random variables, independent of the Poisson random measure $M$. Let $t_1, t_2, \ldots$ be some fixed times; and suppose that $V_t = \sum_i W_i 1_{(0,t]}(t_i)$ is almost surely finite for every $t$. Set

$$Z_t = k(t) + V_t + X_t$$

Then, $t \to Z_t$ is again an increasing right continuous process with independent increments. Conversely, every increasing right continuous process $(Z_t)$ with independent increments can be obtained as above. The process $Z$ has stationary increments if and only if $k(t) = kt$, $V_t = 0$, and $m(dt, dy) = dt\, \mu(dy)$ for some measure $\mu$ on $(0, \infty)$.

### A.3.7 Gamma processes

Let the process $(X_t)$ be constructed as in Section 3.3.6 and assume that the mean measure is given by

$$m(dt, dy) = p e^{-y/c} \frac{1}{y} dt\, dy$$

for some positive numbers $p$ and $c$. Then, by the results in Section 3.3.6, $(X_t)$ has stationary and independent increments with (see Section 3.3.4)

$$\mathbb{E}[\exp(i\lambda X_t)] = \exp\left(t \int_0^\infty \frac{p e^{-c/y}}{y} (e^{i\lambda y} - 1)\, dy\right) = (1 - ic\lambda)^{-pt}$$

Comparing this with Section 3.1.6, we see that $X_t$ has the gamma distribution with shape parameter $pt$ and scale factor $c$.

Accordingly, this process $(X_t)$ is called a gamma process with shape rate $p$ and scale factor $c$.

### A.3.8 Local gamma processes

Let $(X_t)$ be as in Section 3.3.6 again but now assume

$$m(dt, dy) = p(dt) e^{-y/c(t)} \frac{1}{y} dy$$

(If $p(dt) = p\, dt$ and $c(t) = c$ for all $t$ we get the gamma process of the preceding paragraph.)

Then, $(X_t)$ is called a local gamma process with parameter *functions* $p(t)$ and $c(t)$ for shape and scale factor. Here $t \to p(t)$ is increasing right continuous, and $c(t)$ is strictly positive. Clearly, $(X_t)$ has independent increments and is increasing. We have, using the results in Section 3.3.4,

$$\mathbb{E}[X_t] = \int_0^t p(ds)c(s) \qquad \mathrm{Var}\,[X_t] = \int_0^t p(ds)c(s)^2$$

and

$$\mathbb{E}[\exp{(i\lambda X_t)}] = \exp\left(\int_0^t p(ds)\int_0^\infty \frac{e^{-y/c(t)}}{y}\,(e^{i\lambda y} - 1)\,dy\right)$$

$$= \exp\left(\int_0^t p(ds)\log{(ic(s)\lambda - 1)}\right)$$

In the special case where $c(s) = c$, constant, the Fourier transform above becomes $(1 - ic\lambda)^{-p(t)}$, which is that of the gamma distribution with scale factor $c$ and shape parameter $p(t)$. Otherwise, if $c(s)$ is not constant, the distribution of $X_t$ has no known shape.

The local gamma processes were introduced in Çinlar[4]; we refer to it for further results. Please note that, in Çinlar[4], we are talking of scale *parameters*, which are different from the scale *factors* that are being used here. If $c_t$ is the scale factor then the scale parameter is $b_t = 1/c_t$. There is no uniform convention upon this among mathematicians; scale parameters look better in expressing distributions and mean measures, scale factors look better in expressing Fourier transforms and expectations.

## REFERENCES

1. Bazant, Z. P., (1974), 'Theory of creep and shrinkage in concrete structures: a precis of recent developments,' *Mechanics Today*, **2**, 1–93.
2. Çinlar, E. (1972), 'Superposition of point processes', in P. A. W. Lewis (Ed.), *Stochastic Point Processes*, Wiley, New York, pp. 549–606.
3. Çinlar, E., (1979), 'On increasing continuous processes', *Stoch. Proc. & Their Appl.*, **9**, 147–54.
4. Çinlar, E. (1980), 'On a generalization of gamma processes', *J. Appl. Prob.*, **17**, 467–480.
5. Çinlar, E., Bazant Z. P., and Osman E. (1977), 'Stochastic process for extrapolating concrete creep', *J. Eng. Mech. Div., Am. Soc. Civ. Eng.*, **103**, 1069–88.
6. Sze, D. Y., (1980), 'Mixing properties and point processes', *Doctoral Dissertation*, Dept. Mathematics, University of Chicago.

Creep and Shrinkage in Concrete Structures
Edited by Z. P. Bažant and F. H. Wittmann
© 1982 John Wiley & Sons Ltd

Chapter 4

# Estimation of Drying of Concrete at Different Relative Humidities and Temperatures of Ambient Air with Special Discussion about Fundamental Features of Drying and Shrinkage

*S. E. Pihlajavaara*

## 4.1  INTRODUCTION

Generally speaking, the drying of porous solids is a complicated physico-chemical hygrothermal phenomenon. As is often the case with technological problems, there are two chief approaches to the study of the phenomenon of drying: a strictly scientific mathematical approach and a practical technical approach tending to ignore minor effects. Although the ultimate goal of both approaches may often be practical application, the presentation and thus the intelligibility to non-specialized scientists and to engineers, differs greatly in the context of the two different doctrines. Two recently published technology doctoral theses on drying elucidate this situation very well—a situation which worried the author when preparing this chapter. Dr M. Lampinen,[7] Helsinki University of Technology, presented a theoretical doctoral thesis in which he stated that its contents cover 'a general theory of the mechanics and thermodynamics of drying'. He stressed further that 'many separate fields of physical research, such as mechanics, thermodynamics, physical surface chemistry, and rheology are associated with the drying problem . . .' and that 'of course there are many problems where some of the equations may be considerably simplified'. On the contrary, Dr L-O. Nilsson,[9] Lund Institute of Technology, Sweden, in his thesis on drying combined the detailed existing data of a material (concrete) with a practically applicable computerized theory. He stated in a special abstract of his thesis that 'a great number of theories were available but could not be used because the necessary material properties regarding moisture were missing'.

These two theses, which do not share a single literature reference, indicate that even the drying specialists have very little in common. Consequently, plenty of demanding and fruitful research is still to be done in the practical field of the drying of concrete.

In the following presentation, I will try to describe the fundamental features of the drying of concrete after sealed precuring in a simple way starting from fundamental ideas. The theoretical treatment is based mainly on the analogy of the simple constant-conductivity (constant-diffusivity) diffusion theory,[17] although it is known that moisture conductivity and, thus, the drying is dependent on internal and external moisture conditions. With carefully chosen parameters, I believe that this simple theory still yields technically applicable, if not scientifically accurate, estimation of the drying of concrete. The value of some more complex theories is reduced by the limited possibility of the reliable estimation of main moisture and transfer data (especially moisture conductivity) for different concretes. However, the system of simple figures, nomograms, and tables presented here for practical estimation of drying could be refined or revised in the near future with the aid of more advanced theories and more precise relevant data.

The excellent, detailed and comprehensive, but also simplified presentations by Professor H. K. Hilsdorf[3,4] of the practical estimation of the water and moisture content of concrete structures under various conditions during their lifetime deserve special mention.

To elucidate the interrelationship between the properties of concrete and drying, the principles of shrinkage will be dealt with as an example in the next section, using the basic principles and definitions of moisture content and drying.

## 4.2   FUNDAMENTAL FEATURES OF DRYING AND SHRINKAGE

### 4.2.1   General

Changes in moisture content alter the properties of concrete. The diminution of moisture content, which is one of the main phenomena in the first phase of the history of concrete, brings about a significant increase in strength and creep (drying creep), and is the chief cause of shrinkage[14] (drying shrinkage). These changes are partly irreversible, which means that concrete is a history-dependent material.[13]

Furthermore, there is a slow uptake of $CO_2$ from the ambient air, which results in microstructural and strength changes as well as in carbonation shrinkage in the surface layer. This phenomenon, together with drying, is dependent on the size of the concrete body or specimen. As drying and concurrent effects, including changes in various properties, are at least partly interrelated, it is not surprising that test results are sometimes contradictory. This is especially true when one or more relevant factors are omitted in testing and in discussion of the significance of the test results. It is to be hoped that some steps forward in the elucidation of these interrelated effects will be reached in this Lausanne Symposium.

To clarify the role of drying phenomena in deformations of concrete, I shall try to outline the basic principles of drying and shrinkage of concrete as I understand them. Unfortunately, I do not have at my disposal enough information (and it is probable that this kind of comprehensive information does not exist) to be able to reach very reliable statements. That is why I call them 'hypotheses'. I shall start my presentation from basic definitions.

### 4.2.2 Drying and basic definitions of water and moisture contents. Effect of carbonation

Figure 4.1 illustrates the general behaviour of drying concrete and the general definitions of water contents and moisture contents of concrete.

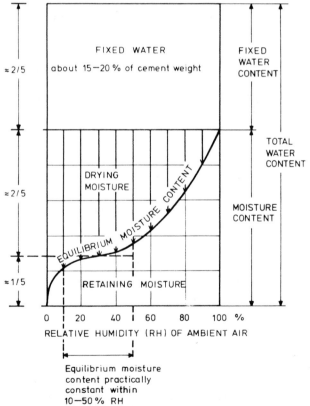

**Figure 4.1** Figure of principle indicating the definitions of water contents and moisture contents of concrete including equilibrium moisture content curve or desorption isotherm at room temperature. Sealed precuring prior to drying

The curve of equilibrium moisture content (or desorption isotherm) at room temperatures divides the moisture content (which will totally evaporate at zero RH or, by the usual definition in drying studies of porous media, in oven drying at 105 °C) into two categories, namely 'drying moisture' and 'retaining moisture'. The latter will remain in pores at a certain RH of ambient air. It is worth noting that for technical purposes the change in equilibrium moisture content, which concrete slowly reaches, is practically constant within the RH range 10–50% (or still better between 20 and 40% RH) (a good rule of thumb).[17] In the determination of desorption isotherms of concrete there are at least three main sources of error, which result in a different shape of the isotherm to that presented in Figure 4.1; these are: (1) carbonation; (2) no forced air circulation in the testing cabinet or desiccator, and (3) too short a testing time.

If the concrete body or specimen is relatively thin, less that 5–10 cm thick, it will be slowly carbonated (uptake of $CO_2$ from the ambient air), and, consequently, its fixed water content[11] and equilibrium moisture content[12,22] will be lower due to microstructural changes (Figure 4.2). Quantitative

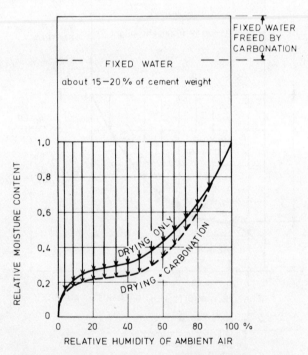

**Figure 4.2**  Figure of principle indicating the effect of carbonation on the fixed water content and on the equilibrium moisture content based on scattered data. Sealed precuring prior to drying

information on the decrease of fixed water content and equilibrium moisture content due to carbonation is scarce and the values of these entities are therefore here roughly estimated. Carbonation is extremely slow in very dry or very wet conditions.

### 4.2.3 Drying and drying shrinkage. Carbonation shrinkage

Figure 4.3 illustrates a technical hypothesis concerning the interrelation of equilibrium moisture content and final drying shrinkage at equilibrium. Although the hypothesis is based on my own experimental results,[14] it seems that other test results are also in agreement.[8,2] This kind of interrelation clearly shows the interdependence of drying shrinkage (starting from sealed precuring) and drying. Further, it is indicated in Figure 4.3 that for practical purposes the final first drying shrinkage is constant within the RH range 10–50%. Experimental results on the effect of the water–cement ratio on shrinkage are presented in Appendix 1.

Unfortunately, the clear conception depicted in Figure 4.3 must be changed to that shown in Figure 4.4 if the carbonation effect is significant in the drying and shrinkage processes.[18,15] This means that non-carbonated and

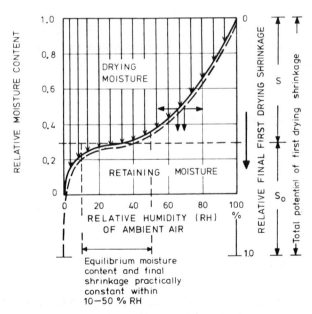

**Figure 4.3** A technical hypothesis: An illustration of the similarity of the dependence of equilibrium moisture content and of corresponding final first drying shrinkage on the RH. $S_0$ varies due to crack formation in low humidities: in thin cement paste specimens $S_0/S$ has been found to be nearly three[2]. Sealed precuring prior to drying

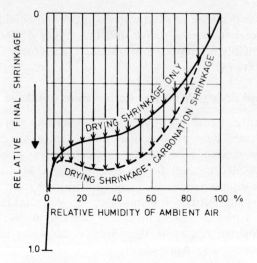

**Figure 4.4**   A technical hypothesis: An illustration
of drying shrinkage and drying shrinkage+
carbonation shrinkage. Sealed precuring prior to
drying

carbonated concrete do not have the same final shrinkage at middle
humidities. Figure 4.4 is based on my own findings, but other results support
the hypothesis. More research is needed for reliable quantitative conclusions.[19]

### 4.2.4   Composition factors of drying

Different types of concrete have different moisture conductivities and thus
different features of drying. The basic composition factors which exert an
influence are:[10]

| (a) Cement paste | (b) Aggregate (dense) | (c) Cement paste–aggregate ratio |
|---|---|---|
| brand of cement | grading | |
| water–cement ratio | | |
| air content | | |
| hydration conditions | | |
| degree of hydration | | |

Generally, the same factors which affect permeability also affect drying,
although it can be anticipated that the effect of cracks on drying is not quite
so drastic as it is on water and gas permeability through a concrete wall, in
which the flow is proportional to $a^3$ ($a$ = crack width). The chief composition
factor, if a possibly very significant effect of the brand of cement is
neglected, is the water–cement ratio (w/c): relative moisture conductivity

increased about 20 times when w/c increased from 0.4 to 1.0[9] and about 2 times when w/c increased from 0.4 to 0.6.[9,23] Next in order of importance are age of hydration (if less than about one week), air content (if particularly high), and modulus of fineness of aggregate (if different from normal values).[9] For normal concretes these effects seem to be small and can be neglected in many technical estimations. Calculations by Nilsson[9] with a composite model gave a decrease in moisture conductivity (diffusivity) of only 5% when aggregate content increased from 1450 to 1750 kg/m$^3$.

The range of moisture conductivity (diffusivity) values due to variations in the composition of good, ordinary Portland cement concretes (w/c = 0.6–0.4) with good precuring before drying can be estimated to be about $10^{-10}$ to $10^{-11}$ m$^2$/s at room temperature.

### 4.2.5 Equilibrium points, if any? Dependence of drying on ambient humidity

Figure 4.5 shows the possible weight or mass equilibrium points of carbonating and non-carbonating concrete.[19,20,21] There are hardly any final mass equilibrium points for a slowly carbonating concrete body if it is relatively thick. Due to difficulties in gravimetric determination of moisture entities as depicted in Figure 4.5, the pore humidity measurement with a moisture probe inserted into concrete has also been used in drying studies.[1,9] However, other difficulties are involved in this type of measurement and it must also be performed with extreme care and with continuous calibration.[16]

Figure 4.6(a) and (b) indicate the difficulties involved in finding a final weight or mass equilibrium point of drying in non-carbonating concrete

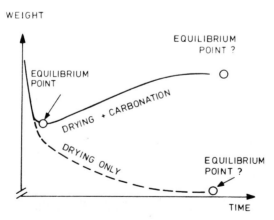

**Figure 4.5** Three equilibrium points to be taken into consideration in drying studies utilizing the gravimetric method

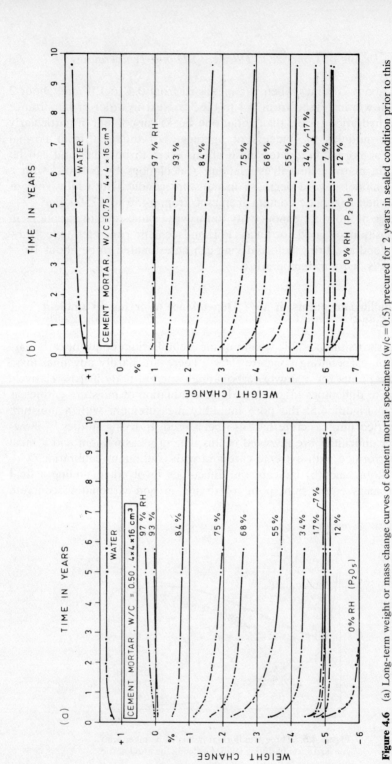

**Figure 4.6** (a) Long-term weight or mass change curves of cement mortar specimens (w/c = 0.5) precured for 2 years in sealed condition prior to this 10-year 'drying' experiment

(b) Long-term weight or mass change curves of cement mortar specimens (w/c = 0.75) precured for 2 years in sealed condition prior to this 10-year 'drying' experiment

bodies. An earlier study[11] verified that there is always a practical moisture equilibrium point in cement paste. However, it is generally accepted that continuous long-term microstructural changes take place in cement paste during its lifetime. Therefore, it is very probable that these alterations bring about continuous changes in fixed water content and moisture content (pore moisture content is dependent on porous microstructure), as Figures 4.6(a) and (b) indicate. In the middle-range humidities relatively thin cement mortar specimens seem to continue to dry out for over 10 years. The highest curves in the figures, which rise slightly, probably indicate slow continuous hydration. The middle curves, which descend continuously, probably reflect internal microstructural changes needing water, but very little, if any, hydration. The lower curves, which have reached an apparent equilibrium, probably indicate that the internal structure has stabilized due to the lack of water for continuous physico-chemical microstructural modifications. The specimens were precured for two years in sealed conditions before this ten-year test. The drying test was performed above salt solutions in desiccators furnished with small electric fans for forced air circulation. Figures 4.6(a) and (b) verify that moisture conductivity or the drying rate of non-carbonating concrete increases when the ambient humidity decreases, as was shown by Yuan, Hilsdorf, and Kesler[23] experimentally at 75, 50, and 25% RH of ambient air. Their tests also indicated that the diffusion theory with a constant diffusion coefficient (= moisture conductivity = diffusivity) selected in accordance with ambient humidity adequately describes the drying of mortars with water–cement ratios ranging from 0.40 to 0.70 and for exposure conditions with a relative humidity, of between 25 and 75% and temperature between 5 and 60 °C, as long as no carbonation takes place during the drying.

An analysis of our drying tests on non-carbonating concrete indicated that the diffusion theory with constant moisture conductivity correlated very well in low humidities of ambient air (5–20% RH), but only 'satisfactorily' in the middle humidities (20–80% RH), and rather well in high humidities above 80% RH. The dependence of moisture conductivity on the relative humidity of ambient air was roughly the same as in Yuan, Hilsdorf, and Kesler.[23] Nevertheless, the results of our drying tests exposed some unsolved peculiarities. Consequently, more experimental data are needed.

### 4.2.6 Effect of internal moisture (=pore humidity) and carbonation on drying

In the preceding section the effect of external moisture or the relative humidity of air (RH) was dealt with on the basis of experimental evidence. The effects of internal moisture and carbonation on drying, and their interplay, have not been sufficiently studied experimentally. In other words, it has been experimentally demonstrated numerous times that in carbonating concrete the moisture conductivity (diffusivity) reduces as the drying proceeds (for example, Pihlajavaara[10,11,20]), but it has not been adequately verified whether the reason for this is a gradually thickening carbonated surface layer or gradually reducing pore humidity (including changes in water transfer in the form of vapour and liquid as well as alterations in microstructure), or both. Unfortunately the mathematical proof based on reduction of pore humidity and, consequently, of moisture conductivity[1,9] cannot be considered as final definite experimental evidence due to the lack of consideration of the carbonation effect and other possible basic phenomena involved. It is well known that the addition of one or two variables decreasing or increasing conveniently but not necessarily having any reliable experimental foundation, can sometimes improve the correlation between a theory and experimental data. Because data from systematic experiments on the effect of the carbonated layer are almost non-existent, a research group headed by H. J. Kropp and H. K. Hilsdorf has undertaken a study on 'the effect of carbonation on drying properties of concrete'.[5] As regards information on the permeability of carbonated concrete, the data available seem to be somewhat contradictory. Some literature sources indicate that the effect of carbonation on the permeability is relatively minor. However, other evidence indicates that carbonated concrete (i.e. of smaller porosity) has a smaller permeability than non-carbonated.

### 4.2.7  Effect of painting on drying, shrinkage, and carbonation

The painting of concrete structures is a common procedure, the effect of which on the behaviour and performance of concrete structures has not been thoroughly studied. A six-year investigation[21] into the effect of painting on the drying, shrinkage, and carbonation of $4 \times 4 \times 16$ cm specimens indicated that painting, as well as slowing down drying and carbonation, diminished considerably drying shrinkage and carbonation shrinkage, and, consequently, reduced surface stresses and possible surface cracking.

## 4.3  SIMPLIFIED METHOD OF ESTIMATION OF DRYING OF PORTLAND CEMENT CONCRETE STRUCTURES PRECURED IN PRACTICALLY SEALED CONDITIONS

### 4.3.1  Background information

Concrete is a very complex material, and systematic experimental research on the drying of concrete is relatively scarce. In particular, large-scale long-term statistically valid research data including all relevant variables, such as composition, thickness, external and internal humidity, carbonation, and temperature, are almost non-existent. Various studies have indicated that transfer entities of concrete governing the drying process, for example moisture conductivities, have a great range and consequently the error in predicting the drying time of a concrete body can easily be of the order of a decade. In addition, the experimental methods, which often vary between different investigations, are complex and must be performed with extreme care and patience. If the analogy of diffusion theory is applied to the description of the drying of different concretes the diffusivity or moisture conductivity varies, let us say, from $10^{-9}$ to $10^{-12}$ m$^2$/s, depending on internal and external moisture conditions and particularly on temperature.[23,17] In normal concretes at room temperature the moisture conductivity can be estimated to be of the order of $10^{-10}$–$10^{-11}$ m$^2$/s. It has been demonstrated that, at least in principle, the application of the analogy of the non-linear diffusion theory with more variables may give or gives better agreement with experimental data than application of the analogy of the simple constant-diffusivity diffusion theory for example.[11,1,6] However, at the present stage of development it is, strictly speaking, hardly possible to obtain more than a passable or a satisfactory quantitative estimation of a drying process, particularly as regards the world-wide validity of the estimation, due to the reasons described above.

### 4.3.2 Charts and tables of estimation

In spite of difficulties in reliable prediction, a simplified method for estimating the drying of concrete, based on some charts (Figures 4.7–4.10) and tables (Tables 4.1–4.2), is presented here. The basic principles of the method have been described earlier.[17] Those interested in a more sophisticated presentation are referred to the latest 'state-of-the-art report'.[9] The estimation system to be presented is based on the following aspects and simplifying assumptions:

(1) The estimation system gives a satisfactory overall view of the drying process of concrete and concrete structures.

(2) The accuracy and range of the values of the entities in the charts and tables are believed to be satisfactory for the preliminary prediction of cases encountered in practice. The range of error in the estimated duration of drying, $t$, can be great, my 'intellectual guess' of the range is $(0.25–4)t$, but at least the range $(0.5–2)t$ should be used.

(3) The entities of the composition, chiefly w/c, cement content, and water content, in the mixes normally used in practice do not differ greatly from the values presented in the charts. The range of error in the initial moisture content at the start of drying, $C_0$, in the equilibrium or surface moisture content at the end of drying, $C_e$, and in the average moisture

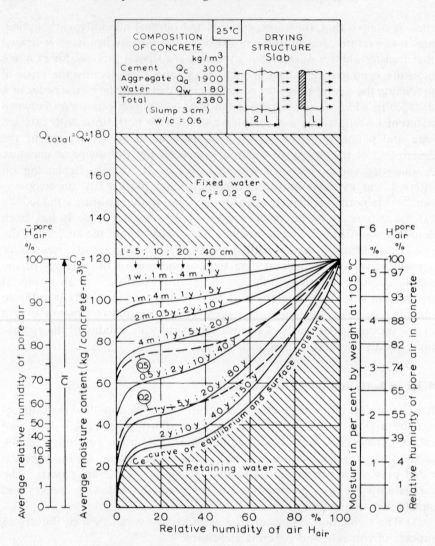

**Figure 4.7**  Estimation of drying of concrete: chart no. 1

content, $\bar{C}$, after a certain fixed drying time is of the order of ±15%, if the chart of nearest w/c to the actual case is used.

(4) Charts 1–4 (Figures 4.7–4.10) give a general idea of the drying process of Portland cement concrete at room temperature and of the values of the entities involved. With simple corrections, interpolation, and extrapolation, these charts can be used for estimation of the drying process in cases other than those presented in the charts.

(5) The charts and tables presented for columns can be used for the rough prediction of the drying of beams (also for drying from three sides only).

(6) The fixed water content of aggregate and the moisture content of aggregate is neglected.

(7) The effect of carbonation is neglected.

The use of charts 1–4 (Figures 4.7–4.10) and Tables 4.1 and 4.2 will be

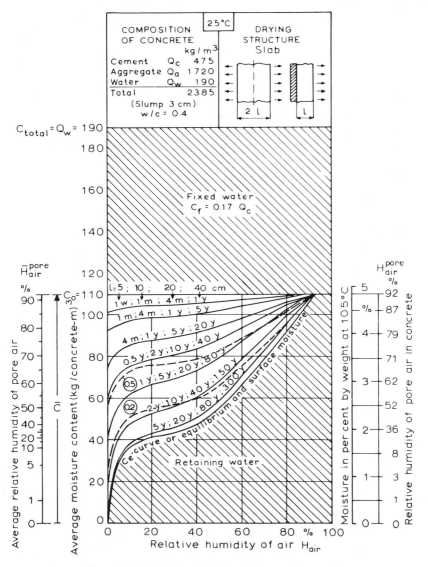

**Figure 4.8** Estimation of drying of concrete: chart no. 2

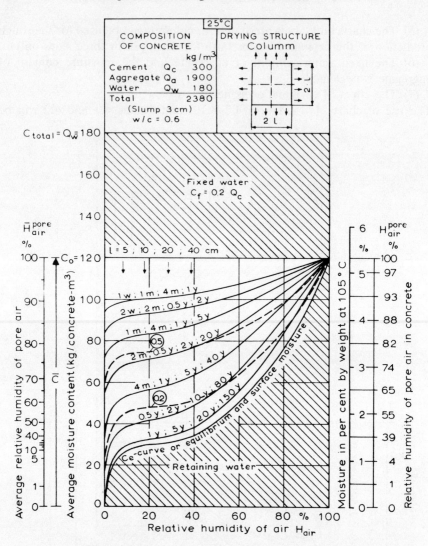

**Figure 4.9** Estimation of drying of concrete: chart no. 3

elucidated in the following paragraphs with some examples. The principles of the construction of the charts and tables are described in *Appendix 2*.

### 4.3.3 Estimation of duration of drying and pertaining moisture content of ordinary concrete structures at room temperature (25 °C) with the aid of charts 1, 2, 3, and 4 (Figures 4.7–4.10)

For slabs and columns with concrete compositions as depicted in charts or nomograms 1, 2, 3, and 4 (Figures 4.7–4.10), immediate answers can be

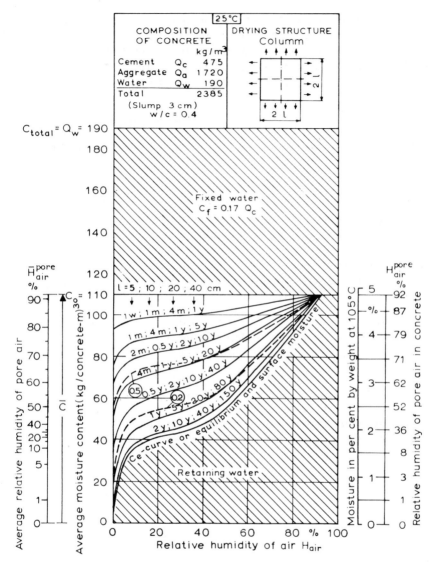

**Figure 4.10**    Estimation of drying of concrete: chart no. 4

obtained to the following questions:

(a)   What is the time lapse required to attain an average moisture content of concrete, $\bar{C}$, or average relative humidity of pore air in the concrete, $\bar{H}$?

(b)   What is the average moisture content of concrete, $\bar{C}$, or average relative humidity of pore air in concrete, $\bar{H}$, which will be attained during a certain period of time?

In addition, it is simple to estimate the time to attain a certain relative content of drying moisture $(1 \to 0)$, the values of which, 0.5 and 0.2 $(\doteq (\bar{C} - C_e)/(C_0 - C_e))$, can easily be read with the aid of the broken curves in the charts. The duration of drying for the structures has been indicated for characteristic thicknesses ($l$) of 5, 10, 20, and 40 cm.

*Example*

Consider a concrete slab as in Chart 1 (Figure 4.7), which has characteristic thickness $l = 10$ cm, is at room temperature, and the relative humidity of the ambient air $H_{air} = 40\%$. Also:

(1) Total water content, $Q_w = 180$ kg/m$^3$.
(2) Initial moisture content at the start of drying, $C_0 = 120$ kg/m$^3$.
(3) Equilibrium moisture content at the end of drying, and surface moisture content, $C_e = 35$ kg/m$^3$ (humidity in pore air $H_{air}^{pore} = \bar{H}_{air}^{pore} = H_{air} = 40\%$).
(4) Duration of drying, $t$, to attain relative average content of drying moisture of 0.5 or $\bar{C} = 77$ kg/m$^3$ ($\bar{H} = 80\%$) is $t_{0.5} = 1.5$ years, and of 0.2 or $\bar{C} = 50$ kg/m$^3$ ($\bar{H} = 60\%$) is $t_{0.2} = 5$ years.

A brief error analysis: If w/c = 0.6–0.5, but the grading of the aggregate and the consistency of the mix are different from the design values of the chart, the roughly estimated ranges of error are

$$C_0 = (120 \pm 20) \text{ kg/m}^3 \qquad C_e = (35 \pm 5) \text{ kg/m}^3$$

$$t_a = (0.5-2)t_{chart}, \quad \text{or} \quad t_{0.5} = 1-3 \text{ years}, \quad \text{and} \quad t_{0.2} = 3-10 \text{ years}$$

### 4.3.4  Correction of duration of drying when the actual characteristic thickness $l_a$ differs from the characteristic thickness $l$ of the chart

$$t_a = \left(\frac{l_a}{l_{chart}}\right)^2 t_{chart} \tag{4.1}$$

If the structure is not a slab or a column with equal sides, $l_a$ can be calculated from

$$l_a = 2\, V/S$$

where $V$ is the volume of the member and $S$ is the drying surface area of the member.

*Example*

Consider a long beam ($a \times h \times \text{length} = 30 \text{ cm} \times 50 \text{ cm} \times \text{length}$) drying from three sides:

$$l_a = 2 \times 30 \times 50 \times \text{length}/(30 \times \text{length} + 2 \times 50 \times \text{length})$$

$$= 2 \times 30 \times 50/(30 + 2 \times 50) = 3000/130 = 23 \text{ cm}$$

Consequently

$$t_a = (l_a/l_{chart}^{column})^2 t_{chart} = (23 \text{ cm}/20 \text{ cm})^2 t_{chart}$$

$$= 1.3 t_{chart}$$

### 4.3.5 Correction of equilibrium moisture content, $C_e$, and of duration of drying when the actual temperature differs from the temperature of the charts (25 °C)

Table 4.1 shows these corrections. $t_{chart}$ and $t_a$ correspond the same relative average moisture content $\bar{u}$:

$$\bar{u} = \frac{\bar{C}^{chart} - C_e^{chart}}{C_0^{chart} - C_e^{chart}} = \frac{\bar{C}^{T\,°C} - C_e^{T\,°C}}{C_0^{chart} - C_e^{T\,°C}}$$

from which average moisture content $\bar{C}^{T\,°C}$ kg/m$^3$ can be calculated easily.

Table 4.1   Temperature correction of moisture, $C_e^{chart}$, and of duration, $t_{chart}$

| | | Temperature, $T$ (°C) | | | | | |
|---|---|---|---|---|---|---|---|
| | | 5 | 25 | 45 | 65 | 85 | 105 |
| Equilibrium moisture or surface moisture, $C_e^{T\,°C}$ | w/c 0.4 | $\dfrac{C_e^{chart}}{0.6}$ | $\dfrac{C_e^{chart}}{1}$ | $\dfrac{C_e^{chart}}{2}$ | $\dfrac{C_e^{chart}}{5}$ | $\dfrac{C_e^{chart}}{20}$ | $c_e = 0$ |
| | w/c 0.6 | $\dfrac{C_e^{chart}}{0.7}$ | $\dfrac{C_e^{chart}}{1}$ | $\dfrac{C_e^{chart}}{1.5}$ | $\dfrac{C_e^{chart}}{2.5}$ | $\dfrac{C_e^{chart}}{6}$ | $C_e = 0$ |
| Duration of drying, $t_a$ | | $\dfrac{t_{chart}}{0.4}$ | $\dfrac{t_{chart}}{1}$ | $\dfrac{t_{chart}}{2.5}$ | $\dfrac{t_{chart}}{6}$ | $\dfrac{t_{chart}}{15}$ | $\dfrac{t_{chart}}{40}$ |

### 4.3.6 Review of duration of drying of concrete structures at different temperatures and humidities

A review of the estimation of duration of drying of different concrete structures at different temperatures and humidities is presented in Table 4.2. It is to be anticipated that the range of error of the durations, $t$, taken from the table is of the order of $(0.5-2)t$.

## 4.4 CONCLUDING REMARKS

Space and time did not allow full treatment of many of the issues broached in this chapter. The utilization of the fundamental principles, theories, and information to yield applicable rules needs thoroughly verified, reliable experimental data and long-term work, the lack of which is too often evident. At the present stage of development it is hardly possible to obtain more than a fair, or sometimes satisfactory, quantitative estimation of the drying process of concrete, especially with regard to the estimated duration of drying, without direct relevant experimental knowledge of the moisture conductivity (moisture permeability) of the case being studied. This means in practice that in cases in which good reliable information on the drying process of a concrete structure is vital: (1) the testing of carefully selected specimen series in relevant conditions; or (2) reliable measurement of the humidity of the pore air with the aid of inserted moisture probes and some

Table 4.2  Examples indicating estimated durations of drying of concrete structures (d = day, m = month, y = year)

| Temperature (°C) | Strength and impermeability of concrete, and RH of ambient air | | Moisture conductivity ($k$ diffusivity) (m²/s) | Column or prism | | | Slab or wall | | |
|---|---|---|---|---|---|---|---|---|---|
| | | | | 5 cm | 10 cm | 20 cm | 5 cm | 10 cm | 20 cm |
| 5 | low | high | $2 \times 10^{-10}$ | 1 m | 5 m | 1.5 y | 3 m | 1 y | 4 y |
| | average | average | $2 \times 10^{-11}$ | 1 y | 4 y | 15 y | 2 y | 10 y | 35 y |
| | high | low | $5 \times 10^{-12}$ | 4 y | 15 y | 60 y | 10 y | 35 y | 140 y |
| 20 | low | high | $5 \times 10^{-10}$ | 0.5 m | 2 m | 7 m | 1 m | 4 m | 1.5 y |
| | average | average | $5 \times 10^{-11}$ | 5 m | 1.5 y | 6 y | 1 y | 4 y | 15 y |
| | high | low | $10^{-11}$ | 2 y | 8 y | 30 y | 5 y | 20 y | 70 y |
| 50 | low | high | $2 \times 10^{-9}$ | 4 d | 0.5 m | 2 m | 8 d | 1 m | 4 m |
| | average | average | $2 \times 10^{-10}$ | 1 m | 5 m | 1.5 y | 3 m | 1 y | 4 y |
| | high | low | $2 \times 10^{-11}$ | 1 y | 4 y | 15 y | 2 y | 10 y | 35 y |
| 100 | low | high | $2 \times 10^{-8}$ | 0.5 d | 1.5 d | 6 d | 1 d | 3 d | 0.5 m |
| | average | average | $2 \times 10^{-9}$ | 4 d | 0.5 m | 2 m | 8 d | 1 m | 4 m |
| | high | low | $10^{-10}$ | 2 m | 9 m | 3 y | 6 m | 2 y | 7 y |

Duration of drying from relative average moisture content $(\bar{C} - C_e)/(C_0 - C_e) = 1$ to 0.2

other relevant experiments, are necessary. In addition, it is evident that good research on shrinkage and creep of concrete also demands simultaneous experiments and documentation on the drying and carbonation processes.

## APPENDIX 1

Experimental results on the effect of the water–cement ratio on shrinkage are presented in Figure 4.11.

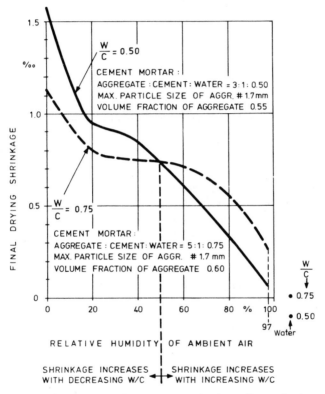

**Figure 4.11** Experimental results of the effect of the water/cement ratio on shrinkage of cement mortar[14]

## APPENDIX 2

Information on the Basic Principles of the Design of Charts 1, 2, 3, and 4 (Figures 4.7–4.10) and Tables 4.1 and 4.2

Figure 4.12 shows the dependence of the moisture conductivities $k$ (diffusivities) of concrete on the water–cement ratio of concrete and on the relative humidity of the ambient air. These interrelations have been used in the calculation of the decrease in average moisture content with time using

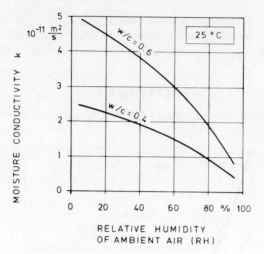

**Figure 4.12**  Dependence of moisture conductivity
$k$ on w/c ratio and RH of ambient air

the constant-diffusivity diffusion theory. Accordingly, diffusivity or moisture conductivity $k$ was constant during the whole drying process but was selected from the figure in accordance with the prevailing humidity and w/c of concrete. Thus, a possible decrease of moisture conductivity during drying has not been taken into account, although such a feature has been demonstrated in some investigations. This effect needs more experimental research before definite conclusions can be reached.

In Figures 4.7–4.10 on the right the dependence of relative humidity of air in pores of concrete, $H_{air}^{pore}$, on the moisture content of the concrete is in accordance with the $C_e$ curve or sorption isotherm and with the pertaining relative humidities of ambient air, $H_{air}$. The numbers in the $H_{air}^{pore}$ scale are related to the moisture content scale expressed in percentage by weight.

The average relative humidity of the pore air, $\bar{H}^{pore}$, in the concrete structure examined is presented in Figures 4.7–4.10 on the left, corresponding to the average relative moisture content, $\bar{C}$, after a certain duration of drying. Our calculations indicated that in the cases presented in the charts the $\bar{H}^{pore}$ values are close to the $H_{air}^{pore}$ values, as the scales demonstrate.

The information in Tables 4.1 and 4.2 has been obtained from the literature as described in Pihlajavaara,[17] with some later modifications.

## REFERENCES

1. Bazant, Z. B., and Najjar, L. J. (1971), 'Drying of concrete as a nonlinear diffusion problem', *Cem. Concr. Res.*, **1**, 461–73.
2. Feldman, R. F., and Swenson, E. G. (1975), 'Volume change on first drying of

hydrated Portland cement with and without admixtures', *Cem. Concr. Res.*, **5**, 25–35.
3. Hilsdorf, H. K. (1967a), 'A method to estimate the water content of concrete shields', *Nucl. Eng. Des.*, **6**, 251–63.
4. Hilsdorf, H. K. (1967b), 'The water content of hardened concrete', *Nucl. Rad. Shield Stud. Rep. No. 4, DASA*-1875, *Univ. Illinois.*
5. Hilsdorf, H. K., and Kropp, H. J. (1979), A private communication (April 1980), and information in 'Research in progress 1979' in *J. Am. Concr. Inst.*, Dec.
6. Kasperkiewics, J. (1972), 'Some aspects on water diffusion process in concrete', *Mater. Constr.*, **5**, 209–14.
7. Lampinen, M. (1979), 'Mechanics and thermodynamics of drying', *Acta Polytech. Scand., Mech. Eng. Ser. No.* 77, *Helsinki.*
8. Mills, R. H. (1968), 'Dependence of drying shrinkage on dimensions of specimens', *Univ. Calgary, Tech. Publ.*
9. Nilsson, L-O. (1980), 'Hygroscopic moisture in concrete—drying, measurement & related material properties', *Lund Inst. Technol., Rep.* TVBM-1003, *Lund, Sweden.*
10. Pihlajavaara, S. E. (1963), 'Notes of drying of concrete', *St. Inst. Tech. Res. Finland, Rep. Ser.* III-*Bldg* 74, *Helsinki.*
11. Pihlajavaara, S. E. (1965), 'On the main features and methods of investigation of drying and related phenomena in concrete', *St. Inst. Tech. Res. Finland, Publ.* 100.
12. Pihlajavaara, S. E. (1968), 'Some results of the effect of carbonation on the porosity and pore size distribution of cement paste', *Mater. Constr.*, **1**, 521–6.
13. Pihlajavaara, S. E. (1971), 'History-dependence, ageing, and reversibility of properties of concrete', in M. Te'eni (Ed.), *Conf. on Structure, Solid Mech. Eng. Des., Southampton 1969*, Part I, Wiley-Interscience, Chichester, pp. 719–41.
14. Pihlajavaara, S. E. (1974a), 'A review of some of the main results of a research on the ageing phenomena of concrete: effect of moisture conditions on strength, shrinkage and creep of mature concrete', *Cem. Concr. Res.*, **4**, 761–71.
15. Pihlajavaara, S. E. (1974b), A discussion of the paper 'A theoretical method for predicting the shrinkage of concrete' authored by N. K. Becker and C. MacInnes, *J. Am. Concr. Inst.*, **71**, 146.
16. Pihlajavaara, S. E., and Paroll, H. (1974c), 'Effect of carbonation on the measurement of humidity in concrete', *2nd Int. CIB/RILEM Symp. on Moisture Problems in Buildings, Rep.* 5.1.7, *Rotterdam.*
17. Pihlajavaara, S. E. (1976a), 'On practical estimation of moisture content of drying concrete structures', *Il Cemento*, **73**, 129–38.
18. Pihlajavaara, S. E. (1976b), 'On the effect of carbonation on shrinkage and weight changes of concrete', *Int. RILEM Symp. on Carbonation of Concrete, 5–6 April, Fulmer, Slough, UK*, Cem. and Concr. Assoc.
19. Pihlajavaara, S. E. (1976c), 'Carbonation, engineering properties, and effects of carbonation on concrete structures', *Int. RILEM Symp. on Carbonation of Concrete, 5–6 April, Fulmer, Slough, UK*, Cem. and Concr. Assoc.
20. Pihlajavaara, S. E. (1977a), 'Carbonation—an important effect on the surfaces of cement based materials', *RILEM/ASTM/CIB Symp. on Evaluation of the Performance of External Vertical Surfaces of Buildings*, Vol. 1, *Tech. Res. Centre Finland, Espoo, Finland.*
21. Pihlajavaara, S. E., and Pihlman, E. (1977b), 'Effect of painting on drying, shrinkage and carbonation of cement-based materials', *RILEM/ASTM/CIB*

*Symp. on Evaluation of the Performance of External Vertical Surfaces of Buildings*, Vol. 1, *Tech. Res. Centre Finland, Espoo, Finland.*

22. Rozental, N. K., and Alekseev, S. N. (1976), 'Change in concrete porosity during carbonation', *Int. RILEM Symp. on Carbonation of Concrete, 5–6 April, Fulmer, Slough, UK*, Cem. and Concr. Assoc.

23. Yuan, R. L., Hilsdorf, H. K., and Kesler, C. E. (1968), 'Effect of temperature on the drying of concrete', *Univ. Illinois, T & A. M. Rep. No. 316.*

# PART II
# CREEP AND SHRINKAGE— ITS MEASUREMENT AND MODELLING

Chapter 5

# Experimental Techniques and Results

*C. D. Pomeroy*

## 5.1 INTRODUCTION

The time-dependent deformation of concrete depends on many factors and not solely upon the applied loads. In particular it depends upon the strength and maturity of the concrete, on the moisture and temperature histories experienced by the concrete, on the aggregate properties, and on the size of the concrete structural member. It follows, therefore, that experimental techniques that are used in studies of the shrinkage and creep of concrete must be chosen as much for the way the environmental parameters can be controlled and varied as for the methods of load control and strain measurement.

There are widely divergent objectives for experimental programmes and these range from tests on hardened cement paste (HCP) to site measurements of movements in actual structural concrete. In this chapter three different classes of test are discussed. Firstly the measurement of the deformation of HCP specimens, secondly the measurement of the deformation of concretes made and tested in the laboratory, and finally the manufacture and site testing of concrete to simultate the actual behaviour of concrete in a structure.

## 5.2 HARDENED CEMENT PASTE

The constituent in concrete that contributes most to the time-dependent dimensional changes is the HCP so that it is not surprising that considerable effort has been applied to studies on HCP specimens. Concretes are made with water/cement ratios by weight that can exceed 1, and at this and much lower levels considerable segregation will occur if special precautions are not taken. Hence, before any consideration is given to the means of measurement of the dimensional changes consequent upon environmental and loading histories, it is essential to develop acceptable techniques for sample manufacture.

Ideally a cement paste specimen is one that accurately reproduces the properties of the cement paste binder in a real concrete, but in concrete the spacing between the aggregate particles is small and settlement of the cement grains will not be so important as it is in a cement and water mixture. Different methods have been devised to overcome this problem. In one the paste is not cast until it has stiffened sufficiently when it is remixed to ensure a uniform composition in the specimen. An alternative procedure is to cast the freshly mixed paste into a sealed mould that can be rotated slowly so that there is not a preferred direction for settlement to take place. When cement and water react there is initially a reduction in volume so that in a sealed mould it is necessary to provide some means for the contraction to occur and a rubber gasket, suitably ventilated has been found successful.[19]

It may be argued that these methods do not faithfully reproduce the paste in a concrete but they are more uniform than is found even for 0.3 w/c pastes when settlement can occur. The effectiveness of the rotation techniques was checked by Parrott[15] who cast flat slabs $290 \times 150 \times 20$ mm in sealed moulds that were rotated at 10 rpm. After the slabs had been moist cured for seven days they were cut into slices. The measured variation in density was always less than 1% and no systematic pattern of variation was observed. It was concluded that the specimens were homogeneous.

Other techniques have included the dry compaction of cement powder to a known porosity followed by evacuation and subsequent immersion in water[10], and the formation of CSH(I) by reacting silica gel and calcium hydroxide solution for many months, the hydration product being collected by filtration, partially dried in a $CO_2$ free atmosphere and compacted into cylinders which were subsequently sliced into discs.[3]

It can thus be seen that it has been necessary to establish special procedures for the manufacture of representative HCP specimens. It is essential to exclude atmospheric $CO_2$ during manufacture, storage, and testing the specimens as carbonation can occur rapidly and affect the results.

### 5.2.1  Specimen shapes

Most studies of the deformational characteristics of HCP involve conditioning the specimens to different levels of moisture equilibrium or to different temperatures, either prior to the start of dimensional observations or during the course of the experiment. The specimen thickness and porosity largely control the time taken to reach equilibrium and it is generally agreed that a thickness of about 1 mm is required, particularly when a rapid response to a change in the environment is sought.

Three different shapes of specimen are commonly used: flat discs made by compaction, hollow cylinders[12] and rectangular prisms, cut from larger slabs with sawn slots to reduce the drying paths.[16] The manufacture of such thin walled cylinders is particularly difficult and sophisticated methods have been developed.[1,2] Care must be taken to avoid carbonation and curing in lime water is essential.

## 5.2.2 Measuring and loading techniques

As shrinkage and creep measurements may take place over a long period the loading and measuring equipment must be stable. Also if changes in the rate of shrinkage or creep are being studied the sensitivity of the strain measuring equipment should be $5 \times 10^{-6}$ or better.

Most creep measurements are made on specimens loaded in uniaxial compression and both hollow cylinders and rectangular HCP prisms have been used satisfactorily. For these small specimens the load can be maintained by a steel spring and two typical loading arrangements are shown in Figure 5.1. The axial contraction on these samples can be measured with ordinary dial gauges that locate on steel inserts fixed to the ends of the specimens. Alternatively linear variable differential transformers have been used to record length changes continuously.

Rectangular prisms and thin disc compacts can be used for flexural tests. The hollow cylinders can be loaded in torsion or by internal pressure so that a wide variety of stress combinations can be studied.

An advantage found for the small rigs used for loading the rectangular prisms ($110 \times 12.5 \times 12.5$ mm) is that if they are made in stainless steel they may be placed into water baths and the studies conveniently made at different or varying temperatures. These rigs are also small enough to be stored in normal laboratory desiccators over saturated salt solutions for drying and rewetting experiments and light enough for changes in weight of the sample to be measured without removal from the loading frames. The small rigs can be fitted with a plastic tube to minimize the risk of sample

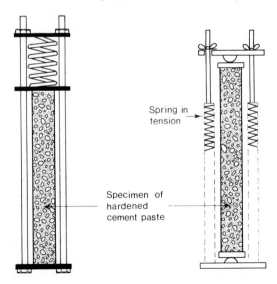

Spring in tension

Specimen of hardened cement paste

**Figure 5.1** Simple creep rigs for hardened cement paste specimens loaded in compression

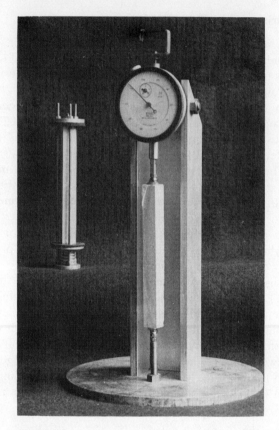

**Figure 5.2**   Loading and measuring frames for creep
measurement

carbonation when they are removed from the desiccators for measurement
in a dial gauge frame (Figure 5.2).

### 5.2.3   Combined stresses

The creep of HCP under different stress systems has also been studied.
Where stresses are applied in more than one direction to a specimen it is
difficult to generate a uniform stress field. In bi-axial compression, for
example, it is possible for some of the load applied in one direction to be
carried by the loading system in the other. Figure 5.3 shows the constant
stress trajectory that could apply if solid loading platens are used in a
bi-axial compression. Various methods have been suggested to minimize this
effect. These include the use of Hilsdorf brush patens,[9] and the insertion of
low friction pads between the steel platen and the surface of the specimen.[7]

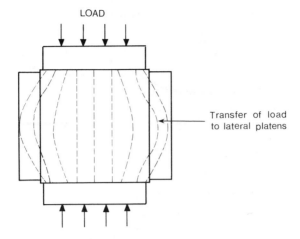

**Figure 5.3** The transfer of stress to solid orthogonal platens in a bi-axial loading rig

Both of these techniques have their limitations and a more satisfactory
system uses 'fluid-platens' that can transmit loads at right angles to a surface
but cannot sustain significant lateral forces. Parrott[14] developed such a
system and checked the uniformity of loading by the measurement of the
surface strain distribution in an aluminium plate using his own design of
vibrating wire strain gauge that was both sensitive and stable over long
periods of time. Details of these fluid platens are shown in Figure 5.4 and in
Figure 5.5 the strain distribution is compared with that when solid platens
are used. It can be seen that the system is satisfactory.

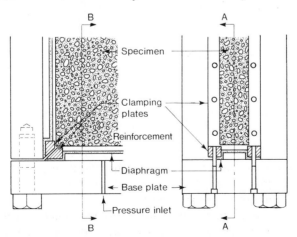

**Figure 5.4** A design of fluid platen to provide uniform
compressive loading

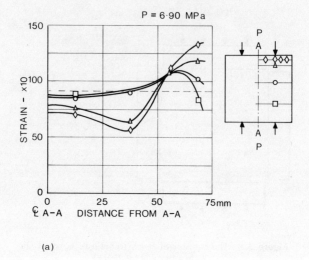

(a)

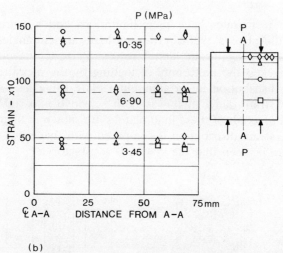

(b)

**Figure 5.5** Strain distribution in aluminium slab loaded
through: (a) solid and (b) fluid platens

Where tension tests have been made, either alone or in combination with
an orthogonal compression, longer slabs were used with a system of levers to
spread the load uniformly (Figure 5.6).

The specimens used in these loading rigs are much larger than those
referred to above and hence alternative methods are needed for tests at
different levels of moisture equilibrium, or when the moisture regime is
changed during the course of the test. In the tests that used HCP slabs the

specimens were conditioned to known moisture levels in a vacuum chamber controlled to a predetermined vapour pressure. When the specimens approach equilibrium they were sealed in a copper foil jacket and the joins were soldered. Before the specimens were loaded they were stored in the sealed state for a period so that any moisture gradients could readjust towards equilibrium. The strains were measured using the surface mounted vibrating wire gauges referred to above. An alternative method of applying composite stresses uses thin hollow cylinders which can be loaded internally or externally by an applied pressure on the curved surface, by an axial compressive load (or even tension) or by torsion. The thin walls of such cylinders are ideal for experiments that involve temperature or humidity changes, but there are difficulties in the manufacture of uniform specimens

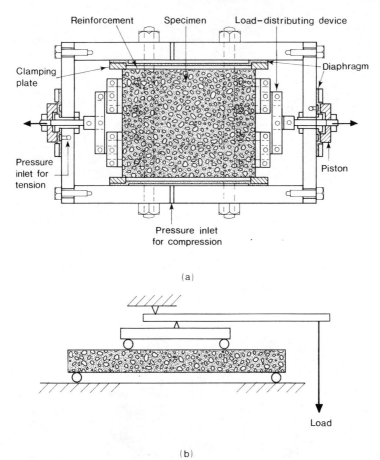

**Figure 5.6** Creep rigs for: (a) tensile (and biaxial compressive) and (b) flexural loading

and in applying uniform loads. Bazant *et al.*[1] have clearly shown that it is
possible to make and test hollow cylinders provided sufficient care is taken.

### 5.2.4  Hydrate compacts

Mention has been made of the use of thin compacts for shrinkage and
deformation studies, the compacts either being formed by dry compaction of

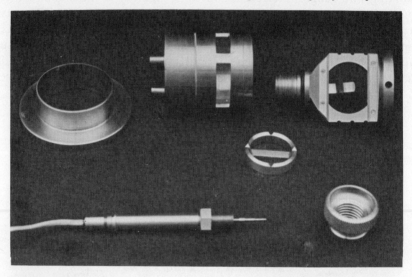

**Figure 5.7**  Equipment for the measurement of overall shrinkage and the change in
interplanar spacing in synthetic CSH(I)

cement powder, that is subsequently hydrated or by the compaction of bottle hydrated material. In particular the calcium silicate hydrate, CSH(I) can be synthesized by reacting silica gel and calcium hydroxide solution over a long period of time (typically a year at 20 °C). Gutteridge and Parrott[4] used this technique and produced CSH(I) with a $CaO:SiO_2$ molar ratio of 0.965:1. This was collected by filtration, partially dried at a relative water vapour pressure of 0.8 and compacted at a pressure of $100 \, N/m^2$ into discs 32 mm in diameter and 1.55 mm thick. The discs were then stored for 18 months in the original mother liquid before being used for a test programme.

The test programme comprised both measurements of shrinkage and expansion as the discs were progressively dried in steps to lower vapour pressures and were subsequently rewetted. Concurrently a comparable specimen was placed in a powder X-ray diffractometer for study of changes in interplanar spacing and length as the moisture content of the specimen changed.

Sophisticated equipment was designed to permit these measurements to be made as shown in Figure 5.7. Gutteridge and Parrott[4] showed that it was possible to measure both gross dimensional changes and changes in the $c$-spacing of the CSH(I) as drying progressed. Observation that only partial recovery in the $c$-spacing occurred upon rewetting provided support to the view that first drying can cause some permanent change in the hydrate structure.

## 5.3 CONCRETE

Creep measurements on concrete are usually made on cylindrical or prismatic specimens loaded in compression, although tests in tension, flexure, and torsion have also been made. In common with most long duration experiments it is necessary to use techniques in which the loads on the concrete can be held approximately constant or can be readjusted and in which the strain measurements are sufficiently precise and accurate and do not drift. An excellent review of loading techniques is given by Neville.[13] The three basic loading methods in compression are: (i) a lever mechanism; (ii) coaxial spring; and (iii) hydraulic. In the ASTM method (C512) for the measurement of creep in compression a spring loading system is described. A sequence of cylinders in series is permitted, thus minimizing the use of space and equipment.

### 5.3.1 Concrete specimens

Concrete specimens are more easily cast than those of cement paste, but even so care must be exercised if subsequent creep tests are to represent structural concrete correctly. As a consequence of aggregate settlement, water migration around aggregate particles and air entrapment concrete can be anisotropic so that its properties may differ in and at right angles to the casting direction. The two types of specimen that are usually used for compressive creep tests are cylinders and rectangular prisms. Normally the cylinders are cast vertically and the prisms horizontally so that there is a difference in the effective loading directions. However these effects are usually of less importance than the curing history of the specimens.

The maximum aggregate sizes relative to the specimen geometry are important, a factor of about 1/5 the minimum dimension providing the practical limit. It follows that hollow cylindrical specimens for torsion tests must either be of very large diameter or have relatively thick walls. If thin wall theory does not apply the creep stresses will be non-uniform and interpretation of creep measurements will be difficult.

It may thus be inferred that care must be taken in selecting a suitable specimen geometry, especially when stresses other than uniaxial compresson are to be applied to the concrete.

### 5.3.2  Strain measurement

Measurements that are continued for a long period that may extend to month or even years require extremely reliable and stable instrumentation. Even if stable equipment is available it is prudent to include in the system standards against which the measurements may be checked.

Mention has already been made of the satisfactory performance obtained from vibrating wire strain gauges and these have been used both for surface mounted instruments or for transducers that can be cast into the concrete for the measurement of internal strains. Ideally a transducer embedded in concrete should match the stiffness of the concrete perfectly so that the stress field is not disturbed. However the ideal cannot be achieved since the concrete properties will change with maturity and the shrinkage and creep characteristics of the transducer and the concrete will differ significantly. Relatively close matching of the properties is possible though and the design of a typical gauge for embedment into concrete is shown in Figure 5.8.

Other forms of measurement have been used satisfactorily and in particular linear variable differential transformers have proved to be reliable and stable for shrinkage measurements. Both vibrating wire and LVDT instruments can be used in data logging systems, an important factor when large numbers of measurements are to be taken over a long period.

It is not always necessary to use such sophisticated equipment and perfectly satisfactory creep and shrinkage measurements have been made with dial gauges or with demountable mechanical extensometers, such as the DEMEC gauge.[11]

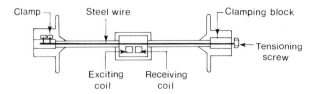

**Figure 5.8**  Vibrating wire strain gauge for embedment in
concrete

### 5.3.3  Alternative loading systems

Mention has been made in Section 5.3.1 of the basic methods that can be
used for the loading of concrete in compression. In this section more
detailed descriptions are given of the systems that have been developed for
load application and load maintenance.

Early research employed either dead-weight loading, sustained through
a compound lever system, or spring loading as indicated in Figure 5.9. More
recently hydraulic systems have become more widely used and preferred,
particularly as load adjustment is both easier and more rapid and the fluid
pressure provides a measure of the applied stress. A typical design for a
hydraulic loading jack is shown in Figure 5.10. Various methods are
available for the applied pressure to be monitored and regulated although,
in practice when the creep measurements are carried out at a closely
controlled temperature it can be adequate to make occasional manual
adjustments to maintain the applied loads within a few per cent.

Either liquid or an inert gas may be used to provide the pressure source.
The latter supplied from a commercial gas cylinder is cheap and convenient,
but care is needed to ensure that the design of the system is safe.

Creep measurements in pure tension are more difficult than in compression, primarily because of the low strength of concrete and hence the low
stress levels that can be applied and the consequent low creep strains. It is
essential to use long specimens, preferably with a reduced cross-sectional
area away from the loading grips, in order to obtain an adequately uniform
stress field. Ross[18] developed a method in which a rubber bag inside a
hollow cylinder was used to induce tensile hoop stresses (Figure 5.11)
although this technique for concrete suffers from the disadvantage already
mentioned with regard to the size of the aggregate and the thickness of the
walls. The system has the potential however for studies under combined
tension and compression, the latter being superimposed by loading in the
axial direction.

Flexure and torsion can be applied to concrete beams or cylindrical
specimens by suitable lever and pulley arrangements, but care should be
taken to ensure that frictional losses are properly taken into account.

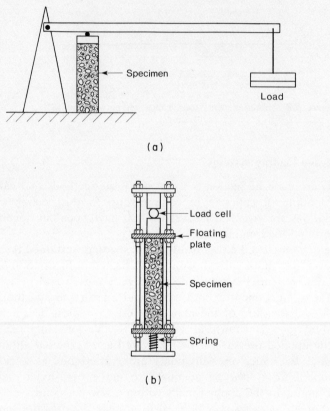

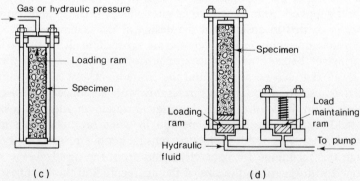

**Figure 5.9**  Compression loading frames for concrete specimens: (a) lever load system; (b) a spring-loaded creep frame with a load cell; (c) a hydraulic load creep frame; (d) a stabilized hydraulic load system for creep tests

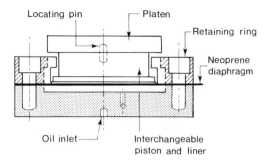

**Figure 5.10** A hydraulic loading jack for creep testing

The deformations of concrete under composite stresses have also been studied. Frequently the loading has been bi-axial or tri-axial compression, with in effect two or three spring or hydraulic compression frames being mounted orthogonally. The difficulty of providing two or three independent stresses has been discussed above in Section 5.2.3, and when solid steel platens are used a significant proportion of the load applied in one direction is carried by the orthogonal loading system. In other words the average stress in the concrete is lower than that applied through the platens. Hilsdorf brush platens partially resolve the situation but the ideal solution is to provide the stresses through fluid platens which cannot carry transverse stresses. One suitable rig is based upon the Hoek triaxial cell[6] in which triaxial compression with $\sigma_1 = \sigma_2 \neq \sigma_3$ can be used, where $\sigma_1$ and $\sigma_2$ are the compressive principal stresses applied through the membrane and $\sigma_3$ the axial principal stress may be compressive, provided by an axial load, or tensile by a hydraulic thrust on shoulders of dumb-bell shaped specimens (Figure 5.12).[5]

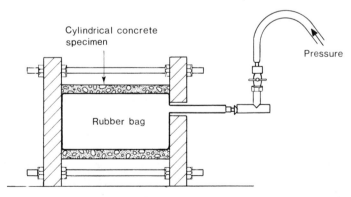

**Figure 5.11** Hoop tensile stresses generated by inflation of an enclosed rubber bag

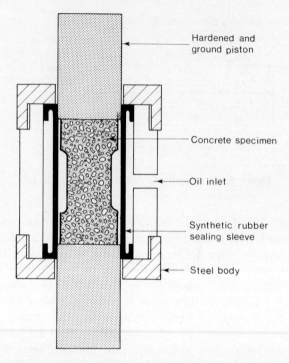

Figure 5.12  Hoek triaxial loading apparatus and dumb-
bell shaped specimen for tensile/compressive stress

It can thus be seen that it is essential to consider loading systems very carefully to be certain that the concrete is stressed in the way desired. It is very easy to employ a system in which some of the loads are transferred to other parts of the system and in which non-uniform and unknown stresses are induced in the specimens. In such a case it is obvious that creep measurements are unsatisfactory.

### 5.3.4  Environmental effects

Creep measurements depend not only upon the prevailing environment but also on the changes that take place while the concrete is under load, particularly the first drying or the first heating to a given temperature. It is not the purpose of this chapter to discuss the effects of such environmental changes but it is pertinent to emphasize the importance of control of the atmosphere around the specimens. The age at which the specimen is loaded and at which it is allowed to dry are also important parameters that must be respected.

Recently recommendations were made for a standard creep test whereby different concretes could be compared.[8] It was suggested that measurements of basic creep, in which drying was prevented by sealing the specimen, was the only way to define a standard test. Tests on three replicates were recommended to allow for specimen variability.

This provides one creep component. When the behaviour of a concrete relative to a known environmental history is required a carefully controlled programme of drying and heating is necessary, proper allowance being made for specimen size. This requirement leads to the section on the measurement of the movement of structural concrete. A good example of the utilization of laboratory measurements to aid the prediction of structural movements is provided by work for concrete pressure vessels in nuclear power engineering. Here the variable parameters included age at stressing and the temperature and loading changes that would subsequently be experienced. Specimens in the laboratory can follow the loading and environmental experience of the structural concrete so that there is a realistic and continuous up-dating of the estimates of the structural concrete.

## 5.4 SITE MEASUREMENTS

The structural movements of concrete will depend upon its maturity when it is loaded and upon the environmental history. These factors are taken into account in the design methods that are used for the estimation of the expected movements. There have been relatively few attempts to compare the actual structural movements that occur with those predicted but such comparisons are an essential stage in the assessment of the validity and usefulness of any design procedure.

When it is possible to make measurements of the structural movements of concrete on site the measurements can be compared with the estimates. In addition specimens of the structural concrete can also be cast and loaded to provide additional data relevant to the assessment.

There are two aspects of site creep work that need emphazing. Firstly the creep rigs must be robust and simple and Tyler[20] developed such unsophisticated equipment. Steel loading frames were made in which the prismatic or cylindrical concrete specimen could be loaded in compression by a small hydraulic jack. The hydraulic jack was powered from a mechanical pump that could be plugged into or disconnected from the jack. This made it possible to load a large number of creep rigs from a single simple power unit and the creep rigs could be stored with the structure under test. The concrete specimen could be loaded at the same age as was the structural concrete so that direct comparison could be made. The creep measurements can be made either from reference studs attached to the surface using a demountable mechanical extensometer or from embedded vibrating wire strain gauges. Unloaded control specimens are used for measurement of the concurrent shrinkage.

The second point that requires emphasizing is the problem of simulation of the drying history of the structural concrete. If a 100 or 150 mm diameter cylinder is used for the creep and shrinkage tests these will dry much more rapidly than would a structural member that was 500 mm or a metre deep. One way to improve the simulation of the drying history of the structural concrete in the smaller test specimen is to seal the specimen partially with a waterproof membrane leaving a restricted drying path. For example a rectangular prism could be sealed on all but one of the larger faces so that drying was from that face alone. It would not be possible to use circular cylinders in this way.

Analysis of *in situ* creep measurements requires some measure of the concurrent unrestrained shrinkage of the concrete. If the strains in the structural concrete are measured with embedded vibrating wire gauges and it is possible to identify a zone subjected to pure compression or tension, two gauges can be mounted orthogonally, one in the compressive or tensile direction. The second gauge then measures the shrinkage and the Poisson creep strain and as creep Poisson's ratio is known, it becomes possible to estimate the shrinkage correction from the structural measurement.[17]

## 5.5  GENERAL DISCUSSION

In this brief chapter it has been possible to highlight a few of the factors that should be considered when a creep investigation is being planned. The main points to be considered are:

  (i)   the purpose of the study;
 (ii)   the precision required;
(iii)   the duration of the experiment;
 (iv)   the curing of the specimens;
  (v)   the environmental changes during the experiment;
 (vi)   the frequency and storage of measurements;
(vii)   the data processing and presentation.

  Items (i) and (ii) affect the choice of loading employed and the instrumentation to be used. They also influence the specimens and the constitution of the specimens (e.g. hardened cement paste or concrete). During the early part of an experiment the creep movement will normally be rapid and hence the frequency of measurement will be higher than is necessary later on. This affects the detailed design of data logging systems that might be used.

  As the drying and temperature histories imposed on the specimens have such an important effect upon their movements it is vital that precise details are provided in every report. Otherwise it becomes impossible to make useful comparisons between the observations by different workers.

For the above reasons it is suggested that a standard creep test for the comparison between different concretes should be made on sealed specimens that undergo solely basic creep.

Further it is recommended that additional specimens should be cast when possible for the measurement of related parameters. These include not only measurements of shrinkage on the unloaded specimens but also of strength, elastic modulus, degree of hydration, and so on. All too often the results of a creep experiment cannot be used by other research workers since some of the pertinent data are missing.

## 5.6 CONCLUSIONS

Creep measurements are of ultimate benefit to the concrete designer and hence the observed data must be relevant to his needs. To achieve this goal it is important that the creep tests are properly selected and carried out. Factors essential to this aim have been discussed.

A basic understanding of the causes of creep is also necessary if concrete is to be used wisely in unusual circumstances. This work is frequently best undertaken on the most active creep component, the hardened cement paste, and some of the additional precautions have also been discussed.

## REFERENCES

1. Bazant, Z. P., Hermann, J. H., Koller, H., and Najjar, L. J. (1973), 'A thin wall cement paste cylinder for creep tests at variable humidity or temperature', *Mater. Struct., RILEM*, **6**, 277–80.
2. Feldman, R. F., and Sereda, P. J. (1963), 'A datum point for estimating the adsorbed water in hydrated Portland cement', *J. Appl. Chem.*, **13**, 375–82.
3. Feldman, R. F. (1972), 'Density and porosity studies of hydrated Portland cement', *J. Cem. Technol.*, **3**, 5–14.
4. Gutteridge, W. A., and Parrott, L. J. (1976), 'A study of the changes in weight, length and interplanar spacing induced by drying and rewetting synthetic CSH(I)', *Cem. Concr. Res.*, **6**, 357–66.
5. Hobbs, D. W. (1970), 'Strength and deformation properties of plain concrete subject to combined stress', *Cem. and Concr. Assoc. (UK)*, Tech. Rep. 42.451, 12 pp.
6. Hoek, E., and Franklin, J. A. (1968), 'Simple triaxial cell for field or laboratory testing of rock', *Trans. Inst. Min. Metall.*, **77A**, 21–6.
7. Hughes, B. P., and Bahramian, B. (1965), 'Cube tests and the uniaxial compressive strength of concrete', *Mag. Concr. Res.*, **17**, 177–82.
8. Illston, J. M., and Pomeroy, C. D. (1975), 'Recommendations for a standard creep test', *Concrete*, **9**, 24–5.
9. Küpfer, H., Hilsdorf, H. K., and Rüsch, H. (1969), 'Behaviour of concrete under biaxial stresses', *Proc. ACI*, **66**, 656–66.

10. Lawrence, C. D. (1969), 'The properties of cement paste compacted under pressure', *Cem. Concr. Assoc. (UK)*, *Res. Rep.* 19, 21 pp.
11. Morice, P. B., and Base, G. D. (1953), 'The design and use of a demountable mechanical strain gauge for concrete structures', *Mag. Concr. Res.*, **5**, 37–42.
12. Mullen, W. G., and Dolch, W. L. (1964), 'Creep of Portland cement paste', *Proc. ASTM*, **64**, 1146–70.
13. Neville, A. M. (1970), *Creep of Concrete: Plain, Reinforced and Prestressed*, North Holland, Amsterdam, Chap. 16.
14. Parrott, L. J. (1970), 'An improved apparatus for biaxial loading of concrete specimens', *J. Strain Anal.*, **5**, 169–76.
15. Parrott, L. J. (1973), 'Load induced dimensional changes of hardened cement paste', *PhD Thesis*, London University.
16. Parrott, L. J. (1975), 'Increase in creep of hardened cement paste due to carbonation under load', *Mag. Concr. Res.*, **27**, 179–81.
17. Parrott, L. J. (1977), 'A study of some long term strains measured in two concrete structures', *Proc. 1st Symp. on Testing in situ of Concrete Structures*, Budapest, Vol. 2, pp. 123–39.
18. Ross, A. D. (1954), 'Experiments on the creep of concrete under two-dimensional stressing', *Mag. Concr. Res.*, **6**, 3–10.
19. Spooner, D. C. (1972), 'The stress–strain relationship for hardened cement pastes in compression', *Mag. Concr. Res.*, **24**, 85–92.
20. Tyler, R. G. (1968), 'Developments on the measurement of strain and stress in concrete bridge structures', *Ministry of Transport (UK)*, *RRL Rep.* LR 189.

*Chapter 6*

# Creep and Shrinkage Mechanisms

*F. H. Wittmann*

## 6.1 INTRODUCTION

There are many papers on creep and shrinkage mechanisms. In the past the problem has been approached from two opposite directions. Some authors tried to formulate deformation mechanisms on the basis of creep and shrinkage measurements carried out on large concrete specimens. Very often special properties have been attributed to the adsorbed water films to explain the macroscopically observed behaviour.

The opposite approach is to investigate directly the physical properties of water near solid surfaces and other details of the microstructure of the xerogel in hardened cement paste.

A concise summary of recent findings on the microstructure has been presented at the 7th International Congress on the Chemistry of Cement[46] and the chapter by J. F. Young in this volume is devoted to this subject.

It will be shown that neither the microstructural approach nor macroscopic observation are sufficient to really understand the complex processes involved in creep and shrinkage. Some physically meaningful mechanisms of creep and shrinkage can be described. The actual behaviour depends, however, to a large extent on so called apparent mechanisms. Apparent mechanisms are phenomena such as crack formation and internally created states of stress which do modify time-dependent deformation.

In this chapter the most important real mechanisms will be outlined and the contribution of apparent mechanisms will be discussed. Most apparent mechanisms can be defined by rigorous mechanical analysis of a drying composite material under load. We will see that any treatise of mechanisms remains confusing and vague without a strict distinction between real and apparent mechanisms.

## 6.2 STRUCTURE OF HARDENED CEMENT PASTE AND CONCRETE

### 6.2.1 Different structural levels of the composite material

Concrete is a heterogeneous material and it has proved to be advantageous to introduce different levels to characterize its structure.[47] The hydration products of cement form a matrix in which the aggregates are embedded. We do not yet understand the structure of the hydration products in all its details. Therefore all available knowledge on the microstructure is summarized in several models. These models represents the microlevel and they are called materials science models.

The behaviour of the composite material cannot be linked directly with mechanisms of the microstructure. The effect of pores and inclusions also has to be taken into consideration. These particularities are introduced on the mesolevel and the resulting models are called materials engineering models.

Finally, by using the results of the models of the micro- and mesolevel, materials laws for the behaviour of concrete can be derived. This is the aim of the work done on the macrolevel and the corresponding engineering models. These models should be presented preferably in such a form that they can be used immediately in advanced numerical analysis. Relevant contributions of the three different structural levels will be discussed briefly.

### 6.2.2  Microlevel: hardened cement paste

It is very difficult to investigate the microstructure of hardened cement paste. Most of the classical methods such as X-ray analysis can hardly be applied. Early sorption measurements and later electron microscopy proved that the hydration products are in a highly dispersed state with an internal surface of some hundred square metres per gram. Such a coherent solid built up essentially by colloidal particles and having a limited water content can suitably be called a xerogel.

All properties of a xerogel are determined by the average coupling strength of individual particles and they are strongly influenced by the interaction of the porous system with water. A lot if not most of the information on the microstructure of hardened cement paste has been gained by means of sorption measurements.

The work of Powers and Brownyard[24] can be looked upon as a real milestone in this area. Later Powers[25] condensed his findings in a model which is essentially based on the thermodynamics of adsorbed water. According to this model creep and shrinkage are caused by the same mechanism, i.e. squeezing out or re-entering of water at the wedge-shaped narrow gaps near points of contact in the xerogel. In the cse of creep the moisture migration is initiated by an external load and in the case of shrinkage by a change of relative humidity. We will come back to this point while discussing real creep and shrinkage mechanisms.

Based on sorption and length-change measurements Feldman and Sereda[11] developed later a different model. They assume that hardened cement paste is built up by a layered irregular structure. But in contrast to

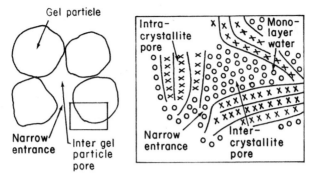

**Figure 6.1** Simplified model for the microstructure of hardened
cement paste as suggested by Kondo and Daimon[18]

Power's model according to Feldman and Sereda the main properties of the
xerogel are not so much influenced by adsorbed water but far more by the
removal or re-entry of interlayer water. Interlayer water in a colloidal
particle might then be squeezed out by an external load just as by a low
water vapour pressure thus causing creep or shrinkage.

In an attempt to unify these diverging views Kondo and Daimon[18]
suggested another model for the microstructure of hardened cement paste.
In this model there is a distinct difference between intergel particle pores,
intercrystallite pores, and intracrystallite pores. A schematic representation
of this model is shown in Figure 6.1. On the surface of gel particles within
intercrystallite pores there is physically adsorbed water which may influence
the properties of the xerogel in a way as suggested by Powers.[25] Moreover,
the interlayer water between layers of crystallites may be removed or it may
re-enter as predicted by Feldman and Sereda.[11]

Taylor[39] and Taylor and Roy[40] combined recent results of different
structural investigations such as trimethylsilylation (TMS) and suggested
another model of the microstructure of type III C–S–H. The schematic
representation of Taylor's model is shown in Figure 6.2. It is suggested that
subcrystalline order can be confined to the immediate neighbourhood of
small fragments of Ca–O layers which are distributed at random and only
rarely parallel to each other. The intervening spaces could then be totally
amorphous.

At this point it is not necessary to discuss chemical and mineralogical
aspects of the microstructure in more detail because this subject is treated
by J. F. Young[52] in Chapter 1 of this volume.

We have, however, to introduce basic elements of another model of the
microstructure which will be used later while discussing mechanisms of
shrinkage and creep, i.e. the Munich model.

Based on a number of different approaches such as the direct observation

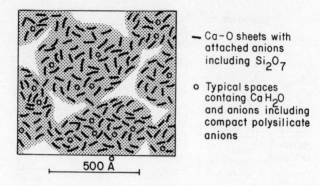

— Ca−O sheets with
attached anions
including $Si_2O_7$

o Typical spaces
containg $Ca H_2O$
and anions including
compact polysilicate
anions

500 Å

**Figure 6.2**  Suggested structure of type III C–S–H according to
Taylor[39]

van der Waals interaction at short distances[36] the study of the coupling of
gel particles within a xerogel by means of the Mössbauer effect[41] or the
determination of the complex permittivity in the range of microwave fre-
quencies[29] and in the range of Hertzian spectroscopy[53] the Munich model
has been developed. It has been described in detail in the literature.[43,44]
Therefore we have to recall relevant aspects only.

It is well known that a liquid droplet having a radius $r$ is under a
hydrostatic pressure $P$:

$$P = 2\gamma/r \tag{6.1}$$

where $\gamma$ represents the surface tension of the liquid. In liquids surface
tension and surface energy are numerically equal. In solids these two values
are at least in the same order of magnitude. In a colloidal system non-
spherical particles may occur. Flood[12] has shown that the mean pressure in
solid particles created by surface tension in such a system can be estimated
with the help of the following equation:

$$P = 2\gamma S/3 \tag{6.2}$$

$S$ stands for the specific surface area and has to be expressed e.g. as $cm^2/cm^3$
in this connection. If the specific surface area is introduced instead of the
radius the actual particle size distribution is neglected or rather replaced by
a mean value. In C–S–H there are particles which are large enough to
ensure that no appreciable internal pressure will be created by surface
tension. Other particles in the same system will experience comparatively
high pressures. The overall response of a system with active and inactive
particles has been calculated by Krasilnikov and co-workers.[19] In their paper
it is pointed out that expansion of gel particles in heterogeneous systems is
not linearly related to the expansion of the total system but a geometrical
magnification factor has to be taken into consideration.

Well aware of the implied simplifications, we may go back to Equation (6.1). Now $r$ has to be looked on as a characteristic value of a given xerogel and $P$ as a mean internal pressure. It is obvious that the resulting internal pressure changes as the surface energy is changed. The surface energy, or rather the interfacial energy of a colloidal system, may be changed by adsorption of gases or vapours. If a film of thickness $\Gamma$ is adsorbed at a given vapour pressure $p$ the interfacial energy measured *in vacuo* decreases by $\Delta\gamma$:[13]

$$\Delta\gamma = \gamma_0 - \gamma = RT \int_0^p \Gamma \, d(\ln p) \qquad (6.3)$$

If $\gamma$ of Equation (6.3) is inserted into Equation (6.1) the change of internal pressure caused by a changing surface energy can be calculated. Each individual gel particle expands as the internal pressure is reduced. Bangham and Fakhoury[3] showed that within certain limits a linear relation exists between the change of interfacial energy and the resulting length change:

$$\Delta l/l = \lambda \, \Delta\gamma \qquad (6.4)$$

Later Hiller[16] expressed $\lambda$ in terms of properties of the colloidal system. It is assumed that in the range of low RH the hydral length change can be described semi-phenomenologically by utilizing Equation (6.4). A more quantitative application of Equation (6.4) is not possible as decisive factors such as the particle size distribution are not sufficiently well known.

As the relative humidity is raised above 50% some surfaces will be separated by disjoining pressure. This leads to additional expansion of the colloidal system. This length change is not caused by a corresponding change of surface energy. Therefore Equation (6.4) cannot be applied in this range.

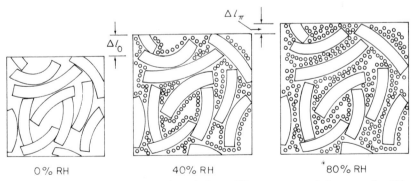

0% RH      40% RH      80% RH

**Figure 6.3** Schematic representation of three different stages of the xerogel within hardened cement paste (Munich model): (a) dry state, all particles are compressed by surface tension (see Equation (1)); (b) surface free energy is reduced by adsorbed water films and therefore the system expands by $\Delta l_0$ (see Equation (3)); and (c) additional swelling ($\Delta l_\pi$) is caused by disjoining pressure and some points of contact are interrupted

Simultaneously the total structure is weakened by the action of disjoining pressure.

A simplifying sketch of the essential aspects of the Munich model is shown in Figure 6.3. Later we will come back to this model.

It has to be pointed out, however, that the xerogel in hardened cement paste is an unstable phase. There is a strong tendency to reach a state of lower energy. Drying and heating both reduce the internal surface thus creating a more stable coarser structure. A rough estimate shows that if the internal surface is reduced from $200 \, \text{m}^2/\text{g}$ to $100 \, \text{m}^2/\text{g}$ an amount of energy of $40 \, \text{J/g}$ is liberated.

### 6.2.3  Mesolevel: pores, inclusions, and cracks

In the preceding section we have discussed some details of the microstructure. Those aspects which can be linked with creep and shrinkage mechanisms have been dealt with in particular so far, however, we have neglected the heterogeneous structure of the material. We have to introduce pores, inclusions, and cracks as the main characteristic features of the mesolevel.

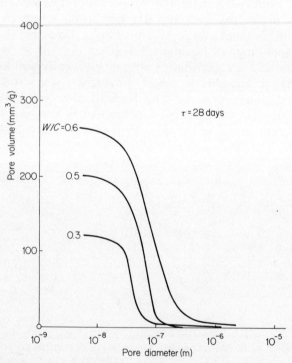

**Figure 6.4**  Influence of water/cement ratio on pore size distribution of hardened cement paste

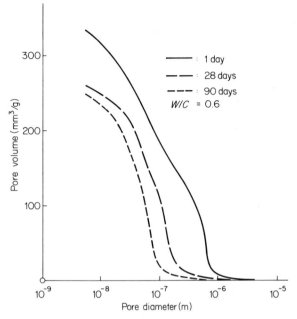

**Figure 6.5** Influence of duration of hydration on pore size distribution of hardened cement paste

Hardened cement paste is a porous material with a wide range of pore sizes. The total porosity as well as the pore size distribution depends on the water/cement ratio and the degree of hydration. In Figures 6.4 and 6.5 some typical examples are given. The drying process and, as a consequence, shrinkage of a porous material are governed essentially by the pore size distribution.

Porosity has a strong influence on creep too. This is mainly due to increasing stress concentrations in the load-bearing solid skeleton as porosity increases.

Most aggregates in concrete can be considered to react in a linear elastic way and to undergo negligible shrinkage. This is one major reason why it is so difficult to relate observed creep and shrinkage strains of concrete with materials mechanisms. The aggregates do not participate in time-dependent deformation but modify the observed behaviour and in particular the internal stress distribution.

If we introduce viscosity $\eta$ as a measure for creep velocity we can apply existing formulae which were derived to describe the flow behaviour of two-phase materials. For low volume concentrations $C_v$ of aggregates the classical Einstein equation can be used:

$$\eta/\eta_0 = 1 + 2.5C_v + 4.4C_v^2 \qquad (6.5)$$

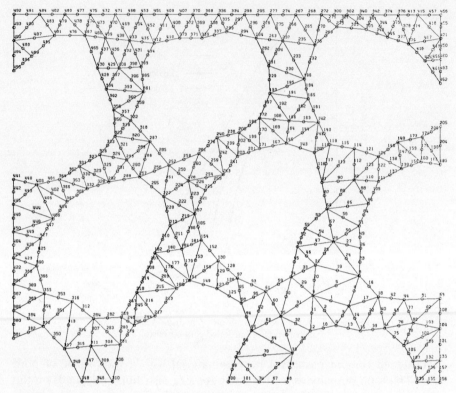

**Figure 6.6** Finite element generated structure of a concrete-like two-phase material (Wittman *et al.*[49])

The limits as well as possible extensions of this approach have been pointed out and discussed earlier.[42]

A totally different possibility to describe creep of composite materials is provided by advanced numerical methods such as finite element analysis. A concrete-like structure can be generated and then the time-dependent behaviour under load can be studied by means of a computer experiment. An example for such a generated structure is given in Figure 6.6. So far most often two-dimensional structures have been studied. It is possible however, to simulate three-dimensional structures as well. In a similar way pores can be taken into consideration.

In principle it is possible to simulate the behaviour of a composite material by means of computer experiments in a realistic way. For the present context it is most important, however, that by finite element analysis it is clearly shown that under usual drying conditions within the composite structure, stresses are created which overcome the tensile strength of the matrix. Cracks formed in this way contribute to the macroscopically observed deformation. We can conclude that crack formation is a major mechanism and attempts to explain shrinkage and creep strains of mortar and concrete by microstructural mechanisms not taking into consideration mechanics of the composite material, are not realistic.

### 6.2.4 Macrolevel: materials laws

Research on creep and shrinkage of concrete has to be oriented towards the application in structural engineering. That means that finally all information available has to be condensed in simplified but reliable materials laws. On the macrolevel the material is considered to be homogeneous and all heterogeneities and structural defects are considered to be smeared out over the total volume.

Empirical materials laws serve to predict materials behaviour under service conditions and examples are relevant national codes. A rather detailed description typical for the macrolevel has been published by Bazant and Panula.[5]

If one looks in more detail even on the macrolevel, materials properties may vary locally. In high columns the lower parts are denser, show less creep, and retarded shrinkage due to hydration under increased hydrostatic pressure. In mass concrete the temperature rise due to hydration in the centre coarsens the structure and thus influences creep and shrinkage.

Surface zones of concrete usually have a lower aggregate volume concentration. In Figure 6.7 a computer generated concrete structure is shown. The spherical aggregates are distributed at random and their size distribution follows a Fuller curve. On the right-hand side the change of the aggregate volume concentration is shown in Figure 6.7(b). It is obvious that the water permeability increases towards the surface and thus the drying process is influenced by this border effect. In addition shrinkage and creep are higher on the surface zones if compared with bulk properties because of the increased content of viscoelastic hardened cement paste (see Figure 6.7(b)).

Although we have seen that, strictly speaking, concrete is inhomogeneous even on the macrolevel, for practical applications simplified materials laws have to be used. In research, however, these effects cannot be neglected and further progress will depend on a balanced systematic study of the various structural details entering on the three different levels.

### 6.3 SHRINKAGE

### 6.3.1 Real mechanisms

#### 6.3.1.1 Capillary shrinkage

It is well known that water in capillaries is under a capillary pressure $P_c$. If the surface tension $\sigma$ of the liquid and the radius of curvature $R$ of the

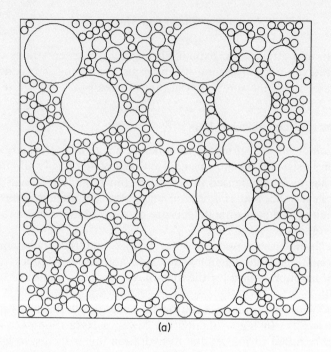

(a)

(b)

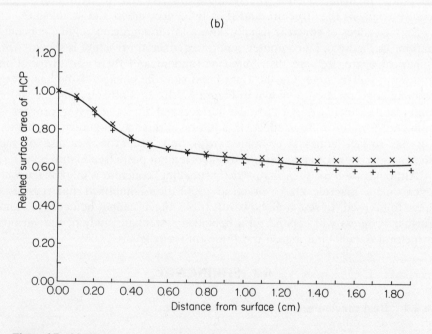

**Figure 6.7** (a) Computer generated structure of concrete with spherical aggregates. The size distribution follows a Fuller curve. (b) The change of the related surface area of hardened cement paste as a function of the distance from the surface. Maximum aggregate size 3 cm (Wittmann *et al.*[49])

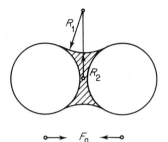

**Figure 6.8**    Capillary pressure and capillary shrinkage of concrete as a function of time

surface of the liquid are known the capillary pressure can be calculated easily. In a general case we have to introduce the main radii $R_1$ and $R_2$ of the curvature of the surface of the liquid instead of the radius $R$. Then the Gauss–Laplace equation reads as follows:

$$P_c = \sigma\left(\frac{1}{R_1} + \frac{1}{R_2}\right) \tag{6.6}$$

It is obvious that in the case of a cylindrical pore the commonly used equation is obtained by putting $R_1 = R_2$. The capillary pressure $P_c$ results in an attractive force $F_a$ between the walls of the capillary or between two particles separated by a liquid filled capillary. This latter situation is shown schematically in Figure 6.8.

In fact in concrete we have two different stages of capillary action. In the fresh state concrete can be thought of as a suspension of cement particles and aggregates in water. In this case the capillary pressure starts to increase if the surface begins to dry. In this case menisci are formed between the particles close to the surface and as consequence all of the pore water is under a capillary pressure. If the process is not excessive and creates cracks this mechanism in fact leads to a beneficial compaction of fresh concrete.

If drying proceeds suddenly the homogeneous water phase disintegrates and the water is trapped in the narrow spaces between neighbouring particles. This case is shown schematically in Figure 6.8.

In Figure 6.9 capillary pressure $P_c$ and capillary shrinkage $\varepsilon$ are plotted as function of duration of hydration under drying conditions.[43] The capillary pressure has been measured directly in fresh concrete. When the water disintegrates capillary pressure cannot be measured with this method. In this second stage one would need a microprobe to observe the capillary pressure in the small capacity spaces between particles. This is the reason why in Figure 6.9 after $3\frac{1}{2}$ hours the observed capillary pressure is suddenly reduced. The peak of capillary pressure in Figure 6.9 marks the end of stage one of capillary shrinkage.

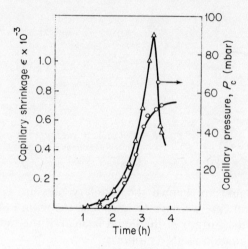

**Figure 6.9** Capillary shrinkage and capillary
pressure of concrete as a function of time

## 6.3.1.2  *Chemical shrinkage*

Chemical shrinkage is a term used to cover a number of distinct shrinkage (or swelling) mechanisms. All these volume changes are caused by chemical reactions. We will mention here the most important mechanisms only. These are:

(a)  hydration shrinkage
(b)  thermal shrinkage
(c)  dehydration shrinkage
(d)  crystallization swelling
(e)  carbonation shrinkage
(f)  conversion shrinkage

Nearly all chemical reactions are followed by a volume change:

$$V_A + V_B \rightarrow V_C + \Delta V$$

If the main constituents of Portland cement (PC) react with water they all undergo a characteristic volume change and a gross volume decrease of about 7% is observed.[10] This is expressed by the following approximate relation:

$$100 \, \text{mm}^3 \, \text{PC} + 125 \, \text{mm}^3 \, \text{H}_2\text{O} \rightarrow 209 \, \text{mm}^3 - \Delta V \qquad (6.7)$$

The volume change due to hydration of cement is proportional to the degree of hydration. In the past hydration shrinkage and capillary shrinkage have not always been separated clearly. Cracking of concrete after two to three hours has sometimes been explained by chemical shrinkage. After two hours, however, the degree of hydration is so small that hydration shrinkage can be neglected and the true reason in most cases is capillary shrinkage. It is important to distinguish between the different shrinkage mechanisms if one is looking for reliable measures to prevent damage.

When Portland cement reacts with water a certain degree of heat is liberated. The specific heat of hydration depends on the chemical composition of the cement. For the main components of Portland cement the following values for the heat of hydration have been found:

$$
\begin{array}{ll}
C_3S & 120\,\text{cal/g} \\
C_2S & 62\,\text{cal/g} \\
C_3A & 207\,\text{cal/g} \\
C_4AF & 100\,\text{cal/g}
\end{array}
$$

From this a value of about 110 cal/g is found for ordinary Portland cement. Some of the heat is liberated while the concrete is still young and easily deformable. In massive elements thermal swelling can be observed. As the rate of hydration slows down the temperature decreases and as a consequence a concrete specimen undergoes thermal shrinkage, which can cause serious cracking.

Some of the hydration products are not stable with respect to a decrease of the relative humidity. Well crystallized phases lose some of their hydrate water at precise values of relative humidity.[34,35] In colloidal particles the transition can be smeared out over a wide range of relative humidity. This loss of hydrate water under drying conditions is always accompanied by a volume change, i.e. the dehydration shrinkage.

During hydration both colloidal products and crystallized phases are formed. The early hydration products can expand in the water filled space between the solid particles. Once the solid skeleton is built up further crystal growth is hindered and thus an internal pressure is created. Crystallization pressure can cause moderate swelling of concrete. Under normal conditions drying shrinkage overcompensates this type of swelling. In swelling or shrinkage compensated cements this mechanism is used. By adding for example sulphates, voluminous sulphoaluminate hydrates are formed during hydration and they may balance normal shrinkage over a certain period if well proportioned.

Finally carbonation shrinkage must be mentioned. Calcium hydroxide formed during hydration of cement is a metastable compound and can react with $CO_2$ from the surrounding air. In a simplified form this reaction can be described by the following equation:

$$Ca(OH)_2 + CO_2 \rightarrow CaCO_3 + H_2O \tag{6.8}$$

Under normal climatic conditions carbonation takes place in the zones near the surface of a concrete member. The liberated water evaporates and in the final state the reaction is accompanied by a decrease of volume. Depending on the quality of concrete the degree of carbonation can vary from several mm to several cm.[28] Therefore carbonation shrinkage is limited to surface zones only.

Some phases in hydrated cement paste, especially aluminate hydrates, undergo slowly a transition (conversion) to a more stable form. This is a major mechanism of chemical shrinkage more precisely called conversion shrinkage in high alumina cements. If the temperature is raised most mechanisms of chemical shrinkage are accelerated.

Not all mechanisms of chemical shrinkage can be linked directly with macroscopically observed dilatation. In some cases the volume change results mainly in an increase (or decrease) of porosity.

### 6.3.1.3  Drying shrinkage

Having introduced the mechanisms of capillary shrinkage and chemical shrinkage we will now discuss mechanisms of drying shrinkage. Sometimes drying shrinkage is supposed to be the only possible hygral length change of concrete. We will see that there are two different mechanisms involved in drying shrinkage.

Drying shrinkage (or wetting swelling) is defined to be the volume change of a colloidal inert system as its moisture content is changed. All chemical reactions are therefore neglected. In the case of hardened cement paste hydration products in the dry state form a xerogel. In Figure 6.10 swelling of

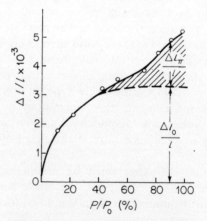

**Figure 6.10**  Swelling of the xerogel in hardened cement paste as a function of relative water vapour pressure (Wittmann[43])

the xerogel in hardened cement paste is shown as function of relative water vapour pressure.[43] Swelling of a dried system is considered because under these conditions chemical shrinkage plays a minor role.

At a relative humidity of about 25% a monolayer of water is adsorbed on the surface. Above 50% RH capillary condensation takes place. In Section 6.2.2 we discussed the interaction of a xerogel with water vapour. It turned out to be helpful to discuss the interaction in the low and the high humidity regions separately.

In the low humidity region the surface energy of the xerogel is changed by adsorbed water films. The Bangham equation (Equation (6.4)) predicts a linear relationship between wetting swelling (or drying shrinkage) and the change of surface energy. If we replot the measured length change of Figure 6.10 as a function of the change of interfacial energy $\Delta\gamma$ we find the data shown in Figure 6.11. Up to a RH of about 50% the Bangham equation is a good approximation and we can conclude that in the low humidity region change of surface energy due to adsorbed water films is the dominant mechanism of drying shrinkage. This result has been verified by many experimental data for different xerogels.

From the results shown in Figures 6.10 and 6.11 it is obvious that in the high humidity region an additional mechanism has to be taken into consideration. If capillary condensation takes place one ought to expect a contraction of the xerogel due to capillary pressure. This is evidently not correct. The reason for this is that water in fine capillaries differs in many respects from bulk water. The chemical potential is changed and the interaction with neighbouring solid surfaces imposes a certain structure to the moving water molecules. In addition an electrical double-layer is formed close to the

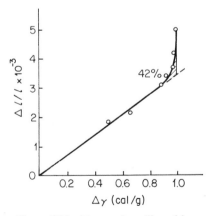

**Figure 6.11** Measured swelling of hardened cement paste as shown in Figure 6.10 as function of change of interfacial energy

surface. In narrow gaps the space charge ($\zeta$-potential) creates a repulsive force.

Because of these peculiarities water in narrow gaps exerts a disjoining pressure on the pore walls.[37] At this moment it is not possible to predict quantitatively the action of disjoining pressure. Further experimental and theoretical investigations are needed. Therefore when we introduce disjoining pressure as the dominant mechanism of drying shrinkage in the high humidity region we have to keep in mind that we can explain the different components of disjoining pressure but that we do not yet fully understand the complex interaction of water with pore walls in fine capillaries.

We have used the Munich model to explain the two basic mechanisms of drying shrinkage. It should be noted at this point that quantitative interpretation of experimental data is hampered by the fact that in drying concrete several mechanisms have to be taken into consideration simultaneously.

### 6.3.2  Apparent mechanisms

#### 6.3.2.1  Influence of geometry

In the literature we find a lot of published data on shrinkage of hardened cement paste, mortar, and concrete. But usually it is not possible to relate the observed total hygral deformation directly with corresponding mechanisms. Therefore shrinkage deformation has been subdivided phenomenologically and sometimes arbitrarily into different components such as reversible and irreversible shrinkage.[15] Being aware that several mechanisms may interact and that they may hardly be separated some authors introduced a distinction between shrinkage due to capillary drying and shrinkage due to gel drying.[7–9,22,51] Bentur *et al.* have also studied changes of the microstructure as a consequence of drying.

The only hygral length change which can be linked directly with mechanisms involved is unrestrained shrinkage. Experimentally, however, it is very difficult to determine unrestrained shrinkage. In a specimen with a given thickness immediately after the drying process begins a hygral gradient builds up. Depending on the geometry and the diffusion coefficient a hygral gradient can exist in concrete for many years. The measured length change under these conditions is the consequence of the resulting internal stress distribution.

In Figure 6.12 the moisture gradient in a prism is shown for three different stages of drying. Unrestrained shrinkage is the related volume change of an infinite volume element. It is in fact the immediate response of a xerogel to a change in moisture content. It is obvious that in a real drying specimen the shrinkage of the outer layers is hindered by the still saturated

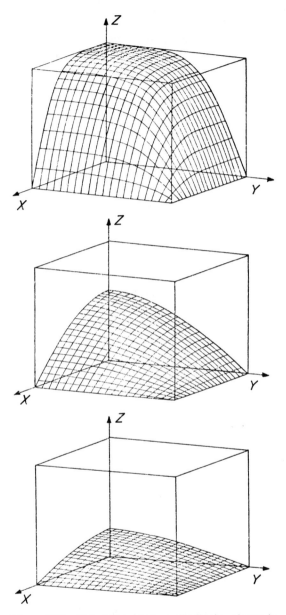

**Figure 6.12** Calculated moisture distribution shown in one quarter of a drying prism. During the initial period the centre part of the concrete specimen is still saturated. From the moisture distribution unrestrained shrinkage of different layers can be calculated. The total hygral length change is then obtained by taking the restraining effect of adjacent layers into account (Wittmann[45])

inner part. In principle there are two ways to determine unrestrained shrinkage approximately. Either the dimensions of moisture movement are chosen sufficiently small (extremely thin specimen) or the humidity is changed by minute steps. It can be shown that in both ways the necessary conditions can hardly be fulfilled in real experiments. The remaining alternative is a rigorous analysis of the internal state of stress by numerical methods. So far this has only been done for comparatively simple geometries.

We can conclude that real mechanisms must be derived from unrestrained shrinkage. The diffusion coefficient depends on the moisture content. As the moisture content is lowered the diffusion coefficient decreases. This leads to a slowed down overall shrinkage but it is a typical apparent mechanism.

Some of the internal state of stress is relaxed by creep of the material. If a specimen is dried from 100% RH to 80% RH the outer shell (tensile zone) will necessarily undergo some creep. If, however, an identical specimen is dried from 100% RH to 30% RH the dry outer shell will show practically no creep during the whole drying process. This change of internal stress results in a modified geometry-dependent shrinkage. This is just another example for an apparent mechanism.

We could easily add some more apparent mechanisms all being caused by geometry, the moisture diffusion process or the change of materials properties as a function of moisture content. But the aim of this section is not to give a complete list of apparent mechanisms but rather to point out that while discussing real mechanisms the apparent mechanisms have to be separated carefully.

### 6.3.2.2   Influence of cracking

So far we have totally neglected crack formation in our discussion. It can be shown easily, that under drying conditions tensile stresses in the drying outer zones usually overcome the tensile strength of the material. This necessarily results in crack formation. Under swelling conditions crack formation takes place in the outer zones of the specimen. It is even possible for swelling cracks to run all through the specimen. Depending on geometry and on the difference of RH imposed during the drying experiment more or less severe cracking takes place. Cracks inevitably change the time dependence as well as the final value of shrinkage and the proportion of reversible and irreversible shrinkage. But crack formation in the tensile zone of a drying (or swelling) specimen is not a real mechanism, it has to be considered to be another apparent mechanism.

Concrete is by no means a homogeneous material and therefore we have to take into consideration its composite structure (see for example Figure 6.6). We realize that shrinkage takes place in the porous matrix only. The hygral length change of most aggregates can be neglected. The resulting differential stresses cause additional cracking in the structure of concrete. Again it is obvious that deformation of the material cannot be linked directly with the shrinkage mechanism. Real shrinkage mechanisms only govern the strains of the porous matrix and crack formation in the heterogeneous matrix is another typical apparent shrinkage mechanism.

## 6.4  CREEP

### 6.4.1  Real mechanisms

#### 6.4.1.1  Short-time creep

We will subdivide the discussion of real creep mechanisms in those termed short-time creep and those termed long-time creep. Ruetz[26] was probably the first to point out that there is a special short-time creep mechanism. Later Sellevold and Richards[31] and Sellevold[32] related low frequency internal friction with short-time creep of hardened cement paste. They determined the loss angle of a small vibrating cantilever of hardened cement paste with one free end. With this method it is possible to observe a distinct transition in the frequency range of about 1 Hz. By using Schwarzl's method[30] this transition can be related directly with short-time creep. This means that in this range hardened cement paste reacts like a linear viscoelastic solid.

Based on these results it can be concluded that short-time creep is caused by a stress induced redistribution of capillary water within the structure of hardened cement paste. Additional evidence for this hypothesis is provided by the measurements of the electromechanical effect.[42] In this case an externally applied alternating electric field forces the capillary water due to the $\zeta$-potential to move. In a small bar of hardened cement paste this water movement creates an alternating bending moment. If the frequency is too high the water cannot follow. As soon as the frequency approaches the characteristic frequency of the transition which corresponds to short-time creep the electrically induced bending moment increases and reaches a maximum value.

Thus we can conclude that the mechanism for short-time creep is water movement and redistribution in the porous structure. The thermodynamic approach developed by Powers[25] seems to be adequate to describe this type of creep.

#### 6.4.1.2  Displacements, rate theory

For short-time creep we could identify one creep mechanism. For long-time creep the situation is far more complex. The xerogel forms a solid skeleton in hardened cement paste. This porous system interacts with adsorbed and capillary condensed water. In the unloaded state all gel particles are fixed to their surroundings partly by primary bonds and partly by secondary bonds. It is impossible to indicate quantitatively and in detail the bonding of an individual gel particle. But we can estimate the average coupling force.[41]

If creep takes place the coupling force of a larger number of gel particles has to be overcome so that these particles can leave their original position in the xerogel. This movement can be looked upon as playing the role of displacement of dislocation, and vacancies in crystalline materials.

Long-time creep in hardened cement paste is in fact the consequence of displacement of gel particles and to some extent creep within particles under high concentrated stress. To describe the creep process realistically we do not necessarily need to know all the different processes involved. If we know well enough the average values and if the number of particles and events is high enough we can apply rate theory. In a colloidal system such as hardened cement paste these two conditions are fulfilled. Therefore we will try to characterize the essential creep mechanisms by introducing basic elements of rate theory.

All gel particles are fixed to their position of rest by a characteristic coupling force. If one wants to separate a given gel particle from its surroundings so that it can be fixed in a new position a certain energy $Q$ is required. This energy is called activation energy of the creep process. The probability $P$ that a particle is removed by thermal energy is given by:

$$P = P_0 \exp\left(-\frac{Q}{RT}\right) \tag{6.9}$$

In Figure 6.13 the coupling of the colloidal particles is shown schematically by a potential trough. It can be seen that all particles which have at least the activation energy $Q$ can leave the trough. Figure 6.13 is a two-dimensional representation of the real three-dimensional process. If there is no load applied the particles which are freed have equal probability to leave their original position to the right or to the left (in the two-dimensional model). That means no net deformation will be observed after many particles have changed their position.

If an external force $F$ is applied the potential trough becomes asymmetrical and now the probability of a jump in the direction of the applied force is increased:

$$P' = P_0 \exp\left[-\left(\frac{Q}{RT} - \frac{V\sigma}{RT}\right)\right] \tag{6.10}$$

and of course the probability of a jump in the opposite direction is

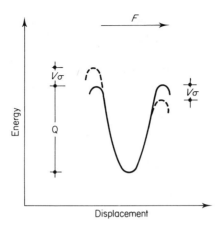

**Figure 6.13** Schematic representation of the potential trough without (solid line) and with an external force $F$ applied (dashed line). The activation energy $Q$ corresponds to the average coupling energy of gel particles in the xerogel of hardened cement paste

decreased in the same way:

$$P'' = P_0 \exp \left[ -\left( \frac{Q}{RT} + \frac{V\sigma}{RT} \right) \right] \qquad (6.11)$$

The asymmetrical trough is shown schematically by dashed lines in Figure 6.13. In the last two equations $\sigma$ stands for the locally existing stress and $V$ has the dimension of a volume. It is therefore called the activation volume of the creep process. If there is a well defined creep mechanism, the activation volume has the meaning of a volume involved in an elementary step. This can be interpreted by the cross section of the moving particle multiplied by the distance of one jump. In a colloidal system $Q$ and $V$ have to be considered to be average values representing a wide spectrum (see e.g. Niklas[23]). Details of rate theory can be found in the book by Krausz and Eyring.[20]

By using Equations (6.10) and (6.11) the following relation for the creep rate can be deduced:[42]

$$\dot{\varepsilon} = \dot{\varepsilon}_0 \exp \left( -\frac{Q}{RT} \right) \sinh \left( \frac{V}{RT} \sigma \right) \qquad (6.12)$$

If correct values for the activation energy and the activation volume are available, Equation (6.12) describes the influence of temperature and of stress level on creep of hardened cement paste and concrete. Corresponding values have been determined experimentally by several authors and they are described in the literature (see e.g. Klug and Wittmann[17], Straub and Wittmann[38], Luijerink[21]).

If particles are moved in the xerogel they are normally fixed more strongly in their new positions. This can be explained by the fact that the weakest links are affected first. Then $\dot{\varepsilon}_0$ from Equation (6.12) can be written so as to describe the time dependence of creep[42]:

$$\dot{\varepsilon}_0 = a_0 (t - \tau)^{n-1} \qquad (6.13)$$

where $(t - \tau)$ is the duration of load with $t$ being the age of the specimen and $\tau$ the age of loading.

In this equation $a_0$ represents the density of creep centres at time $(t - \tau) = 1$, that means the volume concentration of volume elements which contribute to the creep process.

In Figure 6.14 some data taken from Hannant[14] are replotted on a double logarithmic scale. The straight line indicates that the power function theoretically predicted by Equation (6.13) is a good approximation of the experimentally determined creep function. The power function has been found superior to other creep functions by extensive data fitting.[6,50]

From the temperature dependence of creep of concrete as shown in Figure 6.14 the activation energy can be determined. In this case a value of 6.7 kcal/mol has been found.[42]

In concrete technology it is more usual to use creep deformation instead of creep rate, therefore we rewrite Equation (6.13) in an integrated form:

$$\varepsilon_0 = a' (t - \tau)^n \qquad (6.14)$$

So far we have completely neglected the influence of hydration on creep. As the paste matures the xerogel becomes more stable and as a consequence the density of creep centres is reduced. It must be stated at this point that

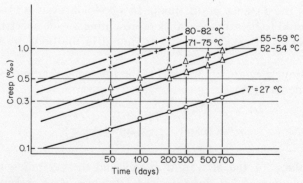

**Figure 6.14** Creep of concrete as function of time. The parameter is the temperature (Hannant[14])

maturing of the xerogel can go on even if hydration has come to an end. In Figure 6.15 the creep deformation after a duration of load $(t - \tau) = 1$ day has been plotted on double logarithmic scale. For the interval chosen a power function represents this relation well:

$$a' = a\tau^{-m} \tag{6.15}$$

By using Equations (6.14) and (6.15) we can rewrite Equation (6.12) in the integrated form:

$$\varepsilon = a\tau^{-m}(t - \tau)^n \exp\left(-\frac{Q}{RT}\right) \sinh\left(\frac{V}{RT}\sigma\right) \tag{6.16}$$

This is the well known double-power law which is now widely used (see e.g. Bazant and Osman[4]). It is evident that all parameters $a$, $m$, $n$, $Q$, and $V$ depend on the actual state of the xerogel.

We have already discussed that fact that the microstructure of hardened cement paste is a unstable phase and that drying especially changes its properties. Therefore it must be expected that a change of moisture content has a significant influence on creep. In Figure 6.16 the parameter $a$ of Equation (6.16) is shown as function of relative humidity.[42] Values of three test series which have been loaded with different levels are plotted in Figure 6.16. All specimens have been equilibrated with the corresponding relative humidity before loading and have been kept in the same relative humidity while under load.

It can be seen that in the low humidity region there is no significant influence of RH on creep. That means that a change of the interfacial energy does not have a great influence on creep. If the disjoining pressure separates some of the gel particles, however, the creep rate increases sharply. This result can be interpreted by the long-time creep mechanism discussed in this section. Disjoining pressure causes loosening of the microstructure and hence increases the density of creep centres. A direct correlation between swelling

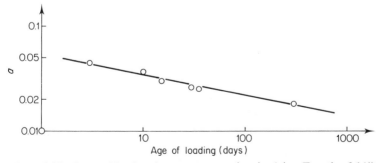

**Figure 6.15**   Creep of hardened cement paste at $(t - \tau) = 1$ (see Equation 6.14)) as a function of age of loading $\tau$ according to Ruetz[26]

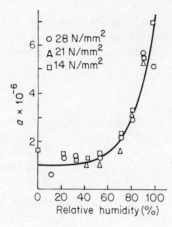

**Figure 6.16** Factor $a$ of equation (6.16) as function of RH. At higher RH the action of disjoining pressure decreases the stability of the xerogel and hence increases the density of creep centres (Wittmann[42])

due to disjoining pressure and change of creep has been pointed out by Setzer.[33]

Thus we have described creep of concrete as a thermally activated process which can be described by rate theory by taking average values for the activation volume, activation energy, and density of creep centres. The influence of age of loading and moisture content on this creep mechanism can be taken into consideration in a reasonable way. In hardened cement paste next to the creep centres we find zones and particles which store energy elastically.

### 6.4.2 Apparent creep mechanisms

The most obvious and probably the most important apparent creep mechanism is drying creep. There are many hypotheses on drying creep mechanisms. However, we will not discuss drying creep here because this is treated in detail in the next section.

If concrete is heated while under load an increased time-dependent deformation is observed. This thermal transient creep is also attributed to special creep mechanisms. But in fact this situation is comparable with drying creep. A temperature change causes firstly a thermal gradient which is followed by a hygral gradient. The resulting internal stress distribution changes the creep rate. At the same time cracking may occur and heat treatment is known to coarsen the fragile microstructure of hardened cement paste. A rigorous analysis of the stress distribution during thermal transient creep has not yet been carried out. Therefore it is not possible at this moment to exclude the existence of a special thermal creep mechanism. It is evident, however, that a major part, if not all, of the observed increase of creep under heating or cooling conditions is due to thermal and hygral gradients, as well as cracking and thus due to apparent creep mechanisms.

In a composite material such as concrete another apparent creep mechanism can be observed. In a simplified way we can say that in the two-phase material the aggregates react in a linear elastic way while the hardened cement paste can be considered to be viscoelastic. If creep deformation in the binding matrix takes place the aggregates are subjected to stress. In this way elastic energy is stored in the two-phase material. This apparent creep mechanism depends essentially on the elastic modulus of the aggregates.

If a concrete is unloaded after a certain period of constant or variable stress, the elastic energy stored in the aggregates causes some reversible creep. This also is an apparent creep mechanism because it is caused purely by stress redistribution within the composite structure. The effect of apparent creep mechanisms caused by the composite structure can be studied in a systematic way by numerical methods. An example is given by the finite element structure shown in Figure 6.6. On a smaller scale the same applies to the heterogeneous structure of hardened cement paste.

This is only a selection of apparent mechanisms. The main reason why they have been introduced is that any discussion on creep mechanisms of concrete is meaningless if not all apparent mechanisms are identified and their contributions to the total observed deformation are not knówn.

## 6.5 SIMULTANEOUS CREEP AND SHRINKAGE

In most cases creep takes place simultaneously with shrinkage. Under these conditions the interpretation of measured data is still a matter of major controversy.

In Figure 6.17 creep and shrinkage deformation is shown schematically as a function of time. If creep and shrinkage take place simultaneously, the observed deformation is always higher than the sum of creep and shrinkage when measured separately at companion specimens. Some authors therefore conclude that creep must be increased by the drying process and consequently different mechanisms of accelerated creep are described in the literature (see e.g. Ali and Kesler[2], Ruetz[27]). Another group states that creep remains unaffected by drying but that shrinkage is increased in a specimen under load (see e.g. Alexandrowsky[1]). The shaded area in Figure 6.17 represents the increase of the total deformation in question. Part of the increased deformation may be explained simply by the non-linear dependence of creep deformation on stress. Here we will try to discuss the possible origins of the increased deformation.

Among the apparent mechanisms of shrinkage we discussed crack formation in the tensile zones of the internal stress field. In Figure 6.18 a typical stress distribution caused by the moisture gradient in a drying specimen is shown schematically. In the example chosen the maximum tensile stress $\sigma_m$

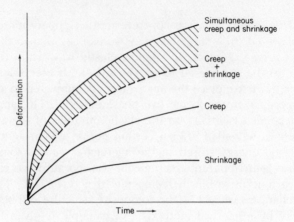

**Figure 6.17**   Schematic representation of creep, shrinkage, and simultanous creep and shrinkage. The increased deformation which is observed when creep and shrinkage take place simultaneously is indicated by the shaded area

overcomes the tensile strength of the material $\beta_t$. Under these conditions the outer drying zones will be cracked. By applying an external compressive load it is possible to reduce the maximum tensile stress so that it remains below the tensile strength. In this way crack formation is avoided.

In fact the internal state of stress as shown in Figure 6.18 is not correct because it will be modified by crack formation. The resulting deformation of a drying specimen depends, however, on the corresponding internal state of stress. Therefore in a realistic analysis it is important to take crack formation into consideration. Shrinkage deformation has been calculated by using the moisture gradient shown in Figure 6.12 and introducing unrestrained shrinkage as function of RH. In addition it was supposed that above a certain value of tensile stress cracks are formed. The result is shown in Figure 6.19

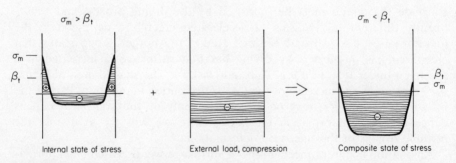

**Figure 6.18**   Internal state of stress due to a moisture gradient. In the example the estimated maximum tensile stres $\sigma_m$ is greater than tensile strength $\beta_t$. If a compressive load is applied the tensile stress in the outer drying zones is reduced and thus crack formation is prevented

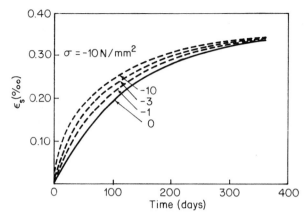

**Figure 6.19** Shrinkage of a concrete as calculated without an external load (0 N/mm$^2$) and with external compressive loads of 1, 3, and 10 N/mm$^2$

by a solid line.[48] If external compressive loads of 1, 3, and 10 N/mm$^2$ are applied crack formation is hindered and therefore the time dependence of shrinkage deformation is influenced. As can be seen from Figure 6.19 the rate of shrinkage is considerably increased by the external compressive load, especially in the initial phase of drying.

Here again we have a typical apparent mechanism. The increase of shrinkage of up to 80% is not due to an accelerated mechanism but solely due to a change of the internal state of stress.

The same numerical analysis can be carried out by assuming the application of an external tensile load. It is evident that cracking under these conditions will be more severe. In Figure 6.20 results are plotted. The solid line corresponds again to shrinkage of an unloaded specimen. A small applied tensile load changes the internal state of stress due to increased cracking in such a way that the initial shrinkage seems to be slowed down. This is just another example of an apparent mechanism. Above a certain level an applied tensile load causes rupture of the specimen. This is also indicated in Figure 6.20 by the shrinkage curves which correspond to tensile loads of 0.8 and 1.0 N/mm$^2$.

In this analysis creep has been neglected to demonstrate clearly the effect of an external load on shrinkage. Under the conditions chosen in this analysis the observed hygral length change may vary by a factor of two if only an external load is added to the internal stress field. This enormous increase is not related to a change of a real shrinkage mechanism. From this fact it follows that shrinkage mechanisms, i.e. processes in the microstructure, cannot be determined by simple observation of macroscopic deformation.

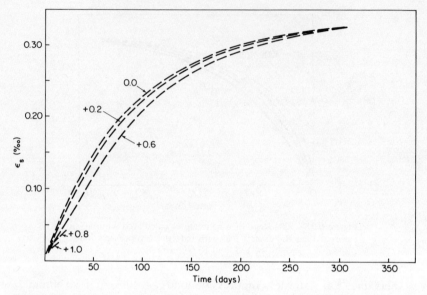

**Figure 6.20**   Shrinkage of an unloaded concrete specimen (solid line) and of companion specimens under a tensile load of $0.2\,\text{N/mm}^2$ and $0.6\,\text{N/mm}^2$. In the case of
external tensile loads of $0.8\,\text{N/mm}^2$ and $1.0\,\text{N/mm}^2$ the specimen fails as the internal
state of stress reaches a critical level

By careful mechanical analysis the unrestrained shrinkage can be obtained. This is the only value with a physical meaning and which can be
explained on the basis of real mechanisms.

Furthermore it is evident that creep and shrinkage of a drying specimen
under load cannot be separated into two components. The resulting total
deformation does not depend on the internal state of stress and the external
load separately but only on the composite state of stress. It follows that
mechanisms which have been defined on the basis of this usual subdivision
have no real meaning.

In a more general way we can say that macroscopically observed total
deformation depends on the external state of stress, and the moisture and
the temperature distribution. In addition the degree of hydration (effective
age) may vary within a specimen. To find the contribution of real mechanisms to macroscopic deformation we have to subdivide the total volume into
$n$ elements. Each volume element $V_i$ (see Figure 6.21) is chosen to be small
enough so that for a given time interval $\Delta t$ it may be looked upon to be in
quasi-equilibrium. Then the deformation of this volume element $V_i$ depends
on the actual existing stress which is the sum of the locally observed part of
the external load $\sigma_e$ and of the internal stress field $\sigma_i$, the temperature $T$, the
relative humidity RH, and the degree of hydration $\alpha$:

$$\varepsilon_i(V_i) = f(\sigma_e, \sigma_i, T, \text{RH}, \alpha) \qquad (6.17)$$

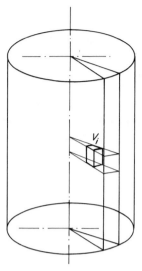

$$\epsilon_i\,(V_i\,) = f(\sigma_e\,,\ \sigma_i,\,T,\ \text{RH},\,\alpha\,)$$

$$\epsilon_{tot} = f\,(\epsilon_i\,)$$

**Figure 6.21** Creep deformation of a small volume element depends on stress, temperature, relative humidities, and degree of hydration. The total deformation of a macroscopic specimen is determined by the unrestrained deformation of all volume elements $V_i$

Within the small volume element we can characterize the time-dependent deformation by real mechanisms. Therefore we call $\epsilon_i(V_i)$ the real creep function. In fact $\epsilon_i(V_i)$ is a complementary value to the unrestrained shrinkage introduced earlier. If we know the real creep function and unrestrained shrinkage as function of RH we can calculate the total deformation.

It has been shown that an increased total deformation of drying concrete under load can be explained on the basis of a mechanical analysis. This part is independent of creep and shrinkage mechanisms. At present it cannot be excluded that under drying conditions accelerated mechanisms contribute to the observed increased total deformation. We can state, however, that if these mechanisms exist they are of minor importance.

## 6.6  CONCLUSIONS

There are several shrinkage mechanisms. They have been subdivided into three distinct groups: capillary shrinkage, chemical shrinkage, and drying shrinkage. Hygral shrinkage can be explained satisfactorily by means of the Munich model.

For creep two different mechanisms are described: short-time creep can be explained by redistribution of water in the microstructure, the long-time creep is caused by displacement of gel particles. Rate theory is a powerful tool to describe long-time creep.

Real shrinkage mechanisms have to be related to unrestrained shrinkage and real creep mechamisms to the real creep function as introduced by Equation (6.17). If creep and shrinkage are separated by definition the resulting functions necessarily have to be inter-related.

Summarizing, the most important real and apparent mechanisms are shown in Table 6.1. These are well defined mechanisms which can be explained by physical and chemical processes and which can be analysed and simulated numerically. In Table 6.1 fictitious mechanisms are also mentioned; they are based on erroneous assumptions.

The actually observed time-dependent deformation cannot be linked directly with creep or shrinkage mechanisms. A number of apparent mechanisms are always involved and they modify the behaviour.

Shrinkage and swelling are governed by the internal state of stress which is caused by the inhomogeneous moisture distribution. Diffusion theory can be applied to predict time-dependent deformations of concrete. The most important factors entering this analysis are diffusion coefficient, unrestrained shrinkage, tensile strength, elastic modulus, and creep. All parameters depend on the moisture content. The only hygral length change which can be interpreted by means of shrinkage mechanisms is unrestrained shrinkage.

Crack formation in the tensile zone has a major influence on the time dependence of hygral length change. The total deformation of a drying specimen under load cannot be subdivided into creep and shrinkage components. All mechanisms which have been based on this subdivision are meaningless.

Total deformation is the only well defined quantity; it is caused by real and apparent mechanisms. In an analysis of creep and shrinkage strains all apparent mechanisms have to be identified.

Table 6.1 Subdivision of processes involved in creep and shrinkage of concrete in real and apparent mechanisms. If this distinction is not made clearly fictitious mechanisms have to be introduced

| | Real mechanisms | Apparent mechanisms | Fictitious mechanisms |
|---|---|---|---|
| Creep (without exchange of moisture) | short-time creep particle displacement | internal stress distribution | |
| Shrinkage (without external load) | capillary shrinkage chemical shrinkage drying shrinkage | hygral gradient crack formation | |
| Simultaneous creep and shrinkage | | hygral gradient crack formation | load-induced shrinkage drying creep |

Numerical methods such as finite element analysis allow us to take the heterogeneous structure of the composite material into consideration. Numerical methods also serve to distinguish between real and apparent mechanisms.

## REFERENCES

1. Aleksandrovsky, C. W. (1966), *Design of concrete and reinforced concrete construction for temperature and moisture action* (with regard to concrete creep) (in Russian), Stroyizdat, Moscow.
2. Ali, I., and Kesler, C. E. (1964), 'Mechanisms of creep in concrete', *Symposium on Creep of Concrete*, American Concrete Institute, SP-9, pp. 35–37.
3. Bangham, D. H., and Fakhoury, N. (1931), 'The swelling of charcoal, Part I, Preliminary experiments with water vapour, carbon dioxide, ammonia and sulphur dioxide', *Proc. R. Soc. A*, **130**, pp. 81–89.
4. Bazant, Z. P., and Osman, E. (1975), 'On the choice of creep function for standard recommendations on practical analysis of structures', *Cem. Concr. Res.*, **5**, 129–38.
5. Bazant, Z. P., and Panula, L. (1978–79), 'Practical prediction of time-dependent deformations of concrete', *Mater. Struct.*, **11**, 307–28, 415–34; **12**, 169–83.
6. Bazant, Z. P., and Thonguthai, W. (1976), 'Optimization check of certain practical formulations for concrete creep', *Mater. Struct.*, **9**, 91–8.
7. Bentur, A., Milestone, N. B., and Young, J. F. (1978), 'Creep and drying shrinkage of calcium silicate pastes II. Induced microstructural and chemical changes', *Cem. Concr. Res.*, **8**, 721–32.
8. Bentur, A., Berger, R. L., Lawrence, F. V., Milestone, N. B., Mindess, S., and Young, J. F. (1979a), 'Creep and drying of calcium silicate pastes III. Hypothesis of irreversible creep and drying shrinkage', *Cem. Concr. Res.*, **9**, 83–6.
9. Bentur, A., Milestone, N. B., Young, J. F., and Mindess, S. (1979b), 'Creep and drying shrinkage of calcium silicate pastes IV. Effect of accelerated curing', *Cem. Concr. Res.*, **9**, 161–70.
10. Czernin, W. (1977), '*Zementchemie für Bauingenieure*, 3rd Edn, Bauverlag, Wiesbaden.
11. Feldman, R. F., and Sereda, P. J. (1968), 'A model for hydrated Portland cement paste as deduced from sorption-length change and mechanical properties'. *Mater. Constr.*, **1**, 509–20.
12. Flood, E. A. (1961), 'Adsorption potentials, adsorbent self-potentials and thermodynamic equilibria, in Solid Surfaces and the Gas–Solid Interface', *Adv. Chem. Ser.*, No. 33, American Chemical Society, Washington, pp. 249–63.
13. Gibbs, J. W. (1957), *Collected Works*, Yale University Press, New Haven.
14. Hannant, D. J. (1967), 'Strain behavior of concrete up to 95 °C under compressive stresses', *Proc. Conf. on Prestressed Concrete Pressure Vessels*, The Institution of Civil Engineers, London, pp. 177–91.
15. Helmuth, R. A., and Turk, D. H. (1967), 'The reversible and irreversible drying shrinkage of hardened cement paste and tricalcium silicate pastes', *J. PCA Res. Dev. Lab.*, **9**, 8–21.
16. Hiller, K. H. (1964), 'Strength reduction and length changes in porous glass caused by water vapour adsorption', *J. Appl. Phys.*, **35**, 1622–8.
17. Klug, P., and Wittmann, F. H. (1974), 'Activation energy and activation volume of creep of hardened cement paste', *Mater. Sci. Ing.*, **15**, 63–6.

18. Kondo, R., and Daimon, M. (1974), 'Phase composition of hardened cement paste', *Proc. 5th Int. Congr. on the Chemistry of Cement*, Moscow. (See also: Daimon, M., Abo-El-Enein, S. A., Hosaka, G., Goto, S., and Kondo, R. (1977). 'Pore structure of calcium silicate hydrated in hydrated tricalcium silicate', *J. Amer. Ceram. Soc.*, **60**, 110–4.)

19. Krasilnikov, K. G., Podvalny, A. M., and Segalov, A. E. (1974), 'On the self-induced deformations in porous bodies', *Kolloidnyi. Zhurnal*, **36**, 266–71.

20. Krausz, A. S., and Eyring, H. (1975), *Deformation Kinetics*, Wiley, New York.

21. Luijerink, J. (1980), 'Deformation kinetics of concrete', in *Fundamental Research on Creep and Shrinkage of Concrete*, ed. by F. H. Wittmann, Martinus Nijhoff Publishers, The Hague, pp. 27–34.

22. Mindess, S., Lawrence, F. V., and Young, J. F. (1978), 'Creep and drying shrinkage of calcium silicate pastes I, Pretreatment and mechanical properties', *Cem. Concr. Res.*, **8**, 591–600.

23. Niklas, (1967), 'Uber den Mechanismus des Andrade-Kriechens von amorphen Polymeren im Glaszustand', *Z. angew. Physik*, **23**, 470–6.

24. Powers, T. C., and Brownyard, T. L. (1974), 'Studies of the physical properties of hardened Portland cement paste', *Portland Cement Association, Res. Bull.*, **22**. (Originally published in *J. Amer. Concr. Inst.*, Oct. 1946–Apr. 1947).

25. Powers, T. C. (1968), 'The thermodynamics of volume change and creep', *Mater. Constr.*, **1**, 487–507.

26. Ruetz, W. (1966), 'Das Kriechen des Zementsteins im Beton und seine Beeinflussung durch gleichzeitiges Schwinden', *Dtsch. Ausschuss Stahlbeton, Schriftenr. Heft* 183, Wilhelm Ernst & Sohn, Berlin.

27. Ruetz, W. (1965), 'The two different physical mechanisms of creep in concrete', *Proc. Int. Conf. on the Structure of Concrete*, Cem. Concr. Assoc., London, pp. 146–53.

28. Schiessl, P. (1976), 'Zur Frage der zulässigen Rissbreite und der erforderlichen Betondeckung im Stahlbetonbau unter besonderer Berïcksichtigung der Karbonatisierung des Betons', *Dtsch. Ausschuss Stahlbeton, Schriftenr. Heft* 255, Wilhelm Ernst & Sohn, Berlin.

29. Schlude, F., and Wittmann, F. H. (1974), 'Ueber ein Verfahren zur raschen Bestimmung der Komplexen DK im Mikrowellenbereich', *Nachrichtentech. Ztg*, **27**, 365–8.

30. Schwarzl, F. R. (1969), 'The numerical calculation of storage and less compliance from creep data for linear viscoelastic materials', *Rheol. Acta*, **8**, 6–17.

31. Sellevold, E. J., and Richard, C. W. (1972), 'Short-time creep transition for hardened cement paste', *J. Am. Ceram. Soc.*, **55**, 284–9.

32. Sellevold, E. J. (1976), 'Low frequency internal friction and short-time creep of hardened cement paste: an experimental correlation', *Proc. Conf. on Hydraulic Cement Pastes: Their Structure and Properties*, Sheffield, pp. 330–4.

33. Setzer, M. J. (1980), 'A model of hardened cement paste for linking shrinkage and creep phenomena', in *Fundamental Research on Creep and Shrinkage of Concrete*, ed. by F. H. Wittmann, Martinus Nijhoff Publishers, The Hague, pp. 3–13.

34. Skoblinskaya, N. N., and Krasilnikov, K. G. (1975a), 'Changes in crystal structure of ettringite on dehydration 1', *Cem. Concr. Res.*, **5**, 381–94.

35. Skoblinskaya, N. N., Krasilnikov, K. G., Nikitina, L. V., and Varlamov, V. P. (1975b), 'Changes in crystal structure of ettringite on dehydration 2', *Cem. Concr. Res.*, **5**, 419–32.

36. Splittgerber, H., and Wittmann, F. H. (1974), 'Einfluss adsorbierter Wasserfilme

auf die van der Waals Kraft zwischen Quarzglasoberflächen', *Surf. Sci.*, **41**, 504–14.
37. Stockhausen, N. (1981), 'Die Dilatation hochporöser Festkörper bei Wasseraufnahme und Eisbildung', *Dissertation*, Techn. Univ. München.
38. Straub, F., and Wittmann, F. H. (1976), 'Activation energy and activation volume of compressive and tensile creep of hardened cement paste', *Proc. Conf. on Hydraulic Cement Pastes: Their Structure and Properties, Sheffield*, pp. 227–30.
39. Taylor, H. F. W. (1979), 'Cement hydration reactions: the silicate phases', *Proc. Eng. Found. Conf. on Cement Production and Use*, Franklin Pierce College, Eng. Found., New York, pp. 107–116.
40. Taylor, H. F. W., and Roy, D. M. (1980), 'Structure and composition of hydrates', *Proc. 7th Int. Congr. on Chemistry of Cement*, Paris, Vol. I, Paper II-2.
41. Übelhack, H., and Wittmann, F. H. (1976), 'Coupling of colloidal particles and recoilless fraction', *J. Phys. Paris.*, **37**, (C6), 269–71.
42. Wittmann, F. H. (1974), 'Bestimmung physikalischer Eigenschaften des Zementsteins', *Dtsch. Ausschuss Stahlbeton, Schriftenr. Heft* 232, Wilhelm Ernst & Sohn, Berlin, pp. 1–63.
43. Wittmann, F. H. (1976), 'The structure of hardened cement paste—a basis for a better understanding of the materials properties', *Proc. Conf. on Hydraulic Cement Pastes: Their Structure and Properties*, Sheffield, pp. 96–117.
44. Wittmann, F. H. (1977), 'Grundlagen eines Modells zur Beschreibung charakteristischer Eigenschaften des Betons', *Dtsch. Ausschuss Stahlbeton Schriftenr. Heft* 290, pp. 43–101.
45. Wittmann, F. H. (1979), 'Trends in research on creep and shrinkage of concrete', *Proc. Eng. Found. Conf. on Cement Production and Use*, Franklin Pierce College, pp. 143–61
46. Wittmann, F. H. (1980), 'Properties of hardened cement paste', *Proc. 7th Int. Congr. on Chemistry of Cement*, Paris, Vol. I, Subtheme VI-2.
47. Wittmann, F. H. (1982), 'Modelling of concrete behaviour', *Proc. Conf. on Contemporary European Concrete Research*, Swedish Cement and Concrete Research Institute, Stockholm. pp. 171–189
48. Wittmann, F. H., and Roelfstra, P. E. (1980), 'Total deformation of loaded drying concrete', *Cem. Concr. Res.*, **10**, 601–10.
49. Wittmann, F. H., Roelfstra, P. E., and Sadouki, H. (1982), 'Simulation and numerical analysis of composite materials, (to be published).
50. Wittmann, F. H., and Setzer, M. J. (1971), 'Vergleich einiger Kriechfunktionen mit Versuchesergebnissen', *Cem. Concr. Res.*, **1**, 679–90.
51. Young, J. F., Berger, R. L., and Bentur, A. (1978), 'Shrinkage of tricalcium silicate pastes: superposition of several mechanisms', *Il Cemento*, **75**, 391–8.
52. Young, J. F. (1982), 'The microstructure of hardened Portland cement paste', Chapter 1 of this volume.
53. Zech, B., and Wittmann, F. H. (1974), 'Studium des dielektrischen Verhaltens von dünnen adsorbierten Wasserfilmen', *Z. Phys. Chem.*, Neue Folge, **92**, 45–62.

Creep and Shrinkage in Concrete Structures
Edited by Z. P. Bažant and F. H. Wittmann
© 1982 John Wiley & Sons Ltd

*Chapter 7*

# Mathematical Models for Creep and Shrinkage of Concrete

*Z. P. Bažant*

## 7.1 INTRODUCTION

Since the advent of the computer era structural analysis capabilities have been advancing at a rapid pace. The large finite element programs to which these advances have led can however serve a useful purpose only if a good mathematical model of the material is available.

Great progress has been achieved in this direction in the field of creep and shrinkage of concrete. Within the linear range, the theory is now reasonably well understood. However, many questions and gaps of knowledge remain, despite the recent vast expansion of the literature on the subject.

This chapter attempts a state-of-art exposition, stating the principal facts, properties, and formulations, and frankly admitting the limitations, uncertainties, and questions. The reader must be warned that the survey which follows does not atempt an exhaustive coverage and is characterized by a certain degree of bias for the contributions made at my home institution with which I am most familiar.

An engineer who merely wants to get a quick information on the models he could use, and not to worry about more subtle or unanswered questions, need not study the whole of this chapter. It suffices for him to look first at Section 7.3.5 for a brief description of the simplest method of analysis, then either at Section 7.3.4 if his structural system is not large and at Sections 7.4.1, 7.4.2, and 7.4.4 if it is large, and finally at Sections 7.2.5–7.2.7, 7.7.1–7.7.4 for the characterization of material properties. Even those sections, however, are not instruction manuals and the appropriate references must be consulted for details.

## 7.2 CREEP AND SHRINKAGE PROPERTIES

### 7.2.1 Definitions

When a load is applied on a concrete specimen, the specimen first shows an instantaneous deformation which is then followed by slow further increase of deformation. This slow increase of deformation, discovered in 1907 by Hatt,[97] is called creep. Concrete specimens slowly deform in time even in the absence of applied loads. These deformations are called shrinkage when temperature is constant.

To define creep one must consider two identical specimens subjected to exactly the same environmental histories, one specimen being loaded and the other load-free (companion specimen). The difference of the deformation of these two specimens defines the instantaneous deformation plus creep.

### 7.2.2 Physical nature of creep and shrinkage

Creep of concrete has its source in the hardened cement paste and, at high stresses, also in failure of the paste–aggregate bond. The paste consists of solid cement gel and contains numerous capillary pores.[72,153,152,188,194] The cement gel contains about 40 to 55% of pores in volume, has an enormous pore surface area (roughly $500 \, m^2/cm^3$), and is made up of sheets of colloidal dimensions (of average thickness about 30 Å, with average gaps about 15 Å between the sheets). The sheets are formed mostly of calcium silicate hydrates and are strongly hydrophylic. Because the pores of cement gel are micropores of subcapillary dimensions they cannot contain liquid water or vapour; but they do contain evaporable water (water that is not chemically bound in the hydrates), which is strongly held by solid surfaces and may be regarded as (hindered) absorbed water or interlayer water. This water can exert on the pore walls a significant pressure called the disjoining pressure[25,151,194] the value of which depends on temperature and the degree of water saturation of capillary pores.

The bonds and contacts between the colloidal sheets in cement gel are highly disordered and unstable. Therefore, creep may be expected to be caused by changes in the solid structure. Although the precise creep mechanism is still debated, bonding and rebonding processes similar to movement of a dislocation may be involved, and it may also be possible that various solid particles displace or migrate (diffuse) from highly stressed zones to stress-free zones such as the surfaces of larger pores. Because of the disjoining pressure, bonds get weakened by the presence of water, and this explains why after drying the creep is less. [25,191–194]

During drying, on the other hand, the creep is higher than in sealed specimens. This effect, called drying creep or Pickett effect,[148] probably has two sources. One may be the fact that as water is diffusing out of the loaded gel micropores it creates disorder, facilitating migrations of solid particles.[16,24,25] Another cause, possibly the major one,[195] is likely to be macroscopic, namely the stresses and microcracking[52,195] produced by drying in the specimen as a whole.

As the solid particles migrate out of the loaded regions, the load on them (or the disjoining pressure) is gradually relaxed, being transferred onto more stable parts of the microstructure. This causes the creep rate to decline. At the same time hydration proceeds, which causes the volume of cement gel to increase at the expense of large (capillary) pores, and the number of bonds in the existing gel to also increase. This reduces creep, too.

Shrinkage results from the increase of solid surface tension and capillary tension due to drying, as well as from the decline of disjoining pressure in the gel.[190,194]

### 7.2.3   Elementary characteristics

The total strain of a uniaxially loaded concrete specimen at time $t$ after the casting of concrete (age) may be subdivided as

$$\varepsilon(t) = \varepsilon_E(t) + \varepsilon_C(t) + \varepsilon_S(t) + \varepsilon_T(t) = \varepsilon_E(t) + \varepsilon''(t)$$
$$= \varepsilon_E(t) + \varepsilon_C(t) + \varepsilon^0(t) = \varepsilon_\sigma(t) + \varepsilon^0(t) \qquad (7.1)$$

in which $\varepsilon_E(t)$ is the instantaneous strain, which is elastic if the stress is small, $\varepsilon_C(t)$ is the creep strain, $\varepsilon_S(t)$ is the shrinkage strain, $\varepsilon_T$ is the thermal dilatation, $\varepsilon^0(t)$ is the stress-independent inelastic strain, $\varepsilon''(t)$ is the inelastic strain, and $\varepsilon_\sigma(t)$ is the stress-produced strain. $\varepsilon_E(t)$ is reversible (i.e. recoverable) upon unloading right after the moment of loading but not later, due, principally, to further hydration.

The thermal strain will not interest us here beyond noting that it is calculated as $\varepsilon_T = \int_{T_0}^{T} \alpha \, dT$ where $T_0$ is the chosen reference temperature and $\alpha$ is the thermal dilatation coefficient, which roughly equals $10^{-5}\,°\mathrm{C}^{-1}$ but actually depends on $T$ and even more on the specific moisture content, $w$.

The dependence of creep on stress may be shown graphically by creep isochrones (Figure 7.1), which are the lines connecting the values of strain $(\varepsilon - \varepsilon^0)$ produced by various constant stresses $\alpha$ during the same time

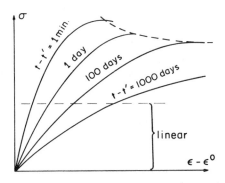

**Figure 7.1**   Creep isochrones

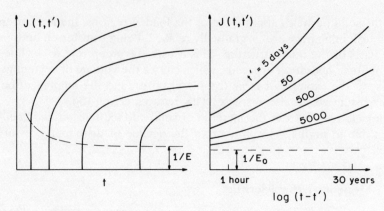

**Figure 7.2**   Typical creep curves for various ages $t'$ at loading

period. Plotting creep isochrones from test results (see Figure 7.1), one finds that for stresses within the service range, or up to about 50% of the strength, the creep is approximately proportional to stress. For constant uniaxial stress $\sigma$ we may then write

$$\varepsilon(t) = \sigma J(t, t') + \varepsilon^0(t) \qquad (7.2)$$

in which $\varepsilon$ is the uniaxial strain, $t$ is the time, which we normally choose to coincide with the age of concrete, and $J(t, t')$ is the compliance function (often also called the creep function), which represents the strain at time $t$ produced by a unit constant stress that has been acting since time $t'$. Due to the proportionality property, the creep is completely characterized by the function $J(t, t')$, the typical shape of which is sketched in Figure 7.2. This function may be expressed as

$$J(t, t') = \frac{1}{E(t')} + C(t, t') = \frac{1 + \phi(t, t')}{E(t')} \qquad (7.3)$$

where $1/E(t')$ represents the instantaneous elastic deformation at age $t'$, $C(t, t')$ is the creep compliance (also called the specific creep), and $\phi(t, t')$ is the ratio of the creep deformation to the elastic deformation, called the creep coefficient. The instantaneous deformation has a large inelastic (irreversible) component at high stresses but in the service stress range (below about $\frac{1}{2}$ of the strength) it is essentially elastic, i.e. reversible immediately after loading.

For long-time loading, the values of creep coefficient are usually between 1.0 and 6.0, with 2.5 as the typical value. So, realizing that creep deformations are normally larger than the elastic ones, we recognize the importance of taking creep into account in calculations of stresses, deformations, cracking, buckling, and failure of structures under sustained loads. Shrinkage typically attains values from 200 to $800 \times 10^{-6}$. Compared to this, a typical creep strain for $E(t') = 3.5 \times 10^6$ psi (24,000 MPa), $\phi = 2.5$, and $\sigma = 2000$ psi (14 Mpa) is $1400 \times 10^{-6}$. Thus, shrinkage is normally somewhat less important than creep, except when the long-time stresses produced by load are small.

The magnitude of the part of deformation that is called instantaneous or elastic, i.e. $\varepsilon_E = \sigma/E(t')$, unfortunately suffers by ambiguity because significant creep exists even for extremely short load durations; see the typical curves at constant stress plotted in Figure 7.2 (on log-time scale) in which the left-hand side horizontal asymptote represents the true instantaneous deformation $J(t', t')$ (since $\log 0 = -\infty$). Its value is very difficult to determine experimentally and is, anyhow, not needed for static structural analysis for long-time loads. For this purpose, the deformation which coresponds to any load duration less than about 1 day ($a$ in Figure 7.3) may serve just as well as the conventional instantaneous (or elastic) strain. The conventional elastic modulus obtained from the formulae of ACI or CEB-FIP recommendations (e.g. $57,000 \sqrt{f'_c}$) corresponds to approximately two hours of stress duration and represents approximately half of the true instantaneous modulus.

In previous works, unfortunately, different definitions of the instantaneous (elastic) deformation have been used. Some authors imply, often tacitly, the instantaneous deformation to be that for 1 to 10 min duration (typical duration of strength tests), others that for 0.001 s. There are great differences among all the definitons of $E(t')$ used in the past. Much confusion and error has been caused by carelessly combining incompatible values of $E(t')$ and $\phi(t, t')$ or $C(t, t')$ ($a$ with $b'$ or $b$ with $a'$ in Figure 7.3).

For short-time loading, quasi-elastic structural analyses based on the effective modulus $E_{\text{eff}} = 1/J(t, t')$ normally give very good results. For loads of more than one-day duration, it makes therefore, little difference whether $E(t')$ corresponds to a duration of 1 s or 2 h, provided that $1/E(t')$ and $C(t, t')$ add up to the correct value of $J(t, t')$.

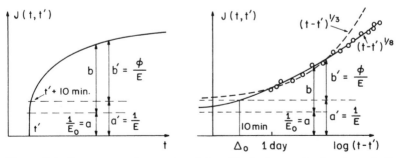

**Figure 7.3** Creep curves in actual and logarithmic time scales ($a$ = true elastic deformation, $b$ = true creep, $a'$ = conventional elastic deformation, $b'$ = conventional creep)

To make these errors impossible, it is preferable to specify initially the creep properties in terms of $J(t, t')$ rather than $C(t, t')$ or $\phi(t, t')$. For the purpose of structural analysis the conventional elastic modulus may then be calculated as

$$E(t') = \frac{1}{J(t' + \Delta, t')} \qquad (7.4)$$

where $\Delta$ is some chosen load duration, less than about 1 day. The creep compliance must then be evaluated as $C(t, t') = J(t, t') - J(t' + \Delta, t')$ and the creep coefficient as $\phi(t, t') = [J(t, t')/J(t' + \Delta, t')] - 1$.

### 7.2.4 Influencing factors

Creep and shrinkage of concrete are influenced by a large number of factors, which may be divided into intrinsic factors and extensive factors. The intrinsic factors are those material characteristics which are fixed once and for all when the concrete is cast. Extensive factors are those which can vary after the casting; they include temperature, pore water content, age at loading, etc..

The main intrinsic factors are the design strength, the elastic modulus of aggregate, the fraction of aggregate in the concrete mix, and the maximum aggregate size.[145,113] Increase of any of these factors causes a decrease of creep as well as shrinkage. This is because the aggregate does not creep appreciably and has a restraining effect on creep and shrinkage. Gap-grading of aggregate further reduces creep and shrinkage. As for shrinkage, it also increases as the water/cement ratio of the concrete mix increases.

Among the extensive factors we must distinguish the local from the external ones. The former, also called the state variables, are those which can be treated as a point property of a continuum. They are the only ones which can legitimately appear in a constitutive equation. Temperature, age, degree of hydration, relative vapour pressure (humidity) in the pores, and pore water content represent state variables affecting creep.

On the other hand, the size of specimen and the environmental humidity are not admissible as state variables in a constitutive equation even though they have a great effect on creep of a concrete specimen. Properly, the environmental humidity must be considered as the boundary condition for the partial differential equation governing pore humidity. It is the pore humidity, not the environmental one, which directly affects creep and can appear in the constitutive equation.

The effects of state variables (documented, e.g., by the text data reported by Neville and Dilger,[145] L'Hermite and Macmillan,[122] Lambotte and Mommens,[117] Hanson,[94] Harboe *et al.*,[93] Troxell *et al.*,[177] Rüsch *et al.*,[16] Wagner[182], Neville[144]) are as follows. Creep decreases as the age of concrete at the instant of loading increases (this is actually the effect of the increase in the degree of hydration). Creep also increases with increasing temperature, but this effect is offset by the fact that a temperature increase also accelerates hydration which in turn reduces creep. Creep at constant pore water content is less for a smaller pore water content or a lower relative humidity in the pores.[160,194] In most practical situations, however, this local effect is overpowered by the effect of the changes in environmental humidity (an extensive factor) upon the overall creep of a specimen or structural member. This effect is opposite—the creep of a specimen is increased, not decreased, by a decrease of environmental humidity.[160,194]

Another important non-local extensive factor which is not a state variable is the size of specimen or structural member. The drying process in a larger specimen is slower, and consequently the creep increase due to drying is less, i.e. creep is less for a larger specimen. Similarly, shrinkage is less for a larger specimen and it is also less for a higher environmental humidity.

In a sealed state, at which (due to hydration) the pore humidity is found to drop gradually to about 97–99%, concrete exhibits a small shrinkage, called the autogeneous shrinkage. It is due to volume changes in the hydration reaction and is about twenty times less than the drying shrinkage. In water immersion (100% humidity) concrete exhibits small swelling (negative shrinkage), which is about ten times less in magnitude than the drying shrinkage.

### 7.2.5 Constitutive properties

Among the simple formulae, the creep of concrete at constant moisture and thermal state (also called the basic creep) may be best described by power curves of load duration $(t - t')$,[194] and by inverse power curves for the effect of age $t'$ at loading. This leads to the double power law[25,41,43,44] (Figure 7.3):

$$J(t, t') = \frac{1}{E_0} + \frac{\phi_1}{E_0} (t'^{-m} + \alpha)(t - t')^n \qquad (7.5)$$

in which, roughly, $n \approx 1/8$, $m \approx 1/3$, $\alpha \approx 0.05$, $\phi_1 \approx 3$ to 6 (if $t'$ and $t$ are in days), and $E_0$ (= asymptotic modulus) $\approx 1.5$ times the conventional elastic modulus for 28-days old concrete. These coefficients can be relatively simply determined from test data; for example, by using the foregoing estimates for $E_0$, and $m$ and $\alpha$ and plotting $y = \log[(E_0 J - 1)/(t'^{-m} + \alpha)]$ versus $\log(t - t')$, one gets a straight-line plot whose slope is $n$ and y-intercept is $\phi_1$. Comparisions with test data are exemplified in Figure 7.4.

Temperature has a major influence on creep. To describe the creep curves at various constant temperatures, Equation (7.5) may be generalized as

$$J(t, t') = \frac{1}{E_0} + \frac{\phi_T}{E_0} (t_e'^{-m} + \alpha)(t - t')^{n_T} \qquad (7.6)$$

in which $t_e' = \int \beta_T(t') \, dt'$ represents the age corrected for the effect of

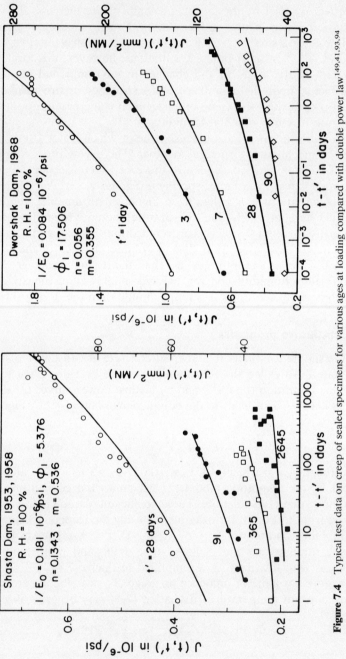

**Figure 7.4**  Typical test data on creep of sealed specimens for various ages at loading compared with double power law[149,41,93,94]

temperature on the rate of hydration (or aging) and is called the reduced age or the equivalent hydration period (or maturity). Coefficients $\phi_T$, $n_T$, and $\beta_T$ are empirical functions of temperature,[27] introduced such that 'at reference (room) temperature $(T_0)$' $\phi_T = \phi_1$, $n_T = n$, and $\beta_T = 1$. For temperature history that equals $T_0$ up to time $t_0$ and then jumps to another constant value $T$, we have $t'_e = t_0 + \beta_T(t - t_0)$.

Since $(t - t')^n = e^{nx}$ where $x = \ln(t - t')$, the power curves of $(t - t')$ appear on a log-time scale as curves of ever-increasing slope and with no bounded final value (Figure 7.2). The question whether there exists a bounded final value of creep (at $t \to \infty$) has been debated for some time and no consensus has been reached. It is, however, clear that if a final value exists it would be reached at times far beyond those of interest. All measurements of creep of sealed or immersed specimens indicate (except for what appears to be statistical scatter) non-decreasing slopes on a log-time scale for the entire test duration. There is no evidence of a final value. For design purposes, however, the question of existence of a bounded final value is not too important because the creep increase from 50 years to 100 years is, according to extrapolations of test data (or Equation (7.5)), anyhow insignificant. Most structures are being designed for 40- or 50-year service lives.

The power law of load duration, first proposed by Straub[173] and Shank,[170] follows theoretically from certain reasonable hypotheses about the microstructural creep mechanism, e.g. rate process theory,[192–194] or a statistical model of creep mechanism.[71] Until recently the power law had been used in conjunction with the conventional elastic modulus for the elastic term ($1/E$ instead of $1/E_0$ in Equation (7.5)). However, this definition of the elastic term greatly restricts the range of applicability. Namely, by choosing the left-hand side of the horizontal asymptote to be too high (Figure 7.3), a higher curvature of the power curve, i.e. a higher exponent (about $n \simeq 1/3$), is required in order to fit the creep data for durations from 3 to 100 days. Then the large curvature due to too high an exponent (1/3 instead of 1/8) causes the curve to pass well above the creep data for longer creep durations (over 100 days); see Figure 7.3(b). It was for this reason that in the older works the power law was deemed to be inapplicable for long-time creep.

It may be of interest to add that an improvement of data fits may be achieved by the so-called log-double power law which asymptotically coincides for short load durations with the double power law and for long load durations tends to straight lines in $\log(t - t')$ which have the same slope for all $t'$ (work in progress by J. C. Cherni, Northampton University).

To be able to fit the creep test data up to many years duration the elastic term ($1/E_0$ in Equation (7.5)) must be taken as the true instantaneous value, i.e. as the left-hand side horizontal asymptote on the log-scale, and the exponent then turns out to be around 1/8. The double power law then acquires an extraordinarily broad range of applicability. It agrees reasonably well with the known data for creep up to 30 years' duration and at the same time it describes very well the test data for load durations under one day and as short as 1 s. It even gives approximately correct values for the dynamic modulus $E_{dyn}$ when one substitutes load duration $t - t' = 10^{-7}$ day. The conventional elastic modulus, along with its age dependence, may be considered as the value of $1/J(t, t')$ for $t - t' = 0.001$ day, for which Equation (7.5) yields:

$$E(t') = \frac{E_0}{1 + \phi_1'(t'^{-m} + \alpha)} \qquad \phi_1' = 10^{-3n}\phi_1 \qquad (7.7)$$

However, the value obtained by substituting $t - t' = 0.1$ day agrees better with the ACI formula ($E = 57,000 \sqrt{f_c'}$).

Since four parameters ($E_0, m, \alpha, \phi_1'$) are needed to describe the age dependence of the elastic modulus, there is only one additional parameter, namely the exponent $n$, which suffices to describe creep. This makes the double power law simpler than any other known formula for creep.

Many other expressions for the compliance function have been proposed.[192,193] Ross and Lorman proposed a hyperbolic expression $C(t, t') = \bar{t}/(a + b\bar{t})$, $\bar{t} = t - t'$, which is convenient for fitting of test data but is unfortunately inapplicable to long creep durations.[31] Hanson[94] proposed a logarithmic law, $C(t, t') = \phi(t') \log(1 + \bar{t})$, which does not approach any final value and gives good predictions for long creep durations but for short durations is not as good as the double power law. Mörsch[137] proposed the expression $C(t, t') = \phi\{1 - \exp[-(bt)^{1/2}]\}^{1/2}$ and Branson et al.[58–60] proposed the expression in Equation (7.13) in the sequel; these exhibit a final value. The expressions of Ross and Mörsch work better for creep at drying which we discuss in the next section. McHenry,[125] Maslov,[133] Arutyunian,[10] Bresler and Selna,[62] Selna,[168,169] and Mukaddam[138,139] used a sum of exponentials of $t - t'$ with coefficients depending on $t'$. Such expressions can be closely adapted to any test data and we will discuss these in Section 7.4.1.

Various expressions have been introduced with the particular purpose of enabling a certain simplified method of creep structural analysis. These include the expressions of Whitney,[185] of Glanville[85] and Dischinger,[77,78] England and Illston,[79] and Illston,[79] and Nielsen,[146,147] which lead to the rate-of-creep (Dischinger's) and rate-of-flow methods for structural analysis and will be mentioned later, and other expressions.[70,10,120]

The double power law exhibits a certain questionable property which was recently discussed in the literature.[27,33] It is the property that the creep curves for different ages $t'$ at loading diverge after a certain creep duration, i.e. there exists a time $t - t' = t_D$ (function of $t'$) after which the difference between these curves increases while up to this time it decreases (Figure 7.4). This property, which is shared with the ACI creep expression but not with that in the CEB-FIP Model Code, is equivalent to the condition that $\partial^2 J(t, t')/\partial t\, \partial t'$ changes sign from positive to negative (it is non-negative if there is no divergence). One objection was that the creep recovery curves obtained by principle of superposition do not have a decreasing slope at all times if the creep curves exhibit the divergence property. This argument is however unrealistic because the principle of superposition cannot be applied to creep recovery (Section 7.3.1). Further it was thought that the divergence property might violate the second law of thermodynamics but it was proven that this is not so.[27,33] So, whether the divergence property is real depends strictly on experimental observations. The evidence from test data is ambivalent; some exhibit the divergence, many do not. It could be that the divergence property is due to some non-linear effect, in which case it would not belong to function $J(t, t')$.

Finally, we should indicate how the thermal strain is determined. Although the deformations due to changes in temperature and moisture content are in reality coupled, we may reasonably well calculate the thermal strain as

$$\varepsilon_T = \int_{T_0}^{T} \alpha(T)\, dT \tag{7.8}$$

where $T_0$ is the initial reference temperature, $T$ the current temperature, and $\alpha(T)$ the coefficient of thermal expansion at temperature $T$. Approximately, $\alpha(T)$ may be considered to be constant (usually $\alpha = 10^{-5}\,°\mathrm{C}^{-1}$), and independent of moisture content and age. Then $\varepsilon_T = \alpha(T - T_0)$.

## 7.2.6 Cross-section behaviour during drying

Concrete as initially cast is wet, with pore humidity 100%. After a certain initial moist treament period, usually 7 to 14 days, most concrete structures (except for those sealed by an impervious liner) are exposed to the environment and dry gradually. The drying process is very slow. If concrete does not crack and is of good quality, it takes over 10 years for the pore humidity at mid-thicknesses of a 6-inch slab to approach that of the environment. For other thicknesses, the drying times are proportional to square of the thickness. This gives a drying time of about 1 year for a 2-inch shell, and of 360 years for a 3 ft slab (if it does not crack). In very thick uncracked structures (mass concrete) there is no significant drying except for up to about 1 foot from the surface. (These times, however, become much shorter if concrete is heated over 100%.)

Drying is the cause of most of the shrinkage and it also profoundly affects creep. Shrinkage is larger for a lower environmental humidity and for a smaller size of cross section. It also decreases as the age of concrete (or the degree of hydration) at the start of drying increases, and as the initial moist period extends. Regarding the intrinsic factors, shrinkage increases with an

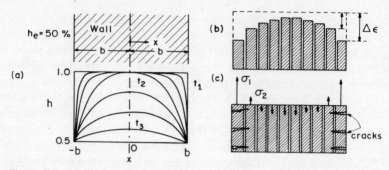

**Figure 7.5** (a) Typical distributions of pore humidity at various times during drying; (b) free shrinkage and creep at various points of cross section; (c) internal stresses

increase in water/cement and cement/aggregate ratios of concrete mix and with a decrease in concrete strength. As for the effect of drying on creep, generally a strong increase of creep, compared to sealed specimens, is observed. This phenomenon is called the drying creep or Pickett effect.[148] (For test data see Neville and Dilger,[145] Ali and Kesler,[4] Troxell,[177] Keeton,[111] L'Hermite and Mamillan[122,123] Weil,[183] Rüsch,[161] Ishai,[105] Wagner,[182] Kesler,[113] Lambotte,[117] LeCamus,[119] and others.)

Certain constitutive relations that govern creep at drying have been proposed on a speculative theoretical basis (e.g. Bažant[25]). However they are considerably more complicated and, thus far, more weakly supported by experiment than those for constant pore humidity, even though the available test data have been successfully fitted.[52] The determination of constitutive relations in the presence of drying is hampered by the fact that for nearly all available creep and shrinkage tests the cross section has been in a highly non-uniform moisture state (Figure 7.5), with totally different shrinkage and creep strains at the core and the surface layer of the specimen. It is certain that this non-uniformity must lead to large internal stresses which cause non-linear triaxial behaviour and microcracking,[202] with microcracks probably so fine that they cannot be seen by the unaided eye. It was shown[52] that these phenomena have a very large effect on observed creep of drying specimens. More recently it has been proposed[195] that nearly all of the increase of creep due to drying might be due to microcracking. We will return to this question in Section 7.6.

What is presently available for long-time deformations at drying are semi-empirical formulae that indicate the overall or mean shrinkage and creep of the cross section of a test specimen. We will outline them in this section. These formulae, however, cannot be considered as a constitutive property, i.e. a point (or local) property of concrete as such. The so-called unrestrained (or free) shrinkage and creep at various points of the cross section is no doubt rather different from the behaviour indicated by such formulae. Nevertheless, in view of the uncertainties and complexities that characterize the presently available constitutive relations at drying, the use of these formulae as constitutive relations may be justified at present for the purpose of cruder calculations aimed at determining just the internal force resultants within the cross section. The distributions of stresses due to drying, however, cannot be determined by such calculations. It should also be realized that applying the same mean creep properties to both bending and axial deformation cannot be correct since the response of concrete in the core, which dries later, has almost no effect on bending while it affects the axial deformation.

Drying is a diffusion process and, as test data confirm, the evolution of pore humidity and water content distributions in time can be reasonably well calculated from a non-linear diffusion equation.[37] Based on this equation it appears (see Section 7.6.4, Equations (7.105)–(7.107)), that the drying times are proportional to the square of the size when geometrically similar bodies are compared (Figure 7.6). The same is true of shrinkage since the free shrinkage strain appears to be a function of pore water content which, in turn, is a function of pore humidity. In practice, the size-square dependence is not exact, being spoiled by the effects of continuing hydration and microcracking, but it agrees with measurements quite well.

Using these results of the diffusion theory, we may express the mean shrinkage of the cross section as[42,43,44]

$$\bar{\varepsilon}_S(t, t_0) = \varepsilon_{sh_\infty} k_h S(\theta) \tag{7.9}$$

where

$$\theta = \frac{t - t_0}{\tau_{sh}} \qquad \tau_{sh} = c_s \frac{(k_s D)^2}{C_1} \tag{7.10}$$

Here $\tau_{sh}$ is the shrinkage square half-time (i.e. the time in which the square of shrinkage strain reaches about 1/2 of its final value); $\varepsilon_{sh_\infty}$ is the final shrinkage at humidity 0%, which depends on the mix ratios and the strength (typically 0.0005 to 0.0013); $k_h$ is a function of environmental humidity $h_e$

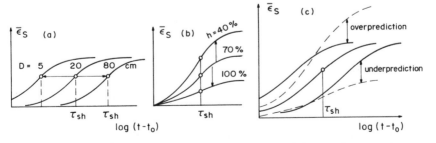

**Figure 7.6**  Effects of size and ambient humidity in mean shrinkage of cross section

(empirically $k_h \simeq 1 - h_e^3$); $S(\theta)$ is a function giving the evolution of shrinkage in non-dimensional time $\theta$; $t_0$ is the age at the start of drying; $C_1$ is drying diffusivity of concrete at the start of drying ($10 \, \text{mm}^2/\text{day}$ in order of magnitude); $k_S$ is the parameter of cross-section shape which can be calculated from the diffusion theory ($k_S = 1$ for slab, 1.15 cylinder, 1.25 square prism 1.30 sphere 1.55 cube); $D$ is the effective thickness of cross section (in mm), defined as $D = 2v/s$, where $v$ is the volume and $s$ is the surface area exposed to drying (for a slab, $D$ represents the actual thickness); and $c_s$ is an empirical constant ($= 0.267 \, \text{mm}^2$). Because non-linear diffusion theory does not permit simple solutions, an empirical expression, namely $S(\theta) = [1 + \tau_{sh}/(t - t_0)]^{-1/2}$, is used.[43,44]

The effects of temperature $T$ and of the age at the start of drying on shrinkage may be described by means of diffusivity and have the form $C_1 = C_0 k_T' k_t$ where $C_0$ is a constant, $k_t$ is an empirical function of age $t_0$; and $k_T'$ is a function of temperature which may be based on activation energy theory.

Two examples of a comparision between calculated shrinkage curves (for different cylinder diameters and different environmental humidities) and test data from the literature are given in Figure 7.7. Figure 7.5 illustrates the effect of a change in environmental humidity, $h_e$, which causes a vertical

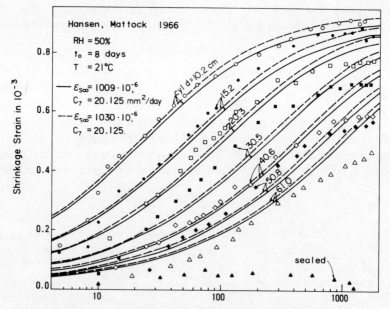

**Figure 7.7** Hansen and Mattock's test data on shrinkage of cylinders of various sizes[92] compared with Equations (7.9) and (7.10) exhibiting the size-square dependence of shrinkage half-time[43]

scaling of shrinkage ordinates, and the effect of changing the size from $D_0$ to $D$, which does not cause a vertical scaling but a horizontal shift of the shrinkage curve in log-time scale (the shift is by the distance $2 \log (D/D_0)$, because $\log \theta = \log (t - t_0) + 2 \log D + \text{constant}$ (see Figure 7.6). In some other practical formulae,[5,60,58,59,68,161] the size effect on shrinkage is handled by scaling the ordinates. This disagrees, however, with the diffusion theory, is not supported by measurements, and leads to underprediction of long-time shrinkage for thick structural members (Figure 7.6(c)). The thickness-square dependence of shrinkage times is the simplest and most essential property of shrinkage.

It must be emphasized that the constitutive relation for the free (unrestrained) shrinkage as a local (point) property is doubtless rather different from Equations (7.9) and (7.10). The specimen size and environmental humidity are not local state variables of a continuum and are therefore inadmissible for a constitutive equation. The free shrinkage is not a function of time but of the specific water content (or pore humidity), the dependence of which on time is not a constitutive property of the material but results from the solution of a boundary value problem. The only dependence on time which is constitutive (local) in nature is that on the degree of hydration (aging).

Direct measurement of the free shrinkage requires lowering the environmental humidity gradually and so slowly that the humidity distribution within the specimen would remain almost uniform.[202] Such tests have been made for cement paste tubular specimens 1 mm thick,[200] but for large specimens the test times become impossibly long. Since microcracking, tensile non-linearity, and creep due to shrinkage-induced stresses reduce the observed shrinkage of a specimen, the free shrinkage is certain to be significantly higher than the final values observed in standard tests (Equation (7.9)).

The mean compliance function $\bar{J}(t, t')$ of the cross section in the presence of drying may be expressed approximately as:[42]

$$\bar{J}(t, t') = J(t, t') + \bar{C}_d(t, t') \qquad (7.11)$$

where $J(t, t')$ is the compliance function for constant pore humidity, as given in Section 7.2 (e.g. Equation (7.5)), and $\bar{C}_d(t, t')$ is the mean additional compliance due to drying (with the indirect effect of simultaneous shrinkage) (Figure 7.8).

For a lower humidity, the drying is more severe, and thus the drying creep term increases as the environmental humidity decreases. When the size tends to infinity, there is no drying in the limit. So the drying creep term $\bar{C}_d$ must decrease with increasing size and approach zero as the size tends to infinity. Some practical models[5,58–60] disregard this condition.

Since drying follows the size-square dependence, the same should be

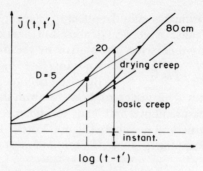

**Figure 7.8** Components of mean creep
of cross section at drying

expected of the drying creep term. So, we may write, in analogy to Equations
(7.9) and (7.10) for shrinkage,

$$\bar{C}_d(t, t') = \frac{f_d(t')}{E_0} k'_h \bar{S}(\theta) \qquad \theta = \frac{t - t'}{\tau_{sh}} \tag{7.12}$$

where $\tau_{sh}$ is the shrinkage-square half-time (same as in Equation (7.10))
$t - t' =$ duration of load; $k'_h$ is an empirical function of environmental humid-
ity ($k'_h \simeq 1 - h_e^{1.5}$); $\bar{S}(\theta)$ is an empirical function of $\theta$ similar to $S(\theta)$; $f_d(t')$ is a
decreasing empirical function of age at loading $t'$. (For detailed expressions
and justification, see Bažant *et al.*[42,43].) An important property of Equation
(7.12) is that the drying creep term is similar to shrinkage, thus reflecting the
fact that shrinkage affects creep and is not simply additive to creep, as the
experimentalists have always been emphasizing.

An essential feature is that the size-square dependence is embodied in
Equation (7.12). A change in size causes a horizontal shift of the curve for
the drying creep term in log-time scale, and superimposing this term on the
basic creep, $J(t, t')$, we may imagine the drying term curve to slide on top of
the basic creep curve as shown in Figure 7.8. A change of environmental
humidity, on the other hand, causes a vertical scaling of the ordinates of this
term. In this manner, many different shapes of the creep curves can be
generated.

This property is not reflected in the older formulae in which both the
humidity and size effects are handled by a multiplicative factor, i.e. vertical
scaling of the creep curve. This then leads to underprediction of long-time
creep for very thick structural members, and overprediction for very thin
ones (like in Figure 7.6).

The fact that the slope of creep curves in log-time, as observed in drying
environment, begins to decrease after a certain period of time (depending on
the size) appears to be due solely to the drying creep term. From this we

may not, however, infer that the creep curves approach a finite value since the basic creep term does not approach one.

### 7.2.7  Practical prediction of creep and shrinkage

To perform a finite element analysis of time-dependent behaviour, suitable analytical expressions must be selected for creep and shrinkage. Since the experimental data for the particular structure to be analysed are usually lacking and are always incomplete, functions $J(t, t')$ (or $\bar{J}(t, t')$) and $\bar{\varepsilon}_s(t)$ to be used in the analysis of a particular structure must be predicted from various influencing factors known in advance. The selection of functions should satisfy the following criteria:

(1) The functions must first of all accurately fit the available experimental data for concretes of the type considered and take into account all important factors (age, temperature and its variations, environmental humidity and its variation, size and shape of cross section, and curing conditions and their duration).

(2) The undetermined coefficients of the functions should be relatively easy to evaluate from the available experimental or empirical data.

(3) The functions should be sufficiently simple to make the numerical evaluation in the program straightforward and efficient.

The last two requirements, i.e. the requirements of simplicity, are certainly not as stringent for finite element analysis as they are for simpler hand calculations. Generally the effort spent on determination of material properties should be commensurate to that devoted to the analysis itself. Since inaccuracies in material characterization usually cause the most serious error in the results of finite element analysis, it clearly makes no sense if the analyst spends, say, only 4 hours on determining the function $J(t, t')$ and then spends one week in getting the finite element solution based on this function. He should spend an equal time on both.

Several practical models for predicting creep and shrinkage properties for a particular concrete and environmental conditions have been developed. They differ in their degree of accuracy and simplicity, and usually one of these must be traded for the other. There exist principally three comprehensive models for the analyst to choose from:

(1)  Model of ACI Committee 209.[58-60]
(2)  Model of CEB-FIP Model Code[68] (Rüsch et al.[162]).
(3)  Bažant and Panula's Model (BP Model), either its complete version[43] or its simplified version.[45]

The ACI Model is the simplest one, while the BP Model is the most comprehensive one, being applicable over the broadest time range (of $t, t'$, and $t_0$) and covering a number of influencing factors neglected by the other two models. It should be remembered that all three models are at least partially empirical (albeit to various extents) and are all based on the fitting of data obtained in certain laboratory controlled tests. Attempts at verification by measurements on structures have so far been inconclusive due to the difficulties in sorting out various influencing factors, which are much more numerous than in laboratory tests.

(I) *ACI Model.*[5] Based on the works of Branson *et al.*[60], ACI Committee 209 recommended the expressions[5]:

$$\bar{J}(t, t') = \frac{1}{E(t')} \left( 1 + \frac{(t-t')^{0.6}}{10+(t-t')^{0.6}} C_u \right) \qquad \bar{\varepsilon}_S(t, t_0) = \frac{t-t_0}{f_c+(t-t_0)} \varepsilon_u^s$$

(7.13)

in which $t'$ is the age at loading in days, $t$ is the current age in days, $t_0$ is the age of concrete in days at the completion of curing; $f_c$ is a constant; $C_u$ is the ultimate creep coefficient, defined as the ratio of the (assumed) creep strain at infinite time to the initial strain at loading; and $\varepsilon_u^s$ is the ultimate shrinkage strain after infinite time. Coefficients $C_u$ and $\varepsilon_u^s$ are defined as functions of environmental humidity, minimum thickness of structural member, slump, cement content, percent fines, and air content; see Section 7.7.1.

(II) *CEB-FIP Model.*[68] According to the CEB-FIP Model Code, the expressions for the mean compliance function and the mean shrinkage of cross section of a structural member have the basic form

$$\bar{J}(t, t') = F_i(t') + \frac{\phi_d \beta_d(t-t')}{E_{c28}} + \frac{\phi_f[\beta_f(t) - \beta_f(t')]}{E_{c28}}$$

(7.14)

$$\bar{\varepsilon}_S(t, t_0) = \varepsilon_{S_0}[\beta_s(t) - \beta_s(t_0)]$$

(7.15)

in which $E_{c28}$ is the elastic modulus of concrete at age 28 days; $\phi_d = 0.4$; $\phi_f$ is a coefficient depending on environmental humidity and effective thickness of member, $\beta_f$ and $\beta_s$ are functions of time and effective thickness, $\beta_d$ is a function of load duration $t - t'$; $F_i(t')$ is a function of age at loading (= sum of instantaneous strain and initial creep strain over a period of several days). These functions are defined by graphs consisting of 16 curves. (The use of graphs is however not too convenient for computer programming.)

(III) *BP Model.* The basic form of this model utilizes Equations (7.5) and (7.9)–(7.12) which ensue from the diffusion theory and activation energy theory, as already explained. The coefficients in these equations were expressed by empirical formulae determined from test results; see Section 7.7.3. For the case of drying, these formulae are relatively complicated, which is, however, at least partly due to consideration of unusually many influencing factors and a very broad range of applicability. A program for computer evaluation of the BP Model or its fitting to given data is available (see full program listing in Ref. 199).

## 7.2.7.1  Comparision of existing models

The BP Model and the ACI Model have in common, for basic creep, the product form of the compliance function, in which a function of the age at loading multiplies a function of the stress duration. In the ACI Model, however, the multiplicative factor $C_u$ introduces not only the effect of age at loading but also the effect of humidity and size. This is very simple but not quite realistic because the diffusion theory leads to a different form of the size effect, as mentioned before (translation in log-time rather than scaling of the ordinates). The same deficiency characterizes the ACI shrinkage formula (Equation (7.13)).

Likewise, the CEB-FIB Model does not follow the size effect of the diffusion theory. In this model, the basic form of $\bar{J}(t, t')$ is based on the idea of reversibility of deformation. The second term in Equation (7.14) is considered to represent the so-called 'reversible' (or delayed elastic) creep, and the last term the so-called 'irreversible' creep. It must be noted, however, that in the case of aging the concept of a reversible creep component lacks theoretical (thermodynamic) justification because this component cannot be defined uniquely (only reversible creep increments can).[27]

The fact that in the CEB-FIP Model the so-called 'reversible' component of $\bar{J}(t, t')$ was calibrated by fitting the creep recovery curves obtained from the superposition principle to recovery test data is also questionable[40,47,45] because linear superposition does not hold in case of unloading, as has been conclusively demonstrated by tests (cf. Section 7.3.1). The domain of approximate validity of the principle of superposition includes only non-decreasing strain histories within the service stress range. Thus, only the creep curves for various ages at loading and the relaxation curves belong to this domain and are suitable for calibrating the compliance function.

The fact that the second term in Equation (7.14) is assumed to be independent of $t'$ and the last one independent of $t-t'$ has also been questioned, on the basis of test data.[40,27] Another aspect which was criticized on the basis of test data is that the humidity and size influences in Equation (7.14) appear only in the irreversible term,[44] and that the size effect in the shrinkage term does not correspond to diffusion theory.

The BP Model is the only one which involves the influence of temperature. It gives this influence for shrinkage, basic creep, and drying creep. It also gives the effect of the load cycling (pulsation), the effects of the delay of the start of loading after the start of drying, the time lag of loading after heating, the decrease of creep after drying, swelling in water, autogeneous shrinkage of sealed concrete,[101] etc. The price paid for this broader range of applicability is larger complexity. The BP Model differs from the ACI and CEB-FIP Models also by the absence of a final (asymptotic) value of creep. We have already commented on this aspect.

The BP Model was obtained by computer analysis and fitting of 80 different data sets on different concretes from different laboratories (over 800 creep and shrinkage curves involving about 10 000 data points). Based on this unusually large (computerized[204]) data bank, the 90% confidence limits (i.e. the relative deviations from the mean having a 5% probability of being exceeded on the plus side, and 5% on the minus side) were found for the BP Model to be $\omega_{90} = \pm 31\%$. For the ACI Model, comparison with the same data indicated $\omega_{90} = \pm 63\%$, and for the CEB-FIP Model $\omega_{90} = \pm 76\%$.[45] This superiority of the BP Model is obtained in spite of the fact that the data set used to obtain these numbers did not include the temperature effects (which the BP Model also describes well), and that the majority of test data used for comparison pertained to small size specimens and to a limited time range (i.e. $t - t'$ from one week to one year and $t'$ from one week to six months), which the ACI and CEB-FIP Models describe better than large specimens or long times. For very long creep durations ($>10$ years), for very high or very small ages at loading ($>10$ years, $<10$ days), for thick specimens ($>30$ cm), and for the final slopes of creep curves, which matter for extrapolation, the comparison is even more favorable to the BP Model.

However, for drying creep, the confidence limits of the BP Model ($\omega_{90} = \pm 29\%$) are not much better than those of ACI Model ($\pm 42\%$) and especially the CEB-FIP Model ($\pm 32\%$). These two models are of course intended mainly for not too massive structures in a drying environment.

The magnitude of error for all existing models is large and there is no doubt much room for improvement. The greatest part of the errors results from the effects of composition of concrete. This is documented by the fact that prediction errors are greatly reduced when the initial elastic deformation or one short-time shrinkage value is measured.[45] Finite element analysis hardly makes sense if the error in $J(t, t')$ exceeds 20%, and so availability of some short-time tests is a requirement for practical applications. Note also that preferable are creep prediction formulae which can be easily calibrated from given short-time values.[45]

At present no consensus on the proper form of the compliance function $J(t, t')$ or $\bar{J}(t, t')$ has yet been reached. Much of the disagreement is due to the great statistical scatter of available test data, and even more perhaps due to the fact that a linear theory is used for a phenomenon which is not really linear, i.e. necessitates a non-linear theory. A linear theory can be adequate only within a limited range, and specialists still disagree as to what is this range, in particular what is the type of tests to be used for determining the compliance function for a linear theory. Some include only creep or relaxation tests for all ages at loading, which alone define the compliance function completely, while others include information from creep recovery tests (without analysing them by a non-linear theory) at the expense of representing creep for high ages at loading and long times. Since a non-linear theory is not considered appropriate for building codes, the question is which form of the compliance function gives the best results in practical problems over the broadest possible range.

Since the long-time creep is of main interest, efforts have been made to compare the creep prediction models with the 'final' creep values obtained directly from creep measurements. Such comparisons suffer, however, by the error which inevitably occurs in determining such 'final' values from the test data and is just as large as the error of the creep model that is supposed to be checked. For example, it has become almost traditional for the experimentalists to use the Ross hyperbola[144,145,31]: $C = \bar{t}/(a + b\bar{t})$ where $\bar{t} = t - t'$ and $C = J(t, t') - [1/E(t')]$. This relation may be written as $1/C = b + a/\bar{t}$. So, if one plots the measured data as $1/C$ versus $1/\bar{t}$ and approximates this plot by a straight regression line, the slope of the line and its intercept with the $1/C$ axis yield coefficients $a$ and $b$, and the value of $b$ at the same time indicates the extrapolated 'final' value of creep ($\bar{t} \to \infty$). The plot looks very satisfactory if the creep data span over a limited time period, such as from $\bar{t} = 1$ week to 1 year, and one is then inclined to think that the final value obtained from this plot is good. Only such limited data were available in the early investigations in the 1930's and so the use of Ross' hyperbola appeared adequate and became standard practice. At present, however, long-range creep measurements of basic creep are available, and then gross deviations from Ross' hyperbola are found; see Bažant and Chern.[31] It is interesting to observe that even when the errors of the Ross hyperbola are not too conspicuous in the plot of $1/\bar{C}$ versus $1/\bar{t}$ they are blatant in the usual plot of $J(t, t')$ versus log $\bar{t}$. Thus we see that the practice of showing only the plot of $1/C$ versus $1/\bar{t}$ is misleading. The inverse scales $1/C$ and $1/\bar{t}$ obscure the errors for long times by crowding together the points for large $C$ and large $\bar{t}$. One other element of error and ambiguity was already discussed, namely the value to be used for $E(t')$ which must be decided before the plot of $1/C$ versus $1/\bar{t}$ can be constructed.

The 'final' values of creep obtained from creep test data on the basis of Ross' hyperbola were recently compared by Müller and Hilsdorf with various models for creep prediction, and it was observed that the agreement was best for the CEB-FIP 1978 Model Code. From the preceding analysis it is, however, clear that such a method of comparision of various models is fallacious.[31] The 'final' value found by extrapolating the test data strongly depends on the choice of the expression for the creep curve, in this case the Ross hyperbola, and the error of the long-time values of Ross' hyperbola compared to more realistic expressions, such as the double power law, is easily 50%. So one tacitly implies the wrong model in determining the 'final' value, and then one concludes that some other model does not agree with this 'final' value. This is a circular argument.

## 7.3  LINEAR CONSTITUTIVE RELATIONS WITH HISTORY INTEGRALS

### 7.3.1  Principle of superposition

Due to creep, shrinkage, and temperature changes, the stresses in structures normally vary significantly in time even if the loads are constant. Thus, the foregoing exposition of material behaviour under constant stress must be extended to formulate a constitutive equation valid for arbitrarily variable stresses or strains. This task is simplified by the fact that, within the range of service stresses (up to about 1/2 of the strength) and with the exception of decreasing strain, concrete may be approximately treated as a linear material, precisely an aging linear viscoelastic material, the theory of which rests on the principle of superposition. The linearity property simplifies structural analysis, while the aging complicates it greatly.

The principle of superposition, which is equivalent to the hypothesis of linearity, states that a response to a sum of two stress (or strain) histories is the sum of the responses to each of them taken separately. According to this principle, the strain due to any stress history $\sigma(t)$ may be obtained by regarding the history as the sum of increments $d\sigma(t')$ applied at times $t' \in (0, t)$ and summing the corresponding strains which equal $d\sigma(t')J(t, t')$ according to Equation (7.2) (Figure 7.9(a)). This yields

$$\varepsilon(t) = \int_0^t J(t, t')\,d\sigma(t') + \varepsilon^0(t) \qquad (7.16)$$

in which $\varepsilon^0(t)$ is the stress-independent strain (shrinkage plus thermal strain). Equation (7.16) represents the uniaxial constitutive equation which relates general histories of uniaxial stress $\sigma$ and strain $\varepsilon$. The integral in this equation should be understood as the Stieltjes integral, the advantage of which is that it is applicable even for discontinuous stress histories. If $\sigma(t)$ is continuous one has $d\sigma(t') = [d\sigma(t')/dt']\,dt'$, which yields the usual (Riemann)

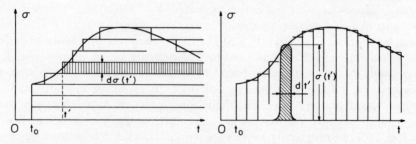

**Figure 7.9** Representation of arbitrary strain history by: (a) stress increments; (b) stress impulses

integral. (The principle of superposition was stated in the works of Boltzmann[57] for non-aging materials and Volterra[181] for aging materials.)

Measurements agree with the principle of superposition very closely under the following conditions:

(1)  The magnitude of stresses is below about 40% of the strength, i.e. within the service stress range.
(2)  The strains do not decrease in magnitude (but the stresses can).
(3)  The specimen undergoes no significant drying during creep.
(4)  There is no large increase of the stress magnitude late after initial loading (cf. Section 7.5.2).

Violation of the last condition causes less error than any of the first three conditions, and so this condition may be dropped in cruder analysis. Condition (2) is very important and excludes, in particular, the creep recovery after unloading, for which the principle of superposition predicts far too much recovery (often twice the observed amount). Some investigators propsed modification of the creep function to predict recovery while keeping the assumption of linearity of creep theory. Such efforts are doomed, however, since one loses more than one gains, sacrificing close approximation of all other behaviour within the linearity range. Prediction of recovery and any response at decreasing strains requires a non-linear theory (Section 7.5.2).

Drying that is simultaneous with creep is a major cause of non-linear dependence on stress. This is probably to a large extent caused by microcracking, cracking, and tensile non-linear behaviour for internal stresses induced by shrinkage and differences in creep. Nevertheless, due to the complexity of non-linear analysis, the principle of superposition is routinely used for structures exposed to a drying environment (such as regular climate conditions) and the mean cross-section compliance $\bar{J}(t, t')$ is substituted for $J(t, t')$. We must however keep in mind that the results of such analysis can be greatly in error. The magnitude of the error is smaller for thicker members and also for prestressed members, since prestress reduces cracking. To eliminate cracking entirely, a large three-dimensional prestress (confinement) is required, which is rarely the case in practice except in spirally reinforced compressed members. However, adequate experimental data on the effects that prestress, confinement, and size have on the deviations from the principle of superposition (linearity) are not available at present.

It is interesting to observe that the proportionality property, which means that if stress history $\sigma(t)$ produces strain history $\varepsilon(t)$ then stress history $k\sigma(t)$ produces strain history $k\varepsilon(t)$, appears to have a broader applicability than the principle of superposition, being verified reasonably well for all loading that meets conditions (1) and (3) but not necessarily (2) and (4). To model this, a non-linear integral-type constitutive relation with a singular kernel seems to work (see Section 7.5.3).

Substituting $d\sigma(t') = [d\sigma(t')/dt']\,dt'$ and integrating by parts, we may transform Equation (7.16) to the following equivalent form:

$$\varepsilon_1(t) = \frac{\sigma(t)}{E(t)} + \int_0^t L(t, t')\sigma(t')\,dt' + \varepsilon^0(t) \tag{7.17}$$

in which $L(t, t') = -\partial J(t, t')/\partial t'$. Geometrically, this equation means that we decompose the stress history into vertical strips and consider each strip as an impulse function (Figure 7.9(b)). The magnitude of each impulse is $\sigma(t')\,dt'$ (area of the strip) and its stress response is $L(t, t')\sigma(t')\,dt'$. Thus, $L(t, t')$ represents the strain at time $t$ caused by a unit stress impulse (Dirac $\delta$-function) at time $t'$ $(t' \leq t)$, and is called the stress impulse memory function.

Another useful relation is obtained by differentiating Equation (7.16):

$$\dot{\varepsilon}(t) = \frac{\dot{\sigma}(t)}{E(t)} + \int_0^t \frac{\partial J(t, t')}{\partial t}\,d\sigma(t') \tag{7.18}$$

where superimposed dots denote time derivatives. The integral gives the creep contribution to the strain rate.

### 7.3.2  Stress relaxation and superposition

The variation of stress at constant strain is called relaxation. It is characterized by the relaxation function, $R(t, t')$ (also called relaxation modulus), which represents the uniaxial stress $\sigma$ at time $t$ (age) caused by a unit constant strain imposed at time $t'$. The typical relaxation function is plotted in Figure 7.10. The response to a general strain history may then be expressed using the principle of superposition. Any strain history may be imagined to consist of small strain increments $d\varepsilon(t')$ introduced at times $t'$, each of which can be regarded as a horizontal strip, similarly to Figure

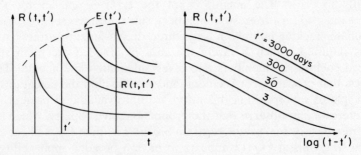

**Figure 7.10**  Typical relaxation curves for various ages $t'$ at strain imposition

7.9(a)). The stress caused by each of them is $R(t, t') \, d\varepsilon(t')$. Summing all these stresses, and subtracting the shrinkage increments $d\varepsilon^0(t')$ since by definition they produce no stress, we have

$$\sigma(t) = \int_0^t R(t, t')[d\varepsilon(t') - d\varepsilon^0(t')] \tag{7.19}$$

For a given stress history, Equation (7.16) represents a Volterra integral equation for the strain history $\varepsilon(t)$. By solving Equation (7.16) for stress at given unit constant strain imposed at various ages $t'$ one may calculate the individual relaxation curves. All these curves together define the relaxation function. Conversely, by solving Equation (7.17) (also a Volterra integral equation) for strain at given unit constant stresses applied at various ages $t'$, one may calculate the individual creep curves, which all together define the compliance function. So, the integral equations, (7.16) and (7.17), are equivalent; one is said to be the resolvent of the other. Of the kernels, $J(t, t')$ and $R(t, t')$, only one may be specified independently and the other one follows.

In absence of drying, the relaxation function obtained by solving Equation (7.19) from the measured compliance function usually agrees with relaxation measurements quite well (Figure 7.11).

The creep function may be converted to the relaxation function by solving the integral equation (Equation (7.16)). We may use for this purpose the step-by-step algorithm given in the next section, for which a simple program was published.[22,199] There also exists a very good approximate formula[34] valid for $t - t' > 1$ day:

$$R(t, t') \simeq \frac{1 - \Delta_0}{J(t, t')} - \frac{0.115}{J(t, t-1)} \left( \frac{J(t-\Delta, t')}{J(t, t'+\Delta)} - 1 \right) \tag{7.20}$$

in which $\Delta = (t - t')/2$, $\Delta_0 \simeq 0.008$, and $t - 1$ means $t$ minus 1 day. Compared to exact solution according to the principle of superposition, the error of this formula is normally (e.g. for double power law) within 1% of the initial value, i.e. within $0.01 R(28 + 0.1, 28)$.

Instead of specifying the compliance function as we did in Section 7.2, the time-dependent behaviour could, alternatively, be described in terms of the relaxation function.[91,158,93,94,115,159,76] The main reason why this is not usually done is that creep tests are somewhat easier to perform than the relaxation tests. Besides, the relaxation function can be determined from the creep function.

This is well applicable, though only for creep without moisture loss, for which the conversion is indeed very accurate. At drying the relaxation function obtained by the principle of superposition from the creep tests may deviate significantly from relaxation measurements. In such cases it might be preferable to use a directly measured relaxation (rather than compliance)

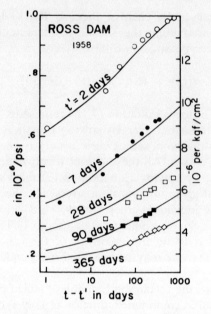

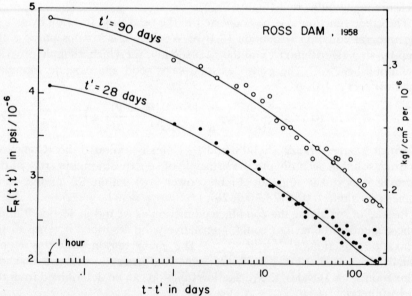

**Figure 7.11** Relaxation curves calculated according to the principle of superposition from measured creep data, and their comparison with measured relaxation data. Calculations from Bažant and Wu[53.]; data from Harboe *et al.*[93] and Hanson[94]

function for the analysis of certain structural problems. These are the relaxation-type problems, such as the effects of a sudden differential settlement or certain types of stress redistribution in the structure for which stresses generally decline and strains are nearly constant.

### 7.3.3 Multiaxial generalization and operator form

The multiaxial generalization of all preceding relations can be easily determined assuming isotropy. It suffices to write stress–strain relations of the same form as the uniaxial ones separately for the volumetric and the deviatoric components of stress and strains. Thus, analogy to Equation (7.16) provides

$$
\left.
\begin{aligned}
3\varepsilon^V(t) &= \int_0^t J^V(t, t')\, d\sigma^V(t') + 3\varepsilon^0(t) \\
2\varepsilon_{ij}^D(t) &= \int_0^t J^D(t, t')\, d\sigma_{ij}^D(t')
\end{aligned}
\right\}
\tag{7.21}
$$

in which $\varepsilon_{ij}^D = \varepsilon_{ij} - \delta_{ij}\varepsilon^V$ is the deviator of the strain tensor $\varepsilon_{ij}$, $\sigma_{ij}^D = \sigma_{ij} - \delta_{ij}\sigma^V$ is the deviator of the stress tensor $\sigma_{ij}$; $\delta_{ij}$ is the Kronecker delta, $\varepsilon^V = \varepsilon_{kk}/3$, $\sigma^V = \sigma_{kk}/3$ (volumetric strain and stress). The subscripts refer to Cartesian coordinates $x_1 \equiv x$, $x_2 \equiv y$, $x_3 \equiv z$ $(i, j = 1, 2, 3)$ (and the summation rule is assumed). Functions $J^V(t, t')$ and $J^D(t, t')$ are the volumetric and deviatoric compliance functions which are related to $J(t, t')$ as follows:

$$
J^V(t, t') = 6(\tfrac{1}{2} - \nu)J(t, t') \qquad J^D(t, t') = 2(1 + \nu)J(t, t') \tag{7.22}
$$

Here $\nu$ is the Poisson ratio, which is in general also a function of $t$ and $t'$, but can be considered as approximately constant ($\nu \approx 0.18$). When, however, drying creep is considered and is described by means of cross-section mean compliance $\bar{J}(t, t')$, then the corresponding mean Poisson ratio is quite variable and can drop to almost zero. Moreover, drying should also cause anisotropy, as a result of microcracking. (For test data on multiaxial creep see McDonald,[124] York,[196] Arthanari and Yu,[9] Neville and Dilger,[145] Meyer,[135] Illston and Jordaan.[103]).

Equation (7.17), based on the relaxation function, and the impulse memory formulations such as Equation (7.21), may be generalized for multiaxial stress similarly.

The linear viscoelastic stress–strain relations of aging material can be also expressed in the form of differential rather than integral equations. This will be outlined in Section 7.4.2.

The constitutive relations may be written in the form $3\varepsilon^V = \mathbf{K}^{-1}\sigma^V + \varepsilon^0$, $2\varepsilon_{ij}^D = \mathbf{G}^{-1}\sigma_{ij}^D$ where $\mathbf{K}^{-1}$ and $\mathbf{G}^{-1}$ are Volterra integral operators which are of non-convolution type and can be manipulated according to the rules of linear algebra. This is exploited in the extension of elastic-viscoelastic analogy to aging materials,[129] which permits converting all equations of linear elasticity to analogous equations for creep.[129,11-14]

### 7.3.4 Application of principle of superposition

A numerical step-by-step solution may be based on the principle of super-position. For this purpose time $t$ is subdivided by discrete times $t_r$ ($r = 0, 1, 2, \ldots$) into time steps $\Delta t_r = t_r - t_{r-1}$. Time $t_0$ coincides with the instant of first loading $t_0$. If there is a sudden change of load at time $t_s$, it is convenient for programming to use a time step of zero duration, i.e. set $t_{s+1} = t_s$ or $t_{s+1} = t_s + 1$ second (Figure 7.12). Often such a sudden load change occurs at the start, i.e. at $t_s = t_0$, and then we choose $t_1 = t_0$ and assume the load to be applied during the first interval $(t_0, t_1)$.

Under constant loads, the strains and stresses vary at a rate which decreases roughly as the inverse of time, and for this reason it must be possible to use increasing time steps $\Delta t_r$ (Figure 7.12). This is also necessary if long times (say $t - t_0 = 50$ years) should be reached in computation, and if the initial time step should be small. Computation is most efficient if we use time steps that are constant in the log $(t - t')$ scale, i.e. $(t_r - t_0)/(t_{r-1} - t_0) =$ constant. Normally about 4 steps per decade in log-time suffice. The first step may usually be chosen as 0.1 day. If there is a sudden change of loading at some later time, one must begin again with small time steps and then increase the steps gradually as long as the load remains constant.

Using the trapezoidal rule, the error of which is proportional to $\Delta\sigma^2$, Equations (7.21) may be approximated[22,38] as

$$3\varepsilon_r^V = \sum_{s=1}^{r} J_{r,s-1/2}^V \Delta\sigma_s^V + 3\Delta\varepsilon_r^0 \qquad 2\varepsilon_{ij_r}^D = \sum_{s=1}^{r} J_{r,s-1/2}^D \Delta\sigma_{ij_s}^D \qquad (7.23)$$

where $\Delta\sigma_s^V = \sigma_s^V - \sigma_{s-1}^V$, $\Delta\sigma_{ij_s}^D = \sigma_{ij_s}^D - \sigma_{ij_{s-1}}^D$, $\sigma_s^V = \sigma^V(t_s)$, etc.; subscripts $r, s$ refer to the discrete times $t_r, t_s$. For $J_{r,s-1/2}^V$ one has two options; either one may take it as $(J_{r,s}^V + J_{r,s-1}^V)/2$ where $J_{r,s}^V$ is a notation for $J^V(t_r, t_s)$, or one may take it as $J^V(t_r, t_{s-1/2})$ where $t_{s-1/2}$ denotes the middle of time interval

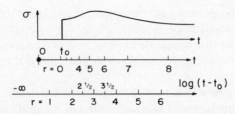

**Figure 7.12** Discrete subdivision of time

$(t_{s-1}, t_s)$ in the log $(t-t_0)$ scale, i.e. $t_{s-1/2} = t_0 + [(t_s - t_0)(t_{s-1} - t_0)]^{1/2}$. A similar notation applies to $J^D_{r,s-1/2}$.

Writing Equation (7.23) also for $\varepsilon^V_{r-1}$ and $\varepsilon^D_{ij_{r-1}}$ and subtracting it then from Equation (7.23), one obtains[22] for volumetric and deviatoric increments:

$$\Delta\sigma^V_r = 3K''_r(\Delta\varepsilon^V_r - \Delta\varepsilon''^V_r) \qquad \Delta\sigma^D_{ij} = 2G''_r(\Delta\varepsilon^D_{ij_r} - \Delta\varepsilon''^D_{ij_r}) \qquad (7.24)$$

in which

$$K''_r = \frac{1}{J^V_{r,r-1/2}} \qquad G''_r = \frac{1}{J^D_{r,r-1/2}} \quad \text{for } r \geq 1 \qquad (7.25)$$

$$3\Delta\varepsilon''^V_r = \sum_{s=1}^{r-1} \Delta J^V_{r,s}\,\Delta\sigma^V_s - 3\Delta\varepsilon^0_r \qquad 2\Delta\varepsilon''^D_{ij} = \sum_{s=1}^{r-1} \Delta J^D_{r,s}\,\Delta\sigma^D_{ij_s} \quad \text{for } r \geq 2$$

$$(7.26)$$

$$\Delta\varepsilon''^V_r = \Delta\varepsilon^0_1 \qquad \Delta\varepsilon''^D_{ij} = 0 \quad \text{for } r = 1$$

$$\Delta J^V_{r,s} = J^V_{r,s-1/2} - J^V_{r-1,s-1/2} \qquad \Delta J^D_{r,s} = J^D_{r,s-1/2} - J^D_{r-1,s-1/2} \qquad (7.27)$$

Since $\Delta\varepsilon''^V_r, \Delta\varepsilon''^D_{ij_r}$ and $K''_r, G''_r$ can be evaluated from the values of the stresses before the current time $t_r$, their values are known before solving $\sigma_r$ and $\varepsilon_r$. Therefore, Equation (7.24) may be regarded as an incremental elastic stress–strain relation,[22,38] with bulk modulus $K''_r$, shear modulus $G''_r$, and inelastic (initial) strains of volumetric and deviatoric components $\Delta\varepsilon''^V_r, \Delta\varepsilon''^D_{ij_r}$. The incremental elastic stiffness matrix can be set up using $K''_r$ and $G''_r$. The creep analysis is thus reduced to a sequence of elastic analyses for the individual time steps. Each of them can be carried out by finite elements. Within each time step, the prescribed increments of loads and displacements must be considered.

Analogous equations may be written for uniaxial behaviour, e.g. for frame analysis.

Algorithms in which the integral from Equation (7.16) is approximated by the rectangle rule, yielding the sum $\sum J_{r,s-1}\,\Delta\sigma_s$ instead of that in Equation (7.23), have been used in practice. However, they are not any simpler. Their error, being of first rather than second order in $\Delta\sigma$, is larger and so more time steps (and more computer storage for the stress history) are required.

To gain an idea of accuracy, the convergence is illustrated in Table 7.1 which gives the values of $R(t, t_0)$ calculated from $J(t, t')$ by the above second-order algorithm (Equations (7.24)–(7.27)) as well as by the first-order algorithm mentioned before. Table 7.1 also gives the values of $J(t, t_0)$ calculated from $R(t, t')$ using a similar second-order algorithm based on Equation (7.19). The calculation of $R(t, t_0)$ was made[22] for ACI compliance function, $t = 1035$ days, $t_0 = 35$ days, and discrete times $t_r = t_0 + [10^{r/m}(0.1 \text{ day})]$; $r = 1, 2, 3, \ldots$; $m = $ number of steps per decade. The calculation of $J(t, t_0)$ was made for $R(t, t')$ obtained by the formula in Equation (7.20)

192       *Creep and Shrinkage in Concrete Structures*

Table 7.1    Convergence of step-by-step method based on superposition

| | $R(t, t_0)$ from $J(t, t')$ | | | | $J(t, t_0)$ from $R(t, t')$ | |
|---|---|---|---|---|---|---|
| Number $m$ of steps per decade | 2nd-order method | | 1st-order method | | 2nd-order method | |
| 1 | | | | ` | 2.3429 | |
| 2 | 0.3532 | 60 | 0.3311 | 157 | 2.3845 | 424 |
| 4 | 0.3592 | 26 | 0.3468 | 88 | 2.4070 | 235 |
| 8 | 0.3618 | 12 | 0.3556 | 43 | 2.4183 | 113 |
| 16 | 0.3630 | 5 | 0.3599 | 20 | 2.4237 | 54 |
| 32 | 0.3635 | 2 | 0.3619 | 10 | 2.4264 | 27 |
| 64 | 0.3637 | | 0.3629 | | — | |

from double power law with $n = 0.14$, $m = 0.3$, $\alpha = 0.04$, $\phi_1 = 1$, and for $t_0 = 28$ days, $t = 10{,}028$ days, and discrete times $t_r = t_0 + [10^{r/m}(0.0001 \text{ day})]$. From the table we see that algorithms based on $J(t, t')$ or $R(t, t')$ are about equally accurate and that in order to keep the error below 1%, about 4 steps per decade suffice for the second-order method and 16 steps per decade for the first-order method.

It is interesting to note that the simple replacement of the impulse memory integral in Equation (7.17) with a sum is computationally less efficient and does not always produce a convergent step-by-step solution. It is partly for this reason that Equation (7.16) has recently been favoured over Equation (7.17) used in earlier works.[1,10,18–20,129]

The foregoing type of algorithm based on the principle of superposition is effective and works well for small to medium size structural systems.[102,201] For large systems with many unknowns, however, the demands for computer storage as well as time become excessive. For each finite element one must store all the preceding values of all stress components, and at each time step one must evaluate long sums from these values. This requires a very large storage capacity, which must normally be met by peripheral storage. Numerous transfers to and from the peripheral storage at each time step then greatly prolong the running time. A decade ago these demands were forbidding and only medium-size structural systems could be solved (and at great expense) by the step-by-step methods based directly on the compliance function.[154,69,164] Today even large systems could be solved in this manner on the largest computers in existence. However, that would be wasteful. Far more effective methods have recently been developed, and we will discuss these in Section 7.4.

### 7.3.5   Age-adjusted effective modulus method

In case the structure is suddenly loaded at time $t_0$ and afterwards the loads are steady, the simplest method is to use a certain quasi-elastic stress–strain relation for the total increment from $t_0$, the time of first loading, to current time $t$. For uniaxial stress $\sigma_{11}$ this relation may be written as:

$$\Delta\varepsilon_{11} = \frac{\Delta\sigma_{11}}{E_{11}} + \Delta\varepsilon_{11}'' \qquad \Delta\varepsilon_{11}'' = \frac{\sigma_{11}(t_0)}{E(t_0)}\phi(t, t_0) + \Delta\varepsilon^0 \qquad (7.28)$$

where $\Delta\sigma_{11} = \sigma_{11}(t) - \sigma_{11}(t_0)$, $\Delta\varepsilon_{11} = \varepsilon_{11}(t) - \varepsilon_{11}(t_0)$, $\varepsilon_{11}(t_0) = \sigma_{11}(t_0)/E(t_0)$; and $E''$ serves as the apparent elastic modulus for the increment, $\Delta\varepsilon_{11}''$ is the inelastic strain increment, $\Delta\varepsilon^0$ is the shrinkage strain increment. One might think at first that a good estimate would be $\Delta\varepsilon_{11} = \Delta\sigma_{11}/E(t_0) + \frac{1}{2}[\sigma(t_0) + \sigma(t)]\,\phi(t, t_0)/E(t_0)$, which is equivalent to Equation (7.28) if we set $E'' = E(t_0)/[1 + \phi(t, t_0)/2]$. However, comparisons with the exact solution of the integral equation shows that this value of $E''$ is usually too high, and far too high when $t_0$ is great. A better estimate, actually far better in case of high $t_0$, is $\varepsilon_{11}(t) = \sigma_{11}(t)/E_{\text{eff}}$ with $E_{\text{eff}} = E(t_0)/[1 + \phi(t, t_0)] =$ effective (sustained) modulus;[126,80] this relation is obtained from Equation (7.28) when $E'' = E_{\text{eff}}$. The best estimate thus seems to be $E'' = E(t_0)/[1 + \chi\phi(t, t_0)]$ where $\chi$ is some coefficient between 0.5 and 1.0.

Calibrating the value of $\chi$ according to approximate but good results for $\Delta\sigma$ for the relaxation test (and neglecting the age dependence of $E(t)$), Trost[176] obtained suitable approximate values of $\chi$, typically $\chi = 0.8$ to 0.9. Subsequently, an exact statement (a theorem) which underlies the effectiveness of this approach and applies also for age-dependent $E(t)$, was found.[23] Namely, if the strain history is linear in $J(t, t')$, i.e. of the form $\varepsilon(t) = \varepsilon_0 + c_1 J(t, t_0)$ (Figure 7.13), then the stress history is linear in $R(t, t_0)$, i.e. of the form $\sigma(t) = \sigma_0 + c_2 R(t, t_0)$, and Equation (7.28) is exact if

$$E'' = \frac{E(t_0) - R(t, t_0)}{\phi(t, t_0)} \qquad (7.29)$$

as proven by Bažant[23] (see Section 7.7.5). The method is effective because the actual histories of stress and strain in structures under a steady load or a load which changes gradually at a decaying rate (shrinkage) are rather close to linear functions of $J(t, t_0)$ and $R(t, t_0)$.

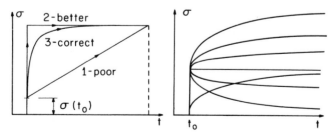

**Figure 7.13** Strain histories expressed as a linear combination of the relaxation function

$E''$ is called the age-adjusted effective modulus. The term comes from the fact that in the absence of aging (or when $t_0$ is high) $E''$ is nearly exactly equal to $E_{eff}$, which reveals that the difference between $E''$ and $E_{eff}$ is almost entirely due to aging. For practical calculations of $E''$, it suffices to use in Equation (7.29) the approximate expression for $R(t, t_0)$ according to Equation (7.20).

In case of multiaxial stress, Equation (7.28) may be generalized as

$$\Delta \varepsilon^V = \frac{\Delta \sigma^V}{3K''} + \Delta \varepsilon''^V \qquad \Delta \varepsilon_{ij}^D = \frac{\Delta \sigma_{ij}^D}{2G''} + \Delta \varepsilon_{ij}''^D \tag{7.30}$$

with

$$\Delta \varepsilon''^V = \frac{\sigma^V(t_0)}{3K(t_0)}\phi(t, t_0) + \Delta \varepsilon^0 \qquad \Delta \varepsilon_{ij}''^D = \frac{\sigma_{ij}^D(t_0)}{2G(t_0)}\phi(t, t_0) \tag{7.31}$$

in which $K''$ and $G''$ are the bulk and shear moduli that correspond to Poisson's ratio $\nu$ and Young's modulus $E''$ given by Equation (7.29); i.e. $3K'' = E''/(1-2\nu)$, $2G'' = E''/(1+\nu)$.

In the case where several steady loads or imposed deformations start at different times $t_0$, the effect of each of them may be analysed separately according to Equations (7.28) or (7.31) and the results may then be superimposed.

For applications see, e.g. Bažant, Carreira and Walser,[30] Bažant and Najjar,[38] Bažant and Panula.[45]

## 7.4  LINEAR CONSTITUTIVE RELATIONS WITHOUT HISTORY INTEGRALS

### 7.4.1  Degenerate kernel

The need for storing and using the complete history of stresses or strains may be eliminated if the integral-type creep law (Equations (7.16) or (7.19) or (7.21)) can be converted to a rate-type creep law, i.e. a creep law given by a system of first-order differential equations. It appears that this can always be done, not exactly but with any desired accuracy. The key is to approximate the kernel of one of the integral equations for the creep law (Equations (7.16) or (7.19) or (7.21)) by the so-called degenerate kernel, the general form of which is a sum of products of functions of $t$ and functions of $t'$. The form may be written as

$$J(t, t') = \sum_{\mu=1}^{N} [1/C_\mu(t')] - \sum_{\mu=1}^{N} [B_\mu(t)/(B_\mu(t')C_\mu(t'))]$$

where $C_\mu$ and $B_\mu$ are functions of time. It is more convenient to denote

$y_\mu(t) = -\ln B_\mu(t)$, in which case $B_\mu(t) = \exp[-y_\mu(t)]$. Thus, the most general form of a degenerate kernel may always be written as

$$J(t, t') = \sum_{\mu=1}^{N} \frac{1}{C_\mu(t')} \{1 - \exp[y_\mu(t') - y_\mu(t)]\} \tag{7.32}$$

In previous works only the special case when

$$y_\mu(t) = t/\tau_\mu \quad (\mu = 1, 2, \ldots, N) \tag{7.33}$$

has been considered. Constants $\tau_\mu$ are called the retardation times. Equation (7.32) then becomes

$$J(t, t') = \sum_{\mu=1}^{N} \frac{1}{C_\mu(t')} \{1 - \exp[-(t - t')/\tau_\mu]\} \tag{7.34}$$

This is a series of real exponentials, called the Dirichlet series (sometimes also called the Prony series).† To represent the instantaneous (elastic) part of the compliance function, we choose very small first retardation time $\tau_1(\mu = 1)$, e.g. $\tau_1 = 10^{-9}$ day, which means that the first term of the series is nearly exactly $1/C_1(t')$ and represents the instantaneous compliance; then we have $C_1(t') = E(t')$. This is more convenient for computer programming than using in Equation (7.3) a separate instantaneous term which is not a part of the sum.

When plotted in $\log(t - t')$ scale, the individual exponential terms in Equation (7.40) look like step functions with the step spread out over the period of about one decade (Figure 7.14(a)). Outside this decade on both left and right, each exponential term gives an almost horizontal curve. The point $t - t' = \tau_\mu$ is located roughly at the centre of the rise. The approximation of a creep curve by a sum of exponential curves (Equation (7.34)) may be imagined as shown in Figure 7.14(b). By passing horizontal strips and picking as $\tau_\mu$ the times at the centre of the rise for each strip, it is possible to obtain graphically a crude Dirichlet series approximation (i.e. the values of $\hat{E}_\mu(t')$ for each chosen $t'$). From this graphical construction several salient properties become evident.

(1) The approximation is not unique, since various divisions in horizontal strips in Figure 7.14 can be used to approximate the same creep curve. In particular, various choices of $\tau_\mu$ must yield equally good results. Therefore, the values of $\tau_\mu$ must be chosen in advance. Attempting to calculate them, e.g. from a least-square condition, leads to an unstable problem characterized by an ill-conditioned equation system and a non-unique solution.[118]

(2) For the sake of simplicity, one can choose the same $\tau_\mu$ values for the

† Hardy and Riesz,[96] Lanczos,[118] Cost,[73] Schapery,[166] Williams.[187]

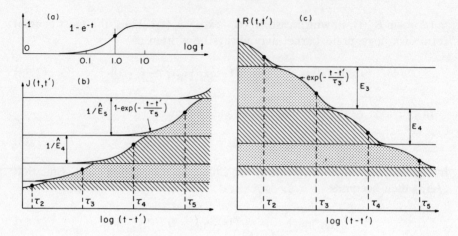

**Figure 7.14**   Representation by Dirichlet series

creep curves for all ages $t'$ at loading, i.e. $\tau_\mu$ can be considered to be constant without any loss in the capability to fit test data.

(3) The choice of $\tau_\mu$ is however not entirely arbitrary. Since the rise of each exponential term spreads roughly over one decade, $\tau_\mu$ values cannot be spaced more than a decade apart in $\log(t-t')$ scale. So, the smallest possible number of exponential terms is obtained with the choice

$$\tau_\mu = 10^{\mu-2}\tau_2 \quad (\mu = 2, 3, \ldots, N) \tag{7.35}$$

although the choice $\tau_\mu = a^{\mu-2}\tau_2$ with $a < 10$ gives a somewhat better accuracy and smoother response curves.

(4) The values of $\tau_\mu$ must cover the entire time range of interest. If we want to calculate the response from time $\tau_{\min}$ until $\tau_{\max}$ after load application, the smallest $\tau_\mu$ (i.e. $\tau_2$) must be such that $\tau_2 \lesssim 3\tau_{\min}$ and the last one such that $\tau_N \gtrsim 0.5\tau_{\max}$.

(5) To take aging into account accurately, the smallest $\tau_\mu$ must be much less than the age of concrete, $t_0$, when the structure is first loaded, i.e. $\tau_2 \lesssim 0.1t_0$.

The functions $C_\mu(t')$ may in general be identified from any given $J(t, t')$ by a computer subroutine based on minimizing a sum of square-deviations from given $J(t, t')$. This subroutine is listed in Ref. 28. (However, do not take from Ref. 28 the subroutine that converts $J(t, t')$ into $R(t, t')$ since it contains two misprints; use the one from Ref. 22 or 199.) When the creep curves are given by $J(t, t')$ as power curves, i.e.

$$J(t, t') = [1 + \Psi(t')(t-t')^n]/E_0, \tag{7.36}$$

there exists explicit formulae [26]

$$\frac{1}{\hat{E}_\mu(t')} = \begin{cases} \dfrac{1}{E_0} + a(n)\left(\dfrac{\tau_1}{0.002}\right)^n \Psi(t') & \text{for } \mu = 1 \\[3mm] b(n)\left(\dfrac{\tau_1}{0.002}\right)^n 10^{n(\mu-1)}\Psi(t') & \text{for } 1 < \mu < N \\[3mm] 1.2b(n)\left(\dfrac{\tau_1}{0.002}\right)^n 10^{n(N-1)}\Psi(t') & \text{for } \mu = N \end{cases} \qquad (7.37)$$

in which the coefficients may be determined from Table 7.2. An explicit formula also exists when $J(t, t') = \Psi(t') \log(t - t' + \text{constant})$; see Bažant and Wu;[51] or Bažant.[26,25]

Although the use of degenerate kernel permits conversion to a rate-type creep law, it does not eliminate all numerical difficulties. The reason is that the smallest retardation time $\tau_\mu$ must be rather short, say three days, in order to represent the initial rapid creep adequately. With the usual step-by-step integration methods, however, the time step $\Delta t$ must not exceed the smallest $\tau_\mu$ for reasons of numerical stability and must be kept much less than this, say $\Delta t = 0.1$ day, to assure sufficient accuracy. Then, however, an enormous number of time steps would be needed to reach the long-time solution for, say, 50 years of load duration. Yet small time steps $\Delta t$ after 10 years of loading should not be required because all variables in case of steady loading vary so slowly that even with a one-year interval their change is small. So, it should be possible to increase the time step gradually from a small initial value, such as 0.1 day, to a large value, such as one year. There exists algorithms which enable this without causing numerical instability and loss of accuracy. These are the recursive exponential algorithms, and we will explain them later.

The choice of reduced times $y_\mu(t)$ for the general degenerate kernel (Equation (7.32)) is still under investigation (J. C. Chern at Northwestern University). It appears that a suitable expression is

$$y_\mu(t) = (t/\tau_\mu)^{q_\mu} \quad (\mu = 1, 2, \dots, N) \qquad (7.38)$$

where $q_\mu = \text{constants}$ ($q_\mu > 0$). Choosing $q_\mu < 1$ obviously helps in representing the decline of the creep rate due to aging. Regarding the choice of

Table 7.2  Coefficients for Dirichlet series expansion of power function of exponent $n$

| $n$ | 0.05 | 0.10 | 0.15 | 0.20 | 0.25 | 0.30 | 0.35 |
|---|---|---|---|---|---|---|---|
| $a(n)$ | 0.6700 | 0.4465 | 0.2929 | 0.1885 | 0.1154 | 0.0611 | 0.0156 |
| $b(n)$ | 0.0819 | 0.1161 | 0.1229 | 0.1152 | 0.1007 | 0.8042 | 0.0681 |

retardation times, we should observe that function $1-\exp[-y_\mu(t)]$
(Equation (7.32)) gives a step that spreads over a width of about one decade
in $\log t^{q_\mu}$ rather than in $\log t$. So it is appropriate to require that the ratio of
$\tau_\mu^{q_\mu}$ to $\tau_{\mu-1}^{q_\mu}$ would not exceed 10, $\bar{q}_\mu$ being the larger of $q_\mu$ and $q_{\mu-1}$. This
suggests the rule

$$\tau_\mu = 10^{1/\bar{q}_\mu}\tau_{\mu-1} \quad (\mu = 3, 4, \ldots, N) \tag{7.39}$$

Now we see that having exponents $q_\mu < 1$ has the advantage that times $\tau_\mu$
can be spread farther apart in $\log t$ scale, the farther apart the smaller is $q_\mu$.
It appears that $q_\mu$ can be as small as 2/3 without impairing the representation of typical test data (as found by J. C. Chern). In that case we would
need about 33% fewer $\tau_\mu$ to cover the same time range, which brings about
a substantial reduction in storage requirements for internal variables $\varepsilon_\mu$ as
well as a reduction in computer time.

Similarly to Equation (7.34), the most general degenerate form of the
relaxation function is

$$R(t, t') = \sum_{\mu=1}^{N} E_\mu(t') \exp[y_\mu(t') - y_\mu(t)] \tag{7.40}$$

where $y_\mu(t)$ are reduced times defined again by Equation (7.33), except that
$\tau_\mu$ are here called the relaxation times (rather than retardation times). So
far, the reduced times have been always considered proportional to actual
time $\tau_\mu$, as in Equation (7.33). Then we have the Dirichlet series expansion

$$R(t, t') = \sum_{\mu=1}^{N} E_\mu(t') \exp[-(t-t')/\tau_\mu] \tag{7.41}$$

For the choice of $\tau_\mu$ the same rules hold as before, except that $\tau_1$ need not
be very small, while $\tau_N = 10^9$ days, if the final creep value should be
bounded. So we may set

$$\tau_\mu = 10^{\mu-1}\tau_1 \quad (\mu = 1, 2, \ldots, N-1) \tag{7.42}$$

Functions $E_\mu(t')$ may be crudely determined by a graphical procedure
based on splitting the individual relaxation curves in horizontal strips (see
Figure 7.14(c)). For accurate results, functions $E_\mu(t)$ may be identified by
the method of least squares, for which an efficient subroutine is listed by
Bažant and Asghari[28] and a refined one is given in the program described by
Bažant, Rossow and Horrigmoe[46] and fully listed in Ref. 199. At first,
however, the function $R(t, t')$ must be obtained from $J(t, t')$, as described in
Section 7.3.4.

The plot of $E_\mu(t)$ versus $\log \tau_\mu$ is called the relaxation spectrum; its
example is shown in Figure 7.15.

All that has been said of the uniaxial compliance or relaxation function also applies separately to volumetric and deviatoric compliance functions and relaxation functions.

A compliance function in the form of Dirichlet series was used by McHenry[125] and by Maslov[133] and Arutyunyan,[10] although for the purpose of converting the structural problem from integral to differential equations in time rather than for the purpose of avoiding the storage of history in a step-by-step solution. The latter advantage of degenerate kernel was first utilized by Selna[168,169] and Bresler and Selna;[62] but their algorithm did not allow increasing the time step beyond a fraction of the smallest retardation time. The exponential algorithm which admits arbitrary time step was first developed for non-aging materials; see Zienkiewicz and Watson,[198] Taylor, Pister and Goudreau[175] and Mukaddam.[138] The exponential algorithms for aging materials, based on degenerate forms of compliance as well as relaxation functions, were developed by Bažant[21] and were applied in a small finite element program by Bažant and Wu.[51,53] Other forms of exponential algorithms which differ in various details were developed by Kabir and Scordelis,[108] Argyris *et al.*,[7,8] Pister *et al.*,[150] and Willam.[186] They used their algorithms in large finite element programs. Smith, Cook and Anderson,[171] Smith and Anderson,[172] and Anderson,[6] implemented Bazant's algorithm[21,51] based on a degenerate form of the compliance function in the general purpose finite element program NONSAP. The same, but for a degenerate form of the relaxation function, was implemented in NONSAP by Bažant, Rossow, and Horrigmoe.[46]

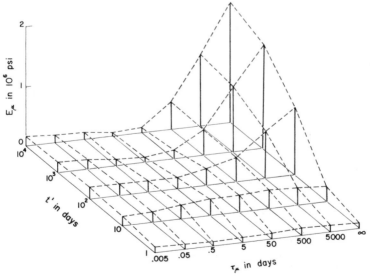

**Figure 7.15**   Example of relaxation spectra for Dworshak Dam concrete
(Corresponding to Figure 7.17(a))

In concluding the subject of Dirichlet series compliance or relaxation functions we should emphasize that they represent merely an approximation to the real creep law (e.g. power law), justified by computational convenience. The Dirichlet series involves too many empirical coefficients to accept it as the creep law *per se*. In organizing a versatile program, the creep properties should be input either in the form of the parameters of some of the creep laws explained in Section 7.2 or as numerical measured values of $J(t, t')$ (and $\varepsilon_S(t, t_0)$). The former involves much fewer parameters for the input than the Dirichlet series. For example, for the double power law only five parameters need to be read on input to characterize all uniaxial creep, instantaneous deformation, and aging. The program should then automatically generate the parameters of the Dirichlet series (or of the rate-type formulation to be described next). The input of the finite element program by Bazant, Rossow, and Horrigmoe,[46] fully listed in Ref. 199, is organized in this manner, with various options (see Section 7.7.4).

### 7.4.2   Rate-type constituve relations

As already mentioned, the degenerate compliance function (Equation (7.32)) can be exploited to convert the integral-type creep law to a differential equation. Substituting Equation (7.32) into Equation (7.16), we may write the resulting expression for $\varepsilon(t)$ in the form

$$\varepsilon(t) = \sum_{\mu=1}^{N} \varepsilon_\mu(t) + \varepsilon^0(t) \tag{7.43}$$

in which

$$\varepsilon_\mu(t) = \int_0^t \frac{d\sigma(t')}{C_\mu(t')} - \gamma_\mu(t) \qquad \gamma_\mu(t) = \exp\left[-y_\mu(t)\right] \int_0^t \exp\left[y_\mu(t')\right] \frac{d\sigma(t')}{dy_\mu(t')} \frac{dy_\mu(t')}{C(t')}$$

$$\tag{7.44a,b}$$

Now, expressing the derivatives $d\varepsilon_\mu/dy_\mu$ and $d^2\varepsilon_\mu/dy_\mu^2$, we may check by substitution that the $\varepsilon_\mu$ always satisfy the following linear differential equations:

$$\frac{d^2\varepsilon_\mu}{dy_\mu^2} + \frac{d\varepsilon_\mu}{dy_\mu} = \frac{1}{C_\mu(t)} \frac{d\sigma}{dy_\mu} \quad (\mu = 1, 2, \ldots, n) \tag{7.45}$$

Furthermore, expressing the derivative $d\gamma_\mu/dy_\mu$ from Equation (7.44), we may check that the $\gamma_\mu$ always satisfy the differential equations:

$$\frac{d\gamma_\mu}{dy_\mu} + \gamma_\mu = \frac{1}{C_\mu(t)} \frac{d\sigma}{dy_\mu} \quad (\mu = 1, 2, \ldots, n) \tag{7.46}$$

$\varepsilon_\mu$ and $\gamma_\mu$ are related to $\varepsilon$ by the differential equations

$$\dot{\varepsilon} = \sum_\mu \dot{\varepsilon}_\mu + \dot{\varepsilon}^0 \qquad \dot{\varepsilon}_\mu = \frac{\dot{\sigma}}{C_\mu(t)} - \dot{\gamma}_\mu \qquad (7.47)$$

One can further check that integration of Equation (7.45) yields Equation (7.44) for $\varepsilon_\mu$, and integration of Equation (7.46) yields Equation (7.44b) for $\gamma_\mu$. Thus the rate-type creep law given either by Equations (7.45) and (7.47) or by Equations (7.46) and (7.47) is equivalent to the integral-type creep law in Equation (7.16) with $J(t, t')$ given by Equation (7.32). In Equations (7.45) or (7.46), functions $y_\mu(t)$ are treated as independent variables, analogous to time; so we may call $y_\mu(t)$ the reduced times. Obviously, $y_\mu(t)$ must be monotonically increasing functions of actual time $t$.

The derivatives with respect to $y_\mu$ may be expressed in terms of time derivatives, substituting $d\varepsilon_\mu/dy_\mu = \dot{\varepsilon}_\mu/\dot{y}_\mu$, $d^2\varepsilon_\mu/dy_\mu^2 = (\ddot{\varepsilon}_\mu\dot{y}_\mu - \dot{\varepsilon}_\mu\ddot{y}_\mu)/\dot{y}_\mu^3$. Thus, Equation (7.45) becomes

$$\ddot{\varepsilon}_\mu + \left(\dot{y}_\mu - \frac{\ddot{y}_\mu}{\dot{y}_\mu}\right)\dot{\varepsilon}_\mu = \frac{\dot{y}_\mu}{C_\mu(t)}\dot{\sigma} \qquad (7.48)$$

The form of a rate-type constitutive relation may be interpreted in terms of a rheologic model consisting of springs and dashpots. For an aging material, the spring moduli $E_\mu$ and dashpot viscosities are functions of age $t$. Consider the Kelvin chain model in Figure 7.16(a), in which $\varepsilon_\mu$ denotes the strain of the $\mu$th Kelvin unit. Now, for an aging elastic response, we must realize that we cannot say that the stress in the spring is $E_\mu(t)\varepsilon_\mu$; but we can say[27,14] that the rate of the stress is $E_\mu(t)\dot{\varepsilon}_\mu$ (see Section 7.5.1). The stress in the dashpot is $\eta_\mu(t)\dot{\varepsilon}_\mu$ and its rate is $[\eta_\mu(t)\dot{\varepsilon}_\mu]^{\cdot}$. So the rate of total stress in the $\mu$th Kelvin unit is $[\eta_\mu(t)\dot{\varepsilon}_\mu] + E_\mu(t)\dot{\varepsilon}_\mu$, which yields:[21]

$$\ddot{\varepsilon}_\mu + \frac{E_\mu(t) + \dot{\eta}_\mu(t)}{\eta_\mu(t)}\dot{\varepsilon}_\mu = \frac{\dot{\sigma}}{\eta_\mu(t)} \qquad (7.49)$$

Note that this differential equation for $\varepsilon_\mu(t)$ is of second order, while for a non-aging material it is of first order. To obtain $J(t, t')$ for the Kelvin chain model, we need to integrate Equation (7.49) for the stress history $\sigma = 1$ for $t \geq t'$ and $\sigma = 0$ for $t < t'$, and using initial conditions $\varepsilon_\mu(t') = 0$

(a)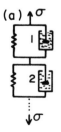

(b)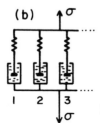

**Figure 7.16** Kelvin and Maxwell chain models

and $\dot{\varepsilon}_\mu(t') = 1/\eta_\mu(t')$, we obtain[14]

$$J(t, t') = \sum_{\mu=1}^{N} \int_0^t \frac{1}{\eta_\mu(\tau)} \exp\left[f_\mu(\tau) - f_\mu(t)\right] d\tau \qquad f_\mu(\xi) = \int_0^\xi \frac{E_\mu(\theta)}{\eta_\mu(\theta)} d\theta$$

(7.49a)

Equating the coefficients of Equations (7.49) and (7.48) we conclude that[14]

$$\eta_\mu(t) = \frac{C_\mu(t)}{\dot{y}_\mu(t)} \qquad E_\mu(t) = C_\mu(t) - \frac{C_\mu(t)}{\dot{y}_\mu(t)}$$

(7.50)

In particular, for $y_\mu = (t/\tau_\mu)^q$ (Equation (7.38)), we have

$$\eta_\mu(t) = \tau_\mu^q C_\mu(t) \frac{t^{1-q}}{q} \qquad E_\mu(t) = C_\mu(t) - \tau_\mu^q \dot{C}_\mu(t) \frac{t^{1-q}}{q}$$

(7.51)

So we can always find a Kelvin chain model which is equivalent to the most general degenerate kernel.[13,14] A more difficult question is whether the values of $\eta_\mu$ obtained from Equation (7.50) are thermodynamically admissible. For example, the minus sign in Equation (7.50) makes us worry lest $E_\mu$ become negative. These questions will be discussed in Section 7.5.1.

An analogous formulation is possible for the degenerate form of the relaxation function (Equations (7.40), (7.41)). Substituting Equation (7.40) into the superposition integral in Equation (7.19) we obtain

$$\sigma(t) = \sum_{\mu=1}^{N} \sigma_\mu(t)$$

(7.52)

$$\sigma_\mu(t) = \exp\left[-y_\mu(t)\right] \int_0^t \exp\left[y_\mu(t')\right] E_\mu(t) [d\varepsilon(t') - d\varepsilon^0(t')]$$

(7.53)

Now expressing the derivative $d\varepsilon_\mu/dy_\mu$, we may check that the $\sigma_\mu$, called the partial stresses, satisfy the differential equations

$$\frac{d\sigma_\mu}{dy_\mu} + \sigma_\mu = E_\mu(t) \frac{d(\varepsilon - \varepsilon^0)}{dy_\mu}$$

(7.54)

Conversely, integration of Equation (7.54) with Equation (7.52) yields Equation (7.53), which implies Equation (7.40). Thus, the rate-type creep law given by Equations (7.52) and (7.54) is equivalent to the integral-type creep law with the degenerate kernel (Equation (7.40)).

Noting that $(d/dy_\mu) = (d/dt)/\dot{y}_\mu$, we may rewrite Equation (7.54) as

$$\dot{\sigma}_\mu + \dot{y}_\mu(t)\sigma_\mu = E_\mu(t)(\dot{\varepsilon} - \dot{\varepsilon}^0)$$

(7.55)

Observe that, in contrast to the aging Kelvin chain (Equation (7.49)), the differential equation for the aging Maxwell chain is of the first order rather than the second order.

Consider now the Maxwell chain model (Figure 7.16b), in which $\sigma_\mu$ denotes the stress in the $\mu$th Maxwell unit. The strain rate in the spring is $\sigma_\mu/E_\mu(t)$ and that in the dashpot is $\sigma_\mu/\eta_\mu(t)$. Summing them, we get the total strain rate $\dot{\varepsilon} - \dot{\varepsilon}^0 = (\dot{\sigma}_\mu/E_\mu) + (\sigma_\mu/\eta_\mu)$, which may be written as

$$\dot{\sigma}_\mu + \frac{E_\mu(t)}{\dot{y}_\mu(t)}\,\sigma_\mu = E_\mu(t)(\dot{\varepsilon} - \dot{\varepsilon}^0) \tag{7.56}$$

Comparing the coefficients of this equation and Equation (7.55), we see that $E_\mu(t)$ is indeed the spring modulus, as we may have anticipated, and that the viscosity of the $\mu$th dashpot is

$$\eta_\mu(t) = E_\mu(t)/\dot{y}_\mu(t) \tag{7.57}$$

In particular, for $y_\mu = (t/\tau_\mu)^a$ (Equation (7.38)) we have

$$\eta_\mu(t) = \tau_\mu^a E_\mu(t)\frac{t^{q-1}}{q} \tag{7.57a}$$

Variables $\sigma_\mu$ or $\varepsilon_\mu$ (or $\gamma_\mu$) represent what is known in continuum thermodynamics as internal variables (i.e. state variables that cannot be directly measured). The current values of these variables characterize the effect of the past history of the material. Thus, we need to store only the current values of, say, about four internal variables ($\mu = 1, \ldots, 4$) to characterize the stress history from, say, $t - t' = 0.1$ day until $10^4$ days. This makes the computations much more efficient. Another term for $\sigma_\mu$ is the hidden stresses or partial stresses, and for $\varepsilon_\mu$ the hidden strains or partial strains.

A comparison of Maxwell chain model predictions with some test data from the literature is shown in Figure 7.17.

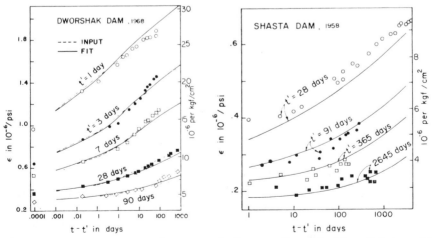

**Figure 7.17** Creep curves for various ages at loading according to the Maxwell chain model, compared with test data[53,149,93,94]

By writing analogous equations for the deviatoric and volumetric components, the foregoing rate-type stress–strain relations are easily generalized to three dimensions.

Since both Maxwell and Kelvin chains can approximate the integral-type creep law with any desired accuracy, these two models must be mutually equivalent, and equivalent also to any other possible spring–dashpot model. For non-aging materials this was rigorously demonstrated long ago by Roscoe.[157] However, certain subtle questions remain in the case of aging material. It appears that not every $J(t, t')$ can be represented by each model unless we relax certain thermodynamic restrictions on the aging process (Section 7.5.1). Thus, the complete mutual equivalence of various spring–dashpot (linear) rheologic models and their equivalence to a general linear integral-type creep law apparently do not hold true in case of aging.

### 7.4.3  Temperature effect

To include the temperature influence, the compliance function of the form of Dirichlet series was generalized by Mukaddam and Bresler,[139] with further refinements by Mukaddam[138] and Kabir and Scordelis.[108] Although a temperature increase always intensifies creep, the use of the integral-type formulation however presents a difficult question: What is the difference between the effects of the current temperature and the past temperatures on the present creep rate? It seems that this question cannot be approached with the integral-type formulation in other than a totally empirical manner.

For the rate-type formulation this difficult question does not arise, since the formulation is history-independent, and so only current temperature matters. Thus, the creep rate must be adjusted only according to the current temperature $T$, and so must be the rate of aging and the rate of change of internal variables such as $\gamma_\mu$ or $\sigma_\mu$. The chief advantage of the rate-type formulation is that a well-founded physical theory, namely the rate-process theory,[86,74] lends itself naturally for describing the rate changes due to temperature.

A temperature increase has two mutually competing effects. Firstly, it accelerates creep, i.e. increases the creep rate. This indicates that the retardation or relaxation times should be reduced as temperature increases. Secondly, a temperature increase further causes an acceleration of hydration or aging, thereby indirectly also reducing creep.

The competition of these two effects explains why rather different temperature influences have been observed in various tests. The creep acceleration (or increase) always prevails. In an old concrete, the creep reduction due to faster aging is small since most of the cement has already been hydrated. However, in a young concrete, in which much hydration still remains to occur, the effect of the acceleration of aging can largely offset the acceleration of creep.

The effect of temperature on the rate of aging may be described by replacing concrete age with a certain equivalent age $t_e$ (also called maturity) representing the hydration period for which at temperature $T$ the same degree of hydration is reached as that reached during actual time period $t$ at reference temperature $T$. Like for all chemical reactions, the rate of hydration depends only on the current temperature and not on the past temperatures. Therefore, for variable temperature we have (cf. Equation (7.6)):

$$t_e = \int \beta_T \, dt \qquad (7.58)$$

where $\beta_T$ is a function of current temperature. Again, like for all chemical reactions, the rate of hydration should follow the activation energy concept[86,74] (Arrhenius's equation), and so

$$\beta_T = \exp\left[ \frac{U_h}{R} \left( \frac{1}{T_0} - \frac{1}{T} \right) \right] \qquad (7.59)$$

Here $T$ is the current temperature (in K), $T_0$ is a reference temperature (in K) (normally 296 K), $R$ is the gas constant, and $U_h$ is the activation energy of hydration; $U_h/R \approx 2700$ K.

The change of creep rate due to temperature may be modelled by accelerating the rate of growth of the reduced times $y_\mu(t)$ as if the retardation or relaxation times $\tau_\mu$ increased. This may be expressed by replacing the previous relation $y_\mu(t) = (t/\tau_\mu)^{q_\mu}$ (Equation (7.38)) by

$$y_\mu(t) = \left( \varphi_T \frac{t}{\tau_\mu} \right)^{q_\mu} \qquad (\mu = 1, 2, \ldots, N) \qquad (7.60)$$

where $\varphi_T$ is a function of current temperature $T(t)$. Since the creep mechanism no doubt consists in breakage and reformation of bonds, which represent thermally activated processes on the molecular scale, coefficient $\varphi_T$ should also follow the activation energy concept. This indicates that[54]

$$\varphi_T = \exp\left[ \frac{U_a}{R} \left( \frac{1}{T_0} - \frac{1}{T} \right) \right] \qquad (7.61)$$

where $U_a$ is the activation energy of creep; $U_a/R \approx 5000$ K. Note also that it makes no difference whether or not the rate coefficient $\varphi_T$ is applied for $\mu = 1$, since the corresponding deformation is almost instantaneous.

It is possible that the activation energies $U_a$ differ for various $\mu = 2, 3, \ldots, N$ but analysis of existing test data[54] did not indicate a need for introducing such a complication. Moreover, it is also possible that more than one activation energy is associated with each $\tau_\mu$, as well as with the aging

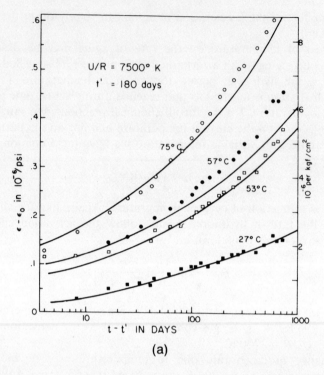

(a)

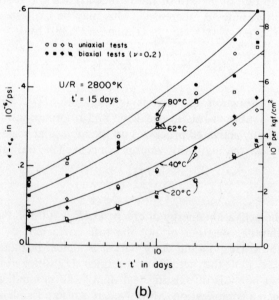

(b)

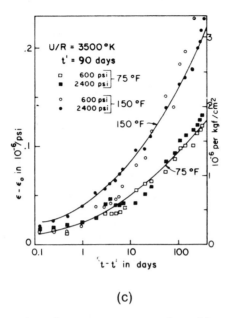

(c)

**Figure 7.18** Creep curves for various temperatures according to Maxwell chain and activation energy models[24], compared with unaxial and biaxial creep measurements[88,89,9,196]

rate (Equation (7.59)); this would make Equations (7.59) and (7.61) inapplicable; but these equations seem to approximate the existing test results reasonably well, as well as can be desired in view of the usual statistical scatter. Fits of some test data from the literature obtained with the use of Equations (7.49)–(7.61) are illustrated in Figure 7.18.

It should be noted that when physical concepts such as activation energy are used, Maxwell chain is often preferable over Kelvin chain (cf. Section 7.5.1). Further, it shoud be noted that since the effective retardation or relaxation times change from $\tau_\mu$ to $\tau_\mu/\varphi_T$, the time range covered by the rate-type model shifts to the left when the temperature increases. Thus, a broader spectrum of relaxation times is necessary to cover the same time range at various temperatures. Denoting as $\tau_{\min} \leqslant t - t_0 \leqslant \tau_{\max}$ the range of load durations for which the creep effects should be accurately calculated, and considering that temperatures vary between $T_{\min}$ and $T_{\max}$, we must use a sufficient number of $\tau_\mu$ (spaced according to Equation (7.34)) such that

$$\tau_2 \leqslant 3\tau_{\min}\varphi_{T_{\min}} \qquad \tau_N \geqslant 0.5\tau_{\max}\varphi_{T_{\max}} \qquad (7.62)$$

This is for Kelvin chain. For Maxwell chain, replace $\tau_2$ and $\tau_N$ with $\tau_1$ and $\tau_{N-1}$.

The compliance function at arbitrarily variable temperature $T(t)$ may be obtained by evaluating the integral in Equations (7.43) and (7.44) for stress history $\sigma = 1$ for $t \geq t'$ and $\sigma = 0$ for $t < t'$. Replacing $C_\mu(t')$ with $C_\mu(t'_e)$, we thus obtain

$$J(t, t') = \sum_{\mu=1}^{N} \frac{1}{C_\mu(t'_e)} \{1 - \exp\left[-y_\mu(t') - y_\mu(t)\right]\} \qquad t'_e = \int_0^{t'} \beta_T \, dt$$

$$y_\mu(t) = \frac{1}{\tau_\mu} \int_0^t \varphi_T(t') \, dt' \qquad (7.63)$$

as the Dirichlet series expansion of the compliance function at variable $T(t)$. If temperature is constant after $t'$, we have $y_\mu(t') - y(t) = (t - t')\tau_\mu\varphi_T$, and if $T$ is constant since time $t = 0$, we further have $t'_e = \beta_T t'$. For generally variable stress, the principle of superposition (Equation (7.16)) based on $J(t, t')$ from Equation (7.63) fully defines the stress–strain relation and is equivalent to the rate-type form (Equations (7.46), (7.47), (7.58)–(7.61)).

Equation (7.63) shows that the compliance function $J(t, t')$ at any temperature kept constant after time $t'$ may be obtained from $J(t, t')$ at reference temperature $T_0$ by the following replacements

$$t' \to t'_e \qquad t - t' \to (t - t')\varphi_T \qquad (7.64)$$

We see that the modification of the double power law which we introduced in Equation (7.6)[26,43] conforms to this rule (which may also be derived from a model of visoelastic porous material in which the volume of the solid grows).[43] Also note that if $\beta_T$ were 1, Equation (7.63) would conform to the time-shift principle for thermorheologically simple materials (unless $\varphi_T$ would depend also on $\mu$). However, $\beta_T$ spoils that.

The formulations of Mukaddam[138] and Mukaddam and Bresler[139] and Kabir[107] are similar to Equation (7.63) but differ in two respects. First, $\varphi_T(t')(t - t')/\tau_\mu$ is used instead of $y_\mu(t) - y_\mu(t')$, which is equivalent only at constant temperature; and, second, $t'$ is used instead of $t'_e$, which means that the acceleration of aging due to temperature increase is neglected. Furthermore, Mukaddam and Bresler[139] consider $C_\mu$ as constants and instead they introduce an empirical 'age-shift' function $\bar{\varphi}(t')$, which is analogous to the formulation used for polymers ('thermorheologically simple' materials).[82,163,187] This approach is convenient for graphical fitting of test data by the time-shift method but does not yield a degenerate form of the compliance function, thus making inapplicable the rate-type formulation. This makes it impossible to find a numerical algorithm that does not need storage of the history, and also precludes the use of the activation energy concept.

The use of $\varphi_T(t')(t - t')/\tau_\mu$ instead of $y(t) - y(t')$ leads in case of variable temperature to certain self-contradictions and non-uniqueness of results. To illustrate it, consider two temperature histories: one for example such that $T = 20\,°C$ all the time, and the other one such that $T = 20\,°C$ all the time except for a rapid rise from 20 °C to 50 °C between 99.99 days and 100 days and rapid drop from 50 °C to 20 °C between 100 days and 100.01 days. Then, for a constant stress applied at $t' = 100$ days, the resulting strains at, say, $t = 1000$ days are rather different for the two temperature histories, while they should be nearly exactly the same. Thus, the response is obtained as a discontinuous functional of the loading history while it obviously should be a continuous functional. As another example, consider two other temperature histories, one such that $T = 20$ °C all the time, and the other one such that $T = 50$ °C all the time except for a drop from 50 °C to 20 °C between 99.9 days and 100 days and a rise from 20 °C to 50 °C between 100 days and 100.1 days. Then one gets the same strains for $t = 1000$ days while the strains should obviously be very different.

Another recently studied effect of temperature is the apparent acceleration of creep shortly after any sudden temperature change, positive or negative. This effect, sometimes called transitional thermal creep,[104] probably has the same physical mechanism as the increase of creep due to a humidity change, and is doubtless also strongly influenced by microcracking.[25,52] Thus it probably does not arise only from the constitutive equation but is influenced by the stress field in the whole specimen. These effects will be more clearly discussed in Section 7.6.3.

Above 100 °C the creep properties are rather different. At constant moisture content, creep continues to increase according to the activation energy.[131,3] Moisture loss however reduces creep significantly,[130-132] even when it happens during creep.[36] The creep in pressurized water over 100 °C is much less than the creep of sealed specimens.[36] The creep Poisson ratio[36] at 200 °C reaches about 0.46. The elastic modulus steadily decreases with increasing temperature.[55,75]

Important experimental data on creep at high temperatures were reported by Browne,[64] Browne and Bamforth,[66] Browne and Blundell,[65] Hannant,[88] Fahmi et al.,[81] Komendant et al.,[116] Maréchal,[131-3] York et al.,[196] Hickey,[99] Nasser and Neville,[142,143] Seki,[167] etc.

### 7.4.4 Application in numerical structural analysis

Although the degenerate form of the kernel allows conversion to differential equations, the usual step-by-step methods for ordinary differential equations cannot be applied. This is either because numerical stability requirements prevent an increase of $\Delta t$ beyond a certain unacceptably small limit or because accuracy requirements do not allow this when the usual unconditionally stable algorithms, such as the central or backward difference methods are used.

Very large time steps, orders of magnitude larger than the shortest retardation or relaxation time, are necessary to reach long times such as 50 years after load application. Such large steps are possible only if we use integration formulae that are exact under certain characteristic conditions, namely for the case when the stress or strain rates and all material stiffness and viscosity prameters are considered constant within each time step although they are allowed to vary by jumps between the steps. Such algorithms were proposed by Bažant.[21] For non-aging materials, a similar algorithm was formulated by Taylor, Pister, and Goudreau,[175] and by Zienkiewicz, Watson, and King.[197]

Let time $t$ be subdivided by discrete times $t_r$ $(r = 1, 2, 3, \ldots)$, and let $\Delta$ refer to the increments from $t_r$ to $t_{r+1}$, e.g. $\Delta y_\mu = y_{\mu_{r+1}} - y_{\mu_r}$, $\Delta\sigma = \sigma_{r+1} - \sigma_r$. Assuming $C_\mu(t)$ and $d\sigma/dy_\mu$ to be constant from $t_r$ to $t_{r+1}$, and setting $C_\mu = C_{\mu_{r+1/2}}$, $d\sigma/dy_\mu = \Delta\sigma/\Delta y_\mu$, the integral of Equation (7.46) then yields exactly,[21] for uniaxial stress,

$$\gamma_{\mu_{r+1}} = \gamma_{\mu_r} \exp\left(-\Delta y_\mu\right) + \frac{\lambda_\mu}{C_{\mu_{r+1/2}}} \Delta\sigma \tag{7.65}$$

in which $C_{\mu_{r+1/2}} = C_\mu(t_{r+1/2})$ and

$$\lambda_\mu = \left[1 - \exp\left(-\Delta y_\mu\right)\right]/\Delta y_\mu \tag{7.66}$$

Substituting Equation (7.65) into Equation (7.44b), Equations (7.43)–(7.44a) may be brought to the form.[21]

$$\Delta\varepsilon = \frac{\Delta\sigma}{E''} + \Delta\varepsilon'' + \Delta\varepsilon^0 \tag{7.67}$$

in which

$$\frac{1}{E''} = \sum_{\mu=1}^{N} \frac{1 - \lambda_\mu}{C_{\mu_{r+1/2}}} \Delta\sigma \qquad \Delta\varepsilon'' = \sum_{\mu=1}^{N} \left[1 - \exp\left(-\Delta y_\mu\right)\right]\gamma_{\mu_r} \tag{7.68}$$

We may now observe that $E''$ and $\Delta\varepsilon''$ can be evaluated if all $\gamma_{\mu_r}$ are known up to the beginning of the current time step, $t_r$. Thus Equation (7.67) may be treated as an elastic stress–strain relation with elastic modulus $E''$ and inelastic strain increment $\Delta\varepsilon''$. Using this relation, the structural problems with prescribed load changes or displacement increments during the step $(t_r, t_{r+1})$ may be solved, yielding the value of $\Delta\sigma$. The internal variables $\gamma_\mu$ at the end of the step, $\gamma_{\mu_{r+1}}$, may then be evaluated[21] from Equation (7.65). Then one can proceed to the next step.

As for the choice of time steps, it is most effective to keep them constant in the scale $\log(t - t_1)$ where $t_1$ is the instant when the first load is applied on the structure or first deformations are imposed. Thus, after choosing the first step $(t_1, t_2)$ we generate subsequent $t_r$ as $t_{r+1} - t_1 = 10^{1/m}(t_r - t_1)$ where $m$ is the chosen number of steps per decade ($m = 2$ to $4$ suffices for good accuracy). The load and imposed deformations must either be constant after $t_1$ or vary gradually at a rate which declines with $t - t_0$ (as, e.g. the shrinkage deformations do). If there is a sudden load (or enforced deformation change) at time $t_s$, one must start again with a small time step $(t_s, t_{s+1})$ and then increase the steps so as to keep them constant in $\log(t - t_s)$, i.e. use $t_{r+1} - t_s = 10^{1/m}(t_r - t_s)$.

It is instructive to explain the role of coefficients $\lambda_\mu$ in Equations (7.68) or (7.69). Among all $\tau_\mu$ there may be one, say $\tau_m$, which is of the same order of magnitude as the current time step $\Delta t$. Then for all $\tau_\mu < \tau_m$ we have $\Delta y_\mu \gg 1$, $\exp(-\Delta y_\mu) \approx 0$, $1 - \exp(-\Delta y_\mu) \approx 1$ and $\lambda_\mu \approx 0$, whereas for all $\tau_\mu > \tau_m$ we have $\Delta y_\mu \ll 1$, $\exp(-\Delta y_\mu) \approx 1$, $1 - \exp(-\Delta y_\mu) \approx 0$, and $\lambda_\mu \approx 1$. Thus, we see from Equation (7.68) that the partial compliances $1/C_\mu$ which contribute to the instantaneous incremental compliance $1/E''$ are only those for which $\tau_\mu < \tau_m$ (or $\tau_\mu \ll \Delta t$). This is intuitively obvious because the stress in the dashpots of Kelvin units for which $\tau_\mu \ll \Delta t$ must almost completely relax within a time less than the step duration. So, the effect of $\lambda_\mu$ as the time step is increased in the step-by-step computation is to gradually 'uncouple' the dashpots as their relaxation time becomes too small compared to $\Delta t$. Furthermore, from the values of $\lambda_\mu$ we see that the inelastic strain increments are negligible and the behaviour becomes elastic for all $\tau_\mu > \tau_m$, i.e. $\tau_\mu \ll \Delta t$.

Equations analogous to Equations (7.65)–(7.68) hold for the volumetric and deviatoric components. When the spatial problem is solved by finite elements, the computational algorithm may be described as follows:

(1) Initiate stresses $\sigma_{ij}$, strains $\varepsilon_{ij}$ and internal variables $\gamma_{ij_{\mu_r}}$ at starting time $t_1$ as zero for all finite elements; set $r$ as 1.

(2) For each finite element (and each integration point of finite element) calculate the volumetric and deviatoric inelastic strain increments and bulk and shear moduli for the step[21]

$$\Delta\varepsilon''^{V} = \sum_{\mu=1}^{N}(1 - e^{-\Delta y_\mu})\gamma_{\mu_r}^{V} \qquad \Delta\varepsilon_{ij}''^{D} = \sum_{\mu=1}^{N}(1 - e^{-\Delta y_\mu})\gamma_{ij_\mu}^{D} \qquad (7.69a)$$

$$K'' = \left(\sum_{\mu=1}^{N}(1 - \lambda_\mu)\hat{K}_{\mu_{r+1/2}}^{-1}\right)^{-1} \qquad G'' = \left(\sum_{\mu=1}^{N}(1 - \lambda_\mu)\hat{G}_{\mu_{r+1/2}}^{-1}\right)^{-1} \qquad (7.69b)$$

Here $\gamma_\mu^{D}, \gamma_{ij_\mu}^{D}$ are the volumetric and deviatoric components of internal variable tensor $\gamma_{ij_\mu}$ (corresponding to $\gamma_\mu$ in uniaxial formulation, Equation (7.44)); $K_{r+1/2}$ and $G_{r+1/2}$ are the bulk modulus and shear modulus corresponding to equivalent age $t_{e_{r+1/2}}$ at the time when the actual age is $t_{r+1/2}$; $\hat{K}_{\mu_{r+1/2}}$ and $\hat{G}_{\mu_{r+1/2}}$ are the bulk and shear moduli for individual terms of Dirichlet series expansion, corresponding to moduli for $t' = t_{e_{r+1/2}}$ in the uniaxial formulation (7.36).

(3) The incremental stress–strain relation for each finite element and each integration point has then the form

$$\Delta\sigma^{V} = 3K''(\Delta\varepsilon^{V} - \Delta\varepsilon''^{V} - \Delta\varepsilon^{0}) \qquad \Delta\sigma_{ij}^{D} = 2G''(\Delta\varepsilon_{ij}^{D} - \Delta\varepsilon_{ij}''^{D}) \qquad (7.70)$$

Since moduli $K''$ and $G''$ and inelastic strain increments $\Delta\varepsilon''^{\mathrm{V}}$ and $\Delta\varepsilon_{ij}''^{\mathrm{D}}$, as well as $\Delta\varepsilon^0$, can be determined in advance, we may treat Equation (7.70) as an elastic stress–strain relation. So we have an elastic problem with general inelastic (or initial) strains, which may be solved by a finite element program in the usual manner. In this elastic analysis we also apply all load increments as prescribed for the current time step $(t_r, t_{r+1})$, if any, and all displacement increments if any are prescribed. The solution yields displacement increments $\Delta u_i$ for all nodes and strain increments $\Delta\varepsilon_{ij}$ for all elements and all integration points in the elements. We then split $\Delta\varepsilon_{ij}$ into the volumetric and deviatoric components $\Delta\varepsilon^{\mathrm{V}}$ znd $\Delta\varepsilon_{ij}^{\mathrm{D}}$, calculate the volumetric and deviatoric stress increments $\Delta\sigma^{\mathrm{V}}$ and $\Delta\sigma_{ij}^{\mathrm{D}}$ from Equation (7.70), and superimpose them to get $\Delta\sigma_{ij}$ for all finite elements and all integration points in the elements.

(4) Then we evaluate the volumetric and deviatoric parts of the internal variables $\gamma_{ij_\mu}$ at the end of the step, $t_{r+1}$, from the recurrent relations:

$$\gamma_{\mu_{r+1}}^{\mathrm{V}} = \gamma_{\mu_r}^{\mathrm{V}} e^{-\Delta y_\mu} + \frac{\lambda_\mu}{3K_\mu''}\Delta\sigma^{\mathrm{V}} \qquad \gamma_{ij_{r+1}}^{\mathrm{D}} = \gamma_{ij_r}^{\mathrm{D}} e^{-\Delta y_\mu} + \frac{\lambda_\mu}{2G_\mu''}\Delta\sigma_{ij}^{\mathrm{D}} \quad (\mu = 1, 2, \ldots, n)$$

$$(7.71)$$

for all $\mu$, all elements and all integration points.

(5) If $t_r <$ final time, go back to step 2 and start the next time step resetting $r$ $(r \leftarrow r+1)$.

The most efficient way for programming is to take an existing elastic finite element program (which can handle arbitrary inelastic strains), place it in a DO loop over discrete times and attach separate subroutines for steps 2 and 4 described above. The foregoing algorithm,[21] along with a model for cracking, has been put in this manner on general purpose finite element program NONSAP by Anderson et al.[6,171,172,207]

To illustrate accuracy, Table 7.3 gives the stress $\sigma(t)$ at 29031 days due to strain $10^{-6}$ enforced at $t_0 = 35$ days, as computed according to Equations (7.65)–(7.71) with Dirichlet series approximation of ACI compliance function for various numbers of steps up to terminal time $t$. The first time step was always $\Delta t = 0.1$ day. We see that this algorithm is even more accurate

Table 7.3 Numerical results for stress relaxation obtained with Bažant's[21] exponential algorithm based on Dirichlet series expansion of compliance function

| No. of time steps | Approx. no. of steps per decade | $\sigma(t)$ |
|---|---|---|
| 13 | 2 | 1.5320 |
| 25 | 4 | 1.5411 |
| 49 | 9 | 1.5438 |
| 97 | 18 | 1.5443 |
| 193 | 35 | 1.5445 |

than the second-order step-by-step method based on superposition (Table 7.1).

The Maxwell chain model offers certain theoretical advantages over the Kelvin chain model (Section 7.5.1), and therefore a similar algorithm was developed[21] for the relaxation function. We assume $E_\mu(t)$, $d\varepsilon^0/dy_\mu$ and $d\varepsilon/dy_\mu$ in (7.54) to be constant from $t_r$ to $t_{r+1}$, setting $E_\mu = E_{\mu_{r+1/2}}$ and $d\varepsilon^0/dy_\mu = \Delta\varepsilon^0/\Delta y_\mu$. Equation (7.54) is then a linear first-order differential equation with constant coefficients and the initial conditions $\sigma_\mu(t) = \sigma_{\mu_r}$. Integration then yields (exactly), for uniaxial stress,

$$\sigma_{\mu_{r+1}} = \sigma_{\mu_r} e^{-\Delta y_\mu} + E_{\mu_{r+1/2}} \lambda_\mu (\Delta\varepsilon - \Delta\varepsilon^0) \qquad (7.72)$$

where $\lambda_\mu$ is given again by Equation (7.66). Substituting this in Equation (7.52), we obtain $\Delta\sigma = E''(\Delta\varepsilon - \Delta\varepsilon'' - \Delta\varepsilon^0)$ where

$$E'' = \sum_{\mu=1}^{N} \lambda_\mu E_{\mu_{r+1/2}} \qquad \Delta\varepsilon'' = E'' \sum_{\mu=1}^{N} (1 - e^{-\Delta y_\mu}) \sigma_{\mu_r} \qquad (7.73)$$

Since $E''$ and $\Delta\varepsilon$ can be evaluated before $\sigma_{r+1}$ and $\varepsilon_{r+1}$ are known, we may determine $\Delta\sigma$ and $\Delta\varepsilon$ by an elastic structural analysis based on elastic moduli $E''$ and inelastic strain increments $(\Delta\varepsilon'' + \Delta\varepsilon^0)$. This algorithm is based on the relations:[21]

$$K'' = \sum_{\mu=1}^{N} \lambda_\mu K_{\mu_{r+1/2}} \qquad G'' = \sum_{\mu=1}^{N} \lambda_\mu G_{\mu_{r+1/2}} \qquad (7.74)$$

$$3K''\Delta\varepsilon''^V = \sum_{\mu=1}^{N} (1 - e^{-\Delta y_\mu}) \sigma_{\mu_r}^V, \qquad 2G''\Delta\varepsilon_{ij}''^D = \sum_{\mu=1}^{N} (1 - e^{-\Delta y_\mu}) \sigma_{ij_{\mu_r}}^D \qquad (7.75)$$

where $\sigma_{ij_\mu}$ are the partial stresses coresponding to $\sigma_\mu$ from Equation (7.54); and $K_\mu$, $G_\mu$ are coefficients of Dirichlet series expansions of $J^V(t, t')$ and $J^D(t, t')$, corresponding to $E_\mu(t')$ from Equation (7.41). The volumetric and deviatoric internal variables (partial stresses) at the end of time step $\Delta t_r$ are determined from the recursive relations:[21]

$$\sigma_{\mu_{r+1}}^V = \sigma_{\mu_r}^V e^{-\Delta y_\mu} + 3K_{\mu_{r-1/2}}(\Delta\varepsilon^V - \Delta\varepsilon^0) \qquad \sigma_{ij_{\mu_{r+1}}}^D = \sigma_{ij_{\mu_r}}^D e^{-\Delta y_\mu} + 2G_{\mu_{r-1/2}} \Delta\varepsilon_{ij}^D \qquad (7.76)$$

The computational algorithm is essentially the same as that described before, except that Equations (7.68)–(7.69) are replaced by Equations (7.74)–(7.75), Equations (7.71) are replaced by (7.76) and $\sigma_\mu$ are used instead of $\gamma_\mu$. Bažant and Wu[52] used this algorithm in a small finite element program, and recently Bažant, Rossow and Horrigmoe[46] put this algorithm on the general purpose finite element program NONSAP.

Figures 7.17 and 7.18 give examples of comparison with tests.

It is again instructive to explain the role of coefficients $\gamma_\mu$ in Equations (7.73) or (7.74). Let $\tau_m$ be that $\tau_\mu$ which is of the same order of magnitude as the correct $\Delta t$. Then for all $\tau_\mu < \tau_m$ we have $\Delta y_\mu \gg 1$, $e^{-\Delta y_\mu} \simeq 0$, $1 - e^{-\Delta y_\mu} \simeq 1$, and $\lambda_\mu \simeq 0$, whereas for all $\tau_\mu > \tau_m$ we have $\Delta y_\mu \ll 1$, $e^{-\Delta y_\mu} \simeq 1$, $1 - e^{-\Delta y_\mu} \simeq 0$ and $\lambda_\mu \simeq 1$. The stiffness of Maxwell chain model for the given step is given by Equation (7.74) and we see that the partial stiffnesses $K_\mu$ or $G_\mu$ which contribute to the overall stiffness are only those for which $\tau_\mu > \tau_m$, i.e. $\tau_\mu \ll \Delta t$. This is intuitively obvious, because the stress in Maxwell units for which $\tau_\mu \ll \Delta t$ must completely relax within a time less than the step duration. Thus the effect of $\tau_\mu$ as the time step increases during the computation is to gradually 'uncouple' the Maxwell units whose relaxation time is too short with regard to the current $\Delta t$. Further, we see that the inelastic strain increments are negligible and the behaviour is elastic for all $\tau_\mu < \tau_m$, i.e. $\tau_\mu \ll \Delta t$.

Another useful temporal step-by-step algorithm which avoids the storage of the previous history by exploiting the Dirichlet series expansion of the compliance function was developed by Kabir and Scordelis[108] and further applied by Van Zyl and Scordelis,[180,179] Van Greunen[178] and Kang.[110,109] This algorithm is similar to Zienkiewicz, Watson and King's[197] algorithm for non-aging materials; it consists of similar formulae involving exponentials $e^{-\Delta y_\mu}$, but it is of lower-order accuracy than the preceding algorithms. Its approximation error is of the first order $O(\Delta t)$ or $O(\Delta \sigma)$ rather than the second order, $O(\Delta t^2)$ or $O(\Delta \sigma^2)$, since the integral in Equation (7.16) is approximated with a rectangle rule. This less accurate approximation has the advantage that the same incremental elastic stiffness matrix of the structure may be used in all time steps if the age of concrete is the same in all finite elements, while in the preceding algorithms the changes in $E''$ (or $G''$, $K''$) cause that this matrix is different in each time step. This advantage is, however, lost if the structure is of non-uniform age or if changes of stiffness due to cracking or other effects are to be considered.

Finally it should be mentioned that, for the analysis of creep effects of composite beams during construction stages, Schade and Haas[165] produced a general finite element program using Euler–Cauchy and Runge–Kutta methods in conjunction with an aging Kelvin chain, and dealt successfully with the stability problems due to the shortest retardation time.

## 7.5   NON-LINEAR EFFECTS

### 7.5.1   Difficulties with aging in linear viscoelasticity

In every constitutive theory it is necessary to check that no thermodynamic restriction is violated. For non-aging materials this is relatively easy and well understood,[56,155] but not so for aging. Obviously, not every function of $t$ and $t'$ is acceptable as a compliance function $J(t, t')$. Certain thermodynamic restrictions, such as $\partial J(t, t')/\partial t \geq 0$, $\partial^2 J(t, t')/\partial t^2 \leq 0$ and $[\partial J(t, t')/\partial t']_{t \to t'} \leq 0$ are intuitively obvious and we will not discuss them. However, some further restrictions are necessary to express certain aspects of the physical mechanism of aging, particularly the thermodynamic restrictions due to the fact that any new bonds produced by a chemical reaction must be without stress when formed.

At present we know how to guarantee fulfillment of these thermodynamic restrictions only if we first convert the constitutive relation to a rate-type form and then make the hypothesis that these restrictions should be applied to internal variables such as the partial strains or partial stresses in the same way they would be applied to strains and stresses. If we did not accept this hypothesis we could not say anything about thermodynamic restrictions. It might be possible that no thermodynamic restrictions are violated by stresses and strains when they are violated by partial stresses or partial strains. But we cannot guarantee it. It is certainly a matter of concern if we have such violations. It has been found[27] that this actually happens for certain existing creep laws used in the past and we will outline the nature of the problem briefly.

If we reduce the compliance function to a rate-type form corresponding to a spring–dashpot model, fulfillment of the second law of thermodynamics can be guaranteed by certain conditions on spring moduli $E_\mu$ and viscosities $\eta_\mu$. (The second law might be satisfied by the compliance function even when some of these condition are violated, but we cannot be certain of it.) Two obvious conditions are $E_\mu \geq 0$ and $\eta_\mu \geq 0$. However, the second law leads to a further condition when the spring moduli are age-dependent:[27]

$$\dot{\sigma}_\mu = E_\mu(t)\dot{\varepsilon}_\mu \quad \text{for } \dot{E}_\mu \geq 0 \tag{7.77}$$

$$\sigma_\mu = E_\mu(t)\varepsilon_\mu \quad \text{for } \dot{E}_\mu \leq 0 \tag{7.78}$$

where $\sigma_\mu$ and $\varepsilon_\mu$ are the stress and the strain in the $\mu$th spring. The first relation pertains to a solidifying material, such as an aging concrete, while the second relation pertains to a disintegrating (or melting) material, such as concrete at high temperatures (over $150\,°C$) which cause dehydration. If Equation (7.78) is used, it can be shown that the expression $\dot{D}_{ch} = -\sigma_\mu^2 \dot{E}_\mu / 2E_\mu^2$ represents the rate of dissipation of strain energy due to the chemical process, particularly due to disappearance of bonds while the material is in a strained state (i.e. a state in which elastic energy exists in addition to the bond energy). Thus, to assure that $\dot{D}_{ch} \geq 0$ we must have $\dot{E}_\mu \leq 0$ if Equation (7.78) is used. So the dissipation inequality is violated if Equation (7.78) is used (or if its use is implied) for an aging (solidifying, hardening) material.

Many different rate-type forms of the creep law are possible. One form which differs from the one that we already analysed, can be obtained by expanding the memory function $L(t, t')$ from Equation (7.17) into Dirichlet series:

$$L(t, t') = \sum_{\mu=1}^{N} \frac{1}{\eta_{\mu}(t')} e^{-(t-t')/\tau_{\mu}} \qquad (7.79)$$

Substituting this into Equation (7.17) we obtain

$$\varepsilon = \sum_{\mu=1}^{N} \varepsilon_{\mu} + \varepsilon^{0} \qquad \varepsilon_{\mu}(t) = \int_{0}^{t} \frac{\sigma(t')}{\eta_{\mu}(t')} e^{-(t-t')/\tau_{\mu}} \, dt' \qquad (7.80)$$

By differentiating $\varepsilon_{\mu}(t)$ and denoting $E_{\mu}(t) = \eta_{\mu}(t)/\tau_{\mu}$, one can readily verify that $\varepsilon_{\mu}(t)$ satisfies the differential equation

$$\sigma = E_{\mu}(t)\varepsilon_{\mu} + \eta_{\mu}(t)\dot{\varepsilon}_{\mu} \qquad (7.81)$$

From this, the non-viscous part of stress $\sigma$ is $\sigma_{\mu} = E_{\mu}(t)\varepsilon_{\mu}$, and we see that this represents an elastic relation that is admissible only for a disintegrating (melting, dehydrating) material. Thus Equation (7.79), which has been used as the basis of one large finite element program for creep of reactor vessels, implies violation of the dissipation inequality.[27] This puts the practical applicability in question, as the critics think. The proponents of the models that imply Equation (7.81) believe however that the problem is not serious since the solidifying process may be counteracted by drying or temperature decrease, and that a separate application of the thermodynamic restrictions to the solidfying behaviour is not required. Maybe, but what about the special case of constant humidity and temperature?

We may note that Equation (7.81) along with $\varepsilon = \sum_{\mu} \varepsilon_{\mu}$ corresponds to a Kelvin chain model[27] the springs of which are, however, governed by an incorrect equation (Equation (7.78)). If the correct equations for the springs are used ($\dot{\sigma}_{\mu} = E_{\mu}\dot{\varepsilon}_{\mu}$), then the Kelvin chain is characterized by second-order rather than first-order differential equations (Equation (7.49)). The violation of the dissipation inequality by Equations (7.79) or (7.80) is actually due to the fact that the equation for partial strains (Equation (7.81)) is of the first order. One can show[27] that even if a non-linear rate-type creep law is considered such that $\varepsilon = \sum_{\mu} \varepsilon_{\mu}$ and $\dot{\varepsilon}_{\mu} = f_{\mu}(\sigma, \varepsilon_{\mu})$, Equation (7.78) which violates the dissipation inequality is implied as long as these are first-order equations. This is one inherent difficulty of using Kelvin chain type models (i.e. decomposing $\varepsilon$ into partial strains $\varepsilon_{\mu}$). By contrast, the differential equations for the aging Maxwell chain must be of the first order for an aging (solidifying) material, which is one advantageous property of Maxwell chains.

Kelvin chain models (which are implied by making a strain summation assumption $\varepsilon = \sum \varepsilon_{\mu}$) have another limitation. Consider the degenerate creep compliance in Equation (7.32). We calculate $\partial^{2}J/\partial t \, \partial t'$ and substitute

$\dot{C}_\mu(t') = [(C(t') - E_\mu(t')]\dot{y}_\mu(t')$ and $C_\mu(t') = \eta_\mu(t')y_\mu(t')$ according to Equation (7.50) for the Kelvin chain. This yields

$$\frac{\partial^2 J(t, t')}{\partial t\, \partial t'} = \sum_{\mu=1}^{N} \frac{\dot{y}_\mu(t)}{\dot{y}_\mu(t')} \frac{E_\mu(t')}{[\eta_\mu(t')]^2} \exp\left[y_\mu(t') - y_\mu(t)\right] \qquad (7.82)$$

Here we must always have $\dot{y}_\mu > 0$ and $E_\mu \geq 0$. Consequently, thermodynamically admissible Kelvin chain models always yield a compliance function such that

$$\frac{\partial^2 J(t, t')}{\partial t\, \partial t'} \geq 0 \qquad (7.83)$$

The same conclusion may be reached from Equation (7.49a).

Now, what is the meaning of this inequality? Geometrically, it means that the slope of a unit creep curve would get greater as $t'$ increases, which means that two creep curves for different $t'$, plotted versus time $t$ (not $t - t'$, see Figure 7.7(a)) would never diverge as $t$ increases. Is this property borne out by experiment? Due to the large scatter of creep data we cannot answer this with complete certainty, but quite a few test data, although not the majority of them, indicate a divergence of adjacent creep curves beginning with a certain creep duration $t - t'$.[33,27] As for the creep formulae, the double power law (Equation (7.5)) as well as the ACI formulation always exhibit divergence after a certain value of $t - t'$, whereas the CEB-FIP formulation does not. Thus, aging Kelvin chain models cannot closely approximate this behaviour without violating the thermodynamic restrictions $E_\mu \geq 0$, $\nu_\mu \geq 0$, $\dot{E}_\mu \geq 0$. Indeed, the previously described algorithms for determining $E_\mu(t)$ yield negative $E_\mu$ for some $\mu$ and some $t$ whenever $J(t, t')$ with divergent creep curves is fitted.

For aging Maxwell chain models, by contrast, it is possible to violate inequality (7.83) without violating any of the thermodynamic restrictions.[27] Therefore, the aging Maxwell chain models are more general and seem theoretically preferable for describing concrete creep. The equivalence of Maxwell and Kelvin chains to each other as well as to any other rheologic model[157] does not quite apply in the case of aging.

The Maxwell chain model is, however, not entirely trouble-free either. When long-time creep data are fitted, the condition $E_\mu \geq 0$ can be satisfied but the condition that $\dot{E}_\mu \geq 0$ for all $\mu$ and all $t$ cannot (Figure 7.16), as numerical experience reveals. (Thus far, however, this question has been studied only for $y_\mu = t/\tau_\mu$; for a general $y_\mu(t)$ a definite answer must await further results.)

Note that we merely evade answering the question if we restrict ourselves to an integral-type creep law, for without its conversion to a rate-type form we cannot know whether our formulation of aging is thermodynamically admissible. We also evade the answer by introducing a rate-type model without recourse to a rheologic spring–dashpot model, since every rate-type model can be visualized by some such rheologic model.

We must admit, however, that the question of uniqueness of representation by a rheologic model remains unsettled. If we find that a certain creep function $J(t, t')$ leads to one unsatisfactory rate-type model, we are not sure that the same creep function might also be represented by some other rate-type model that is satisfactory.

To summarize, we have two kinds of rate-type models: (a) those whose form is fundamentally objectionable (Equation (7.78)) because it always violates the dissipation inequality; and (b) those which are of correct form (Equation (7.77)) and can represent the aging creep curves for various $t'$ over a limited time range but cannot do so for a broad time range without violating the thermodynamic restrictions ($E_\mu \geq 0$ or $\dot{E}_\mu \geq 0$) for some period of time. Although for the second kind of models the problems are less severe, none of the presently known linear rate-type models is entirely satisfactory.

After a long effort it now seems as if the typical shape of concrete creep curves for various $t'$ cannot be completely satisfactorily described with a linear rate-type model. Hence, the difficulties are likely caused by our use of a linear theory for what is actually a non-linear phenomenon. And measurements relative to the principle of superposition suggest that this may indeed be so. On the other hand, the magnitude of the error caused by the shortcomings outlined in this section is not known well and might not be too serious in many cases. Thus, we will continue to use linear rate-type models in the foreseeable future.

To understand the nature of the aging effect in creep, it was attempted to deduce the constitutive relation from an idealized micromechanics model of the solidification process in a porous material.[26] This approach yielded a certain form of creep function (giving in particular support to the double power law); it did not however answer the questions we just discussed.

### 7.5.2 Adaptation and flow

There are basically two kinds of deviations from the principle of superposition:

(1) High-stress non-linearity or flow, which represents an increased creep compared to the principle of superposition;[83,87,114]
(2) Low stress non-linearity or adaptation, which represents a diminished creep compared to the principle of superposition.

The high-stress non-linearity is significant for basic creep only beyond the service stress range, i.e. above approximately 0.5 of the strength. On the other hand, the adaptation non-linearity is quite significant within the service stress range. It is mainly observed when, after a long period of sustained stress (within the allowable stress range), a further sudden load, positive or negative, is superimposed.

A mathematical model for both these types of non-linearity was recently developed.[35] For certain reasons it is appropriate to introduce the non-linearities in an expression for the creep rate $\dot{\varepsilon}$ rather than for total strain $\varepsilon$. Thus, starting with Equation (7.18) for the creep rate, one makes this equation non-linear in the following manner:

$$\dot{\varepsilon}(t) = \frac{\dot{\sigma}(t)}{E(t_e)} + g[\sigma(t)] \int_0^t \frac{\partial J(t, t_e')}{\partial t} \frac{d\sigma(t')}{1 + a(t')} + \dot{\varepsilon}_f(t) \qquad (7.84)$$

The integral term describes what is called the adaptation, which is brought about in two ways. First, the age effect in creep function and elastic modulus is introduced by replacing age $t$ with a more general expression for the equivalent hydration period;

$$t_e = \int \beta_T \beta_\sigma \, dt \qquad (7.85)$$

where coefficient $\beta_\sigma$, for which a formula is given by Bažant and Kim,[35] is added to account for the gradual increase of bonding (or adaptation) caused in cement paste by sustained compressive stress.[98] Second, an additional stiffening of response (adaptation) is obtained through function $a(t')$ which is defined by an evolution equation of the type $\dot{a}(t) = F_1(t)F_2(\sigma)$. In the limit for very rapid loading these non-linear effects vanish.

Functions $g[\sigma(t)]$ and $\dot{\varepsilon}_f(t)$ describe the high-stress non-linearity. The function $\dot{\varepsilon}_f(t)$, called flow, has the form:

$$\dot{\varepsilon}_f(t) = \frac{\sigma(t) - \alpha(t)}{E_0} f[\sigma(t)]\dot{\phi}(t) \qquad (7.86)$$

where $f[\sigma(t)]$ gives an increased creep rate at high stress, and $\alpha(t)$ may be regarded as the location of the centre of a loading surface that gradually moves toward the sustained stress value (similar to kinematic hardening in plasticity). The evolution equation for $\alpha(t)$ is of the type $\dot{\alpha}(t) = f_1[\sigma(t), \alpha(t)]f_2(t)$.

The adaptation and flow non-linearities are illustrated by test data in Figure 7.19, 7.20, and 7.21. For further data see Aleksandrovskii et al.,[2,3] Roll,[156] Freudenthal and Roll,[83] Komendant,[116] Brettle,[63] Meyers and Slate,[136] and others.

### 7.5.3 Singular history integral for non-linear creep

In the foregoing model, the non-linearities at the working stress level are modelled by adjustments to the superposition principle. These non-

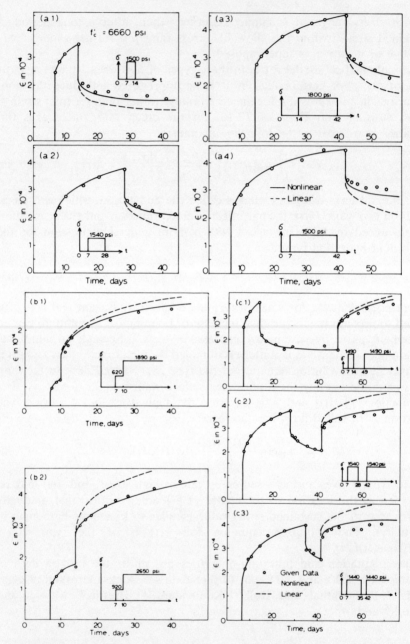

**Figure 7.19** Typical test results[141] showing adaptation non-linearities: solid lines, predictions of nonlinear theory; dashed lines, predictions of linear theory[34,35]

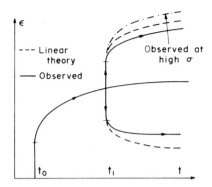

**Figure 7.20**   Deviation from superposition principle for unloading and for step increase of load

linearities may however have a deeper cause in the essential creep mechanism, and so it may be more realistic to abandon the underlying linear superposition principle itself. At the same time it is necessary to preserve the proportionality of the response to an arbitrary load history within the working stress range, a property which is well verified experimentally.[158,141] Such a development has been made recently[50,206] and we will outline it briefly.

We consider a uniaxial creep law of the form

$$\frac{d}{dt}[\gamma(t)^p] = \int_0^t Q(t, \tau)\, d[\sigma(\tau)^r] \qquad (7.87)$$

in which

$$Q(t, \tau) = Ft^{-k}\tau^{-m}(t - \tau)^{-u}[\kappa(t)^s - \kappa(\tau)^s]^{-v} \qquad (7.88)$$

$$\kappa(t) = \int_0^t |d\gamma(\tau)| \qquad (7.89)$$

Here $\sigma$ is uniaxial stress; $\gamma$ is creep strain; $t$ is time ( = age of concrete); $\kappa$ is the path length of creep strain (intrinsic time); $k$, $m$, $p$, $r$, $s$, $u$, and $v$ are non-negative material constants ($s > 0$); $R(t, \tau)$ is the creep kernel; $F$ is a function of $\sigma(t)$ and $\gamma(t)$ which models the creep increase beyond proportionality at high stress. Since we are not interested in this phenomenon at high stress,[116,127,105] we will consider only the case $F$ = constant, which is sufficient for working stress levels. The integral in Equation (7.87) is a Stieltjes integral. For continuous and differentiable $\sigma(t)$ this integral may be replaced by the usual Riemann integral, substituting $d[\sigma(\tau)^r] = r\sigma(\tau)^{r-1}\, d\sigma(\tau)/d\tau$.

If we consider a single-step load history ($\sigma = 0$ for $t < t'$, $\sigma$ = constant $> 0$

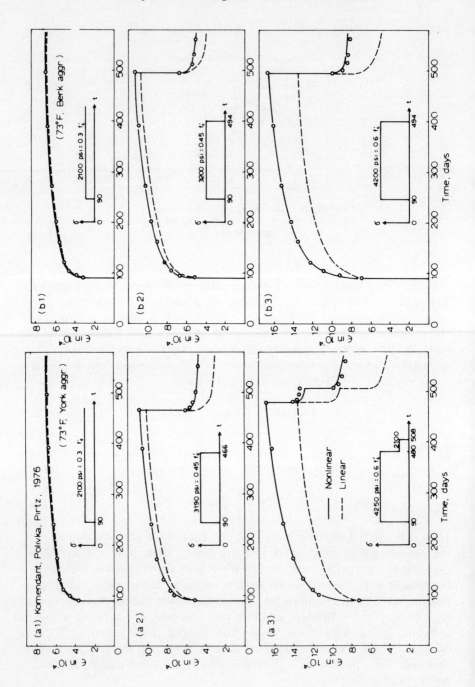

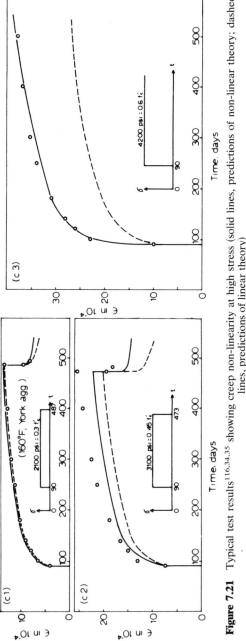

**Figure 7.21** Typical test results[116,34,35] showing creep non-linearity at high stress (solid lines, predictions of non-linear theory; dashed lines, predictions of linear theory)

for $t \geq t'$) under the assumption $r = p + sv$, Equation (7.87) reduces to the form

$$\gamma(t) = \left[ \sigma \frac{F}{(t')^m} \frac{r}{p} B(t, t') \right]^{1/r} \qquad (7.90)$$

in which

$$B(t, t') = \int_{t'}^{t} \frac{d\tau}{\tau^k (\tau - t')^u} \qquad (7.91)$$

For $t - t' \ll t'$, the following asymptotic expression holds:

$$\gamma(t) = \sigma \left( \frac{Fr}{(1-u)p} \frac{(t-t')^{1-u}}{(t')^{m+k}} \right)^{1/r} \qquad (7.92)$$

Equation (7.87) has the following noteworthy features.

(1) For $p = r = 1$ with $v = k = 0$, it reduces to the linear integral-type creep law (superposition principle) based on the double power law and, with $v = 0$ and $k > 0$, it reduces to the one based on the triple power law[26] which has been verified as a slight refinement of the well-substantiated double power law.

(2) If any one of the conditions $p = 1, r = 1, v = 0$ is violated, this creep law ceases to be linear and, therefore, the principle of superposition does not apply. In particular, the stiffening non-linearity is obtained for $p > 1$.

(3) However, if at the same time

$$r = p + sv \qquad (9.93)$$

this non-linear creep law exhibits proportionality in the sense that if $\gamma(t)$ corresponds to history $\sigma(t)$ then $k\gamma(t)$ corresponds to history $k\sigma(t)$. The fact that a non-linear creep law can be obtained without violating proportionality seems useful for modeling experimentally observed properties,[158,141] at working stress levels.

(4) It is also necessary that

$$u + sv < 1 \quad \text{and} \quad u + v < 1 \qquad (7.94)$$

for the creep kernel to be weakly singular and, consequently, integrable. The latter of these conditions must be added with regard to the second and the subsequent steps of a multistep loading history, and prevails when $0 < s < 1$.

(5) As observed in Equation (7.90) for a single-step load history with the proportionality condition (Equation (7.93)) for $k = 0$, Equation (7.87) still leads to the well-verified double power creep law and, for $k > 0$, to the triple power law.[26] Therefore, using the previously obtained results on these power laws, it is possible to estimate some parameters involved in the present model.

(6) A further important property is that not only the term $(t-\tau)^{-u}$, which is present in the previous completely linear integral expressions for the creep rate, but also the term $[\kappa(t)^s - \kappa(\tau)^s]^{-v}$ yields an infinite creep rate $\dot{\gamma}$ (singularity) right after any sudden change in stress $\sigma$. If this term were omitted (i.e. $v = 0$ with $s > 0$), the strength of the singularity of $\dot{\gamma}$ would be given solely by $(t-\tau)^{-u}$, i.e. independent of $\gamma$, and would not contribute to expressing the non-linearity.

(7) With $s = 1$, the strength of the singularity at each stress jump is the same. Analysis of available test data,[158,141] suggests, however, that $s > 1$. This has an interesting consequence for a two-step stress history, i.e. $\sigma = 0$ for $t < t'$, $\sigma = \sigma_1 =$ constant $(>0)$ for $t' < t < t''$ and $\sigma = \sigma_2 =$ constant $(>\sigma_1)$ for $t > t''$. If we let $\sigma_1 \rightarrow 0$ at constant $\sigma_2$, the history approaches a one-step history with a jump of $\sigma_2$ at $t''$, but the singularity strength $(u + v)$ in the limit is not the same as that for a one-step history (i.e. $u + sv$) if $s \neq 1$.

(8) The fact that the integral in Equation (7.87) expresses the creep rate $\dot{\gamma}$ rather than the total creep strain $\gamma$ is appropriate for modelling the non-linear creep properties at high stress as mentioned before (Equation (7.84)).

(9) Asymptotic approximations as well as numerical integration of the creep law have further revealed that at low stress levels the creep law generally gives qualitatively correct deviations from the (linear) superposition principle. For a two-step increasing load the response is after the second step lower than the prediction of the superposition principle. For creep recovery after a period of creep at constant stress, the recovery response is and remains higher than the recovery curve predicted from the superposition principle. In both cases, the deviation vanishes as the duration of the first load step tends to zero. These properties represent the essential non-linear features of concrete creep at low stress levels.

(10) Function $\kappa(t)$ is needed for the case of unloading. This function, which is analogous to the well-known intrinsic time, assures the positiveness of $Q(t, \tau)$. Without excluding the case of creep recovery it is impossible to use $\kappa(t) = \gamma(t)$ because $Q(t, \tau)$ would be negative or undefined for unloading.

From the foregoing discussion, it appears that Equation (7.87) is qualitatively capable of capturing all the significant traits of the non-linear creep behaviour of concrete at working stress levels. It is also encouraging that the proposed creep law is compatible with a realistic picture of the creep mechanism.

We imagine that creep in concrete consists of a vast number of small particle migrations within the cement paste microstructure. Any sudden change of stress, $\Delta\sigma$, is assumed to activate a number of potential migration sites, the number of which, $N_s$, is very large. This points to an infinite strain rate right after any stress jump, which in turn suggests the existence of a singularity in the kernel, resulting from the term $[\kappa(t)^s - \kappa(\tau)^s]^{-v}$. The subsequent growth of this term reflects the gradual exhaustion of potential particle migration sites, thus causing a reduction in creep rate. The exhaustion rate must decrease as the creep strain already caused by stress jump $\Delta\sigma$ increases, i.e. it must decrease as $[\kappa(t)^s - \kappa(\tau)^s]$ and $(t - \tau)$ grow, as reflected in Equations (7.87) and (7.88).

The creep rate must also decrease due to the continuing hydration of the cement paste while it carries the load. The hydration results in formation of further bonds in the microstructure, which reduces the number of potential migration sites. This reduction depends strictly on time and proceeds at a gradually decreasing time rate, as modeled by the term $\tau^{-m}(t - \tau)^{-u}$ in Equations (7.87) and (7.88).

### 7.5.4 Cyclic creep

Another important non-linear phenomenon arises for cyclic (or pulsating) loads with many repetitions. According to the principle of superposition, the creep due to cyclic stress should be approximately the same as the creep due to a constant stress equal to the average of the cyclic stress. In reality, a much higher creep is observed, the excess creep being larger for a larger amplitude of the cyclic component.[184,100,134,84,15,174]

The time-average compliance function for cyclic creep at constant stress amplitude $\Delta$ and constant mean stress may be reasonably well described by an extension of double power law in which $(t - t')^n$ is replaced by the expression $[(t - t')^n + 2.2\phi_\sigma\Delta^2 N^n]$ where $N$ is the number of uniaxial stress cycles of amplitude $\Delta$, $\phi_\sigma$ is a function of $\sigma$, equal to 1.0 when $\sigma = 0.3f'_c$. The cyclic loading does not seem to affect the drying creep component. For details, see Bažant and Panula[43] (Part VI).

### 7.5.5 Multiaxial generalization and operator form

Regarding the multiaxial aspects of non-linear creep, almost no experimental information is available. The multiaxial non-linear behaviour is reasonably explored experimentally only for short-time (rapid, 'instantaneous') loading, and the high-stress non-linearity of creep must approach this behaviour in the limit. This limiting condition is presently just about the only solid information on which generalizations of non-linear creep models to three dimensions can be based.

Based on this scant information, both the endochronic theory and the plastic fracturing theory for non-linear triaxial behaviour have been extended to describe non-linear triaxial creep.[29,39] These models are probably reasonably good for short load durations and large deformations near those for the usual short-time (rapid) tests, but they are entirely hypothetical as far as long load durations are concerned. The adaptation phenomena have not been included in these models. However, an inductive generalization of Equations (7.85) and (7.86), in which $\sigma$ was replaced by certain stress invariants, has been given for the foregoing theory of adaptation (Section 7.5.2).

### 7.5.6 Cracking and tensile non-linear behaviour

The most typical deleterious effect of creep and shrinkage in structures is cracking, both invisible microcracking and, at a later stage, continuous visible cracks. Thus, the calculation of creep and shrinkage effects is complete only if a realistic model of cracking, as well as tensile non-linearity due to microcracking and fracture propagation, is considered.[202]

## 7.6 MOISTURE AND THERMAL EFFECTS

### 7.6.1 Effect of pore humidity and temperature on aging

The moisture effects are much more involved and much less understood than the effects analysed so far, despite considerable research efforts.

The previously indicated expressions for the effect of humidity (Section 7.2.6), as given in current code recommendations and practical creep prediction models, describe only the mean behaviour of the cross section and do not represent constitutive properties and constitutive relations of the material. Thus, they are usable in structural analysis only when the cross section is of single-element width (as is often used for plates and shells); it makes no sense to subdivide the cross section into more than one element.

To determine the distributions of pore humidity and water content within the cross section at various times, it is necessary to solve the moisture diffusion problem. This necessitates a constitutive equation which involves the pore humidity or water content but not the environmental humidity. The latter is inadmissible for use in a constitutive equation and is properly used only as a boundary condition.

One important effect of a decrease in pore humidity $h$ (relative vapour pressure $p/p_{sat}$) is a deceleration and eventual arrest of the hydration process. This may be modelled by extending the previous definition of the equivalent hydration period, $t_e$ (or maturity):

$$t_e = \int \beta_T \beta_h \, dt \tag{7.95}$$

where

$$\beta_h \simeq [1 + 6(1 - h)^4]^{-1} \tag{7.96}$$

Here $\beta_T$ is given by Equation (7.59); $\beta_h$ is an empirical function. Compared to $h = 1.0$, Equation (7.95) gives a reduction of aging rate to 9% at $h = 0.7$ and to 1% at $h = 0.5$.

Still more realistically, taking into account Equation (7.85) one should write $t_e = \int \beta_T \beta_h \beta_\sigma \, dt$.

### 7.6.2  Shrinkage as a constitutive property

Let us now consider shrinkage, $\varepsilon_S$, which is understood as a material property rather than a specimen property, and represents the free (unrestrained) shrinkage at a point of a continuum. As proposed by Carlson[67] and Pickett,[148] the shrinkage is properly modelled as a function of pore water content $w$ (g of water per $cm^3$ of concrete), which is in turn a function of pore humidity $h$. Therefore,

$$\varepsilon_S = \varepsilon_S^0 f_S(h) \qquad (7.97)$$

Here $\varepsilon_S^0$ is the maximum shrinkage (for $h = 0$), which is larger, perhaps much larger, than $\varepsilon_{sh_\infty}$ in Equation (7.9) because the value $\varepsilon_{sh_\infty}$ is reduced by microcracking of the specimen while $\varepsilon_S^0$ is not, by definition. The function $f_S(h)$ is emprical; approximately perhaps $f_S(h) = 1 - h^2$ but this needs to be checked more closely.

Note that, in contrast to Equation (7.9) for the mean cross-section shrinkage $\bar{\varepsilon}_S$, the shrinkage as a material property exhibits no dependence on the duration of drying $t - t_0$ and the age at the start of drying $t_0$. These times affect $\bar{\varepsilon}_S$ only indirectly, through the solution of the diffusion problem which is approximated by Equation (7.9).

There exists, however, some time-dependence in shrinkage, albeit different from that in Equation (7.9). Since the mechanism of shrinkage at least to some extent consists in deformation (compression) of solid particles and solid microstructural framework under the forces caused by changes in solid surface tension, capillary tension, and disjoining pressure in hindered adsorbed water layers, the deformation must depend on the stiffness of the microstructure. This, in turn, depends on the degree of hydration, and thus on the equivalent hydration period $t_e$. Hence, a more accurate expression for shrinkage should be

$$\varepsilon_S = \varepsilon_S^0 f_S(h) g_S(t_e) \qquad (7.98)$$

where the function $g_S$ may approximately be taken as $g_S(t_e) \simeq E_{28}/E(t_e)$, i.e. the inverse ratio of the increase in elastic modulus due to age (hydration).

Another time dependence may exist in shrinkage due to the delay needed to establish thermodynamic equilibrium of water between macropores and micropores. Part of the shrinkage, probably a large part, is due to a change in the disjoining pressure, and since the microdiffusion of water between micropores and macropores through which the thermodynamic equilibrium is established requires some time, the disjoing pressure must respond to humidity $h$ in the macropores with a certain delay. This would mean that for determining the delayed part of $\varepsilon_S(t)$ at time $t$, one would have to substitute the value of $h$ at an earlier time $t - \Delta$, $\Delta$ being the characteristic lag. Alternatively, the delay may be obtained through a formulation exemplified in Equations (7.99)–(7.101) in the sequel.

### 7.6.3  Creep at drying as a constitutive property

The effect of pore humidity on creep is not completely understood at present, chiefly because of the difficulty of determining creep properties from tests on drying specimens which are in a highly non-uniform moisture state during the test and probably undergo significant microcracking.

One effect of pore humidity is however clear; if the pore humidity is constant, then the lower the pore humidity, the smaller the creep.[189,191–194,160] At fully dried state ($h \approx 0$) the creep rate is only about 10% of that at fully saturated state ($h = 1$). This effect may be described by replacing $\tau_\mu$ in the preceding rate-type equations with $\tau_\mu / \phi_h$ where $\phi_h$ is a function of $h$, roughly given as $\phi_h \approx 0.1 + 0.9h^2$.

Another effect is that of a change in humidity. This effect remains rather clouded. Whereas after drying (after $h$ attains a constant value) creep is less at lower humidity, during the drying process the creep is higher than at a sealed state (Hansen,[90] Glucklich and Ishai,[87] Keeton,[111] Kesler *et al.*[113] Kesler and Kung,[112] L'Hermite,[121] L'Hermite and Mamillan,[122] Mamillan,[127] L'Hermite *et al.*,[123] Mamillan and Lelan,[28] Mullen and Dolch,[140] etc.). This phenomenon apparently persists even for some time after the pore humidity has come down and reached equilibrium throughout the specimen. What is uncertain is how much of the creep increase observed in drying specimens (Pickett effect) is due to the non-uniform stress state of the specimen and the inherent microcracking (or tensile non-linearity), and how much of it is due to constitutive properties, e.g. a possible effect of the rate of pore humidity $\dot{h}$ upon the creep rate coefficient $1/\tau_\mu$.

A model which describes both of the aforementioned effects has been developed[16,25,24,52] applying thermodynamics of multiphase systems and of adsorption to obtain a rate-type constitutive model. The effect of $\dot{h}$, if it exists, must be due to a thermodynamic imbalance between macropores and micropores, created by pore humidity changes, and to the resulting local diffusion between these two kinds of pores. In the process of drying (as well as wetting) of a concrete specimen one may distinguish two diffusion processes. One is the macroscopic diffusion in which the water molecules migrate through the pore passages of least resistance, involving the largest (capillary) pores and bypassing most of the micropores (gel pores and interlayer spaces). This diffusion process controls the humidity in the macro-pores, $h$, and is essentially independent of the applied load and deformation. The other diffusion process is the local process of migration of water molecules on the microscale between the macropores and the micropores. This process is driven by a thermodynamic imbalance between these two kinds of pores; more precisely, an imbalance (difference) between the values, $\mu_w$, $\mu_d$, of the specific Gibbs' free energy $\mu$ in these pores. The values $\mu_w$ and $\mu_d$ depend on the water content of pores (as well as temperature) and, for the micropores but not the macropores, also on the stress in the solid gel that is transmitted through water (hindered adsorbed water or interlayer water) in the micropores and is caused by the applied load. The two separate diffusion processes are certain to exist, but at present it is just a hypothesis that the microscopic diffusion of water indeed affects to a significant extent the mobility in the solid microstructure, thereby influencing creep.

The foregoing hypothesis has been applied to the Maxwell chain model in which each partial stress $\sigma_\mu$ ($\mu = 1, 2, \ldots, N$) is separated into two parts, $\sigma_\mu^s$ and $\sigma_\mu^w$, imagined to represent the stresses in solids and in micropore water. The uniaxial version of the constitutive equation then is:[25,52]

$$\dot{\sigma}_\mu^s + \phi_{ss_\mu}\sigma_\mu^s = E_\mu^s(\dot{\varepsilon} - \dot{\varepsilon}_{sh}^0 - \alpha_\mu^s\dot{T}) \tag{7.99}$$

$$\dot{\sigma}_\mu^w + \phi_{ww_\mu}[\sigma_\mu^w - f_\mu(h, T)] = E_\mu^w(\dot{\varepsilon} - \dot{\varepsilon}_{sh}^0 - \alpha_\mu^w\dot{T}) \tag{7.100}$$

$$\sigma = \sum_{\mu=1}^{N} (\sigma_\mu^s + \sigma_\mu^w) \tag{7.101}$$

Here $E_\mu^s(t)$ and $E_\mu^w(t)$ are separate spring moduli for solids and water; $\varepsilon_{sh}^0$ is the part of shrinkage strain that is instantaneous with a change of humidity $h$ (relative vapour pressure) in the capillary pores; $\alpha_\mu^s$, $\alpha_\mu^w$ are coefficients of thermal dilation that are instantaneous with temperature change; $\phi_{ss_\mu}$, $\phi_{ww_\mu}$ are the rate coefficients which replace the role of $1/\eta_\mu$ in Equation (7.56) and reflect the rate of diffusion (or migration) of solids and water (hindered adsorbed water and interlayer water) between the loaded and load-free areas of cement gel microstructure; and $f_\mu(h, T)$ are values of $\sigma_\mu^w$ for which the water in loaded areas (micropores) is in thermodynamic equilibrium with water in the adjacent capillary pores. Coefficient $\phi_{ss_\mu}$ is assumed to increase as $[\sigma_\mu^w - f_\mu(h, T)]^2$ increases; this models the drying creep effect[25,52] and since it expresses the acceleration of creep when thermodynamic equilibrium does not exist between the water in loaded areas and the water in load-free areas. Material functions $f_\mu(h, T)$, $\phi_{ss_\mu}$, $\phi_{ww_\mu}$, $E_\mu^s$, and $E_\mu^w$, which give a good agreement with test data on creep and shrinkage for specimens of various sizes at various regimes of time-variable environmental humidity have been found.[52]

A step-by-step algorithm (of the exponential type) has been developed for Equations (7.99)–(7.101), and their triaxial version was applied to analyse by finite elements the stresses in drying cylinders[52] and in drying floors.[106] Cracking or tensile non-linearity of concrete was considered in these analyses. This rather sophisticated model led to a good agreement with most of the existing test data on creep under various moisture conditions, exceeding by far the results obtained with other constitutive models.

The present test data are however limited in scope, and could perhaps be fitted equally well by different models. At present one cannot even exclude the possibility—an attractive one because of its simplicity—that the pore humidity rate $\dot{h}$ has no significant effect in the constitutive equation *per se* and all of the creep increase due to drying is the consequence of internal stresses and microcracking.[195] More tests and theoretical analyses are urgently needed to check this hypothesis.

Due to random fluctuations of environmental humidity, creep and shrinkage in drying structures should be analysed probabilistically and some steps in this direction have already been taken.[205,203]

### 7.6.4  Calculation of pore humidity

From the preceding exposition we see that calculation of pore humidity as a function of space and time is a necessary part of an analysis of stress distributions in the presence of drying. A satisfactory model already exists for this purpose.

In the early investigations, drying of concrete was considered as a linear diffusion problem, but serious discrepancies were found. It is now reasonably well documented by measurements that the diffusion equation that governs moisture diffusion in concrete at normal temperatures is highly non-linear, due to a strong dependence of permeability $c$ and diffusivity on pore humidity $h$. The governing differential equation may be written as:[37]

$$\frac{\partial h}{\partial t} = k \operatorname{div} \mathbf{J} + \frac{\partial h_s(t_e)}{\partial t} + k\frac{dT}{dt} \qquad \mathbf{J} = -c \operatorname{grad} h \qquad (7.102)$$

where $\mathbf{J}$ is water flux; $k = \partial h/\partial w$ at constant temperature $T$ and constant age ( = slope of desorption isotherm or sorption isotherm); $w$ is the specific pore water content; $\kappa = \partial h/\partial T$ at constant $h$ and constant $t_e$; and $\partial h_s/\partial t$ is the rate of self-desiccation, i.e. of the drop of $h$ due to aging (hydration) at constant $w$ and constant $T$. The function $h_s(t_e)$ is empirical and represents a gradual decrease of $h$ from the initial value 1.00 to about 0.96 to 0.98 after long conservation (without external drying). For desorption at room temperature, coefficient $k$ may be approximately taken as constant, in which case it may be combined with $c$, yielding $C = kc =$ diffusivity.

The graph of $C$ (or $c$) versus $h$ decreases to about 1/20 as $h$ drops from 0.90 to 0.60 (Figure 7.22). This is doubtless due to the fact that the rate of moisture transfer is at room temperature controlled by migration of water

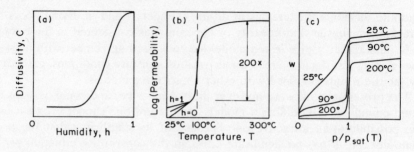

**Figure 7.22**  Dependence of drying diffusivity $C$ of concrete on pore humidity, of permeability on temperature, and sorption isotherms

molecules in adsorbed layers, the rate of migration getting slower as the thickness of the adsorption layers decreases. A suitable empirical expression is[37] (Figure 7.22):

$$C = kc = C_1(T, t_e)\{0.05 + 0.95[1 + 3(1 - h)^4]^{-1}\} \qquad (7.103)$$

where $C_1$ is the diffusivity at $h = 1$, for which an approximate semi-empirical expression based on activation energy is also available:[25]

$$C_1(T, t_e) = C_0\left[0.3 + \left(\frac{13}{t_e}\right)^{1/2}\right]\frac{T}{T_0}\exp\left(\frac{Q}{RT_0} - \frac{Q}{RT}\right) \qquad (7.104)$$

where $Q/R \simeq 4700$ K and $T$ is absolute temperature.

The boundary conditions for $t > t_0$ are: for sealed surface, normal flux $\mathbf{J}_n = 0$; and for perfect moisture transfer, $h = p_{en}/p_{sat}(T)$ where $p_{en}$ is the environmental vapour pressure and $p_{sat}(T)$ is the saturation pressure for temperature $T$ in concrete at its surface.

From the foregoing equations we can determine the size dependence of the drying process. We consider constant temperature, and also neglect the term $\partial h_s/\partial t$ in Equation (7.102) since it is relatively small. We may further neglect the age dependence of slope $k$ and permeability $c$. Then Equation (7.102) becomes

$$\frac{\partial h}{\partial t} = k(h)\frac{\partial}{\partial x_i}\left(c(h)\frac{\partial h}{\partial x_i}\right) \qquad (7.105)$$

where $x_i$ are Cartesian coordinates ($i = 1, 2, 3$). We now introduce the non-dimensional spatial coordinates $\xi_i = x_i/D$, where $D$ is a characteristic dimension of the body, e.g. the effective thickness. We restrict attention to geometrically similar bodies, whose all dimensions are fully characterized by $D$, and we introduce the non-dimensional time

$$\theta = (t - t_0)/\tau_s \quad \text{with } \tau_s = D^2/C_1 \qquad (7.106)$$

where $t_0$ is the age at the start of drying. Then $\partial/\partial t = C_1 D^{-2} \partial/\partial\theta$. Also $\partial/\partial x_i = D^{-1} \partial/\partial\xi_i$. Thus, Equation (7.105) yields:

$$\frac{\partial h}{\partial\theta} = \frac{k(h)}{k_i} \frac{\partial}{\partial\xi_i} \left( \frac{c(h)}{c_1} \frac{\partial h}{\partial\xi_i} \right) \tag{7.107}$$

This diffusion equation is to be solved always for the same region of $\xi_i$. The initial condition consists of prescribed values of $h$. Assuming that the boundary conditions also consist either of prescribed time-constant values of $h$ or of a sealed boundary (normal flux $\mathbf{J}_n = 0$), the corresponding initial and boundary conditions in terms of variables $\xi_i$ and $\theta$ are the same for any $D$. Thus, the solution in terms of $\xi_i$ and $\theta$ is independent of $D$ as well as of $k$, and of $c_1$ (or of $C_1$), and depends only on the coefficients of Equation (7.105), i.e. on the functions $k(h)/k_1$ and $c(h)/c_1$ representing the relative variation of slope $k$ and of permeability $c$. These functions are the same for any $D$. So, the time to reach the same stage of drying (e.g. the shrinkage half-time) is proportional to $t/\tau_S$, i.e. to $D^2/C_1$. This property, which we used in setting up Equation (7.12) for $\tau_s$, is generally known for a linear diffusion equation (constant $k$ and $c$) and here we show that it is also true for the non-linear diffusion equation, provided that the self-desiccation and the age dependence of permeability and of the slope of the sorption diagram are neglected.

When the pore humidity falls below 0.9, the hydration process is nearly arrested. Thus, neglect of the age dependence is well justified for drying at low ambient humidity, such as 0.5, while it is a poor assumption for a high ambient humidity such as 0.9; but this case is of little practical interest. The neglect of aging causes a more severe error for thicker bodies (larger $D$) since pore humidity lingers above 0.9 for a longer time period. Thus, the deviations from a $D^2$ dependence of shrinkage half-time $\tau_s$ are stronger for thicker bodies. On the other hand, in thin bodies another phenomenon may spoil the $D^2$ dependence significantly; it is the cracking (and microcracking) produced by drying, which is more severe for a faster drying because the stresses produced by drying have less time to get relaxed by creep. At present little is known, however, how much the cracking affects permeability.[202] It certainly greatly affects shrinkage and all deformations.

A finite element model for the foregoing diffusion equation (Equation (7.89)) may be developed using the Galerkin procedure, as is well known; see Bažant and Thonguthai[48,49,32] and Figure 7.23.

### 7.6.5 Coupled moisture and heat transfer

Migration of moisture in concrete is produced not only by gradient of moisture concentration $w$ (pore water content) but also by gradient of temperature. It seems that this effect, called thermal moisture transfer, is

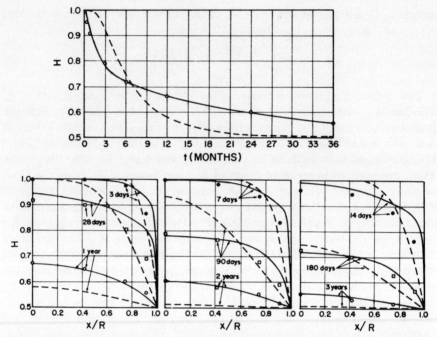

**Figure 7.23** Typical measurements of pore humidity in solid cylinders exposed to drying environment. Test data from Hanson[95]. Solid lines, predictions of nonlinear diffusion theory; dashed lines, predictions of linear diffusion theory[37]

adequately modelled by considering that the driving force of the diffusion flux is not grad $h$ or grad $w$ but grad $p$ where $p$ is the vapour pressure in the pores.

Central to the model are realistic formulations for the moisture diffusivity (or permeability) and for the equation of state of the pore water (sorption isotherms). Both of these properties are rather involved. As already mentioned, the diffusivity at room temperature is found to decrease about twenty times as the pore humidity $h$ decreases from 95% to 60%. Above 100 °C the diffusivity becomes independent of $h$ (i.e. of pore pressure $p$), but another effect is observed (Figure 7.22): The permeability increases about 200 times as we increase the temperature from 90 °C to 120 °C. It seems that this effect may be explained by the enlargement of narrow necks on the flow passages in cement paste, and a transition to a flow that is controlled by viscosity of steam rather than migration of water molecules along adsorption layers which controls the diffusion at room temperature. These phenomena are illustrated in Figure 7.22.

In defining the equation of state, one must take into account the fact that the volume of pores decreases due to dehydration as concrete is heated beyond 100 °C, and that the pressure forces pore water into the microstructure, thereby enlarging the pore volume available to liquid water or vapour.

If these phenomena are taken into account, then the well-known thermodynamic properties of water can be used to calculate pore pressures and moisture transfer and obtain agreement with the scant available measurements. A finite element program, based on Galerkin approach, has been developed for this purpose.[48,49,32]

Due to the sharp rise of permeability, specimens heated over 100 °C lose moisture very rapidly. At room temperature one can almost never expect to deal with fully dried concrete specimens, but at high temperatures the dried condition is typical, except for massive walls as in reactor vessels.

Experimental information on creep and shrinkage under controlled moisture conditions is almost non-existent for high temperatures. Data exist nearly exclusively for uniaxial creep and shrinkage of specimens from which the evaporable water was driven out due to heating, and which probably suffered great non-uniform stresses and microcracking during the heating. Because above 100 °C the escape of water cannot be prevented without significant pressure on all specimen surfaces, triaxial tests are required if the moisture content should be controlled. In fact, uniaxial creep without moisture loss is a meaningless phenomenon above 100 °C, impossible to simulate experimentally.

## 7.7 DETAILS OF SOME MODELS

### 7.7.1 ACI model

The ultimate creep coefficient from Equation (7.13) is specified as follows:[5,61,58,59]

$$C_u = 2.35 K_T^c K_H^c K_T^c K_S^c K_F^c K_A^c \qquad (7.108)$$

where $K_T^c$, $K_H^c$, $K_T^c$, $K_S^c$, $K_F^c$, and $K_A^c$ are called creep correction factors. These factors equal 1.0 (i.e. $C_u = 2.35$) for the following standard conditions: 4 in. or less slump, 40% environmental relative humidity, minimum thickness of member 6 in. or less, loading age 7 days for moist cured concrete and 1–3 days for steam-cured concrete. For other than the standard conditions, one has

$$K_T^c = \begin{cases} 1.25t'^{-0.118} & \text{for moist cured concrete} \\ 1.13t'^{-0.095} & \text{for steam cured concrete} \end{cases}$$

$$K_H^c = 1.27 - 0.0067 h_e \qquad h_e \geqslant 40\%$$

$$K_T^c = \begin{cases} 1.14 - 0.023 T_m & \text{for} \leqslant 1 \text{ year loading} \\ 1.10 - 0.017 T_m & \text{for ultimate value} \end{cases} \qquad (7.109)$$

$$K_S^c = 0.82 + 0.067 S_c \qquad K_F^c = 0.88 + 0.0024 F_a$$

$$K_A^c = \begin{cases} 1.00 & \text{for } A_c \leqslant 6\% \\ 0.46 + 0.090 A_c & \text{for } A_c > 6\% \end{cases}$$

Creep and Shrinkage in Concrete Structures

where $t'$ is the loading age in days, $h_e$ is the environmental relative humidity in percent, $T_m$ is the minimum thickness in inches, $S_c$ is the slump in inches, $F_a$ is the per cent of fine aggregate by weight, and $A_c$ is the air content in per cent of volume of concrete. The initial deformation is defined by

$$E(t') = 33\sqrt{[\rho^3 f'_c(t')]} \qquad f'_c(t') = f'_{c28} \frac{t'}{4 + 0.85t'} \qquad (7.110)$$

where $\rho$ is the unit weight of concrete (normal-weight concretes only). The model is considered applicable only for ages at loading $t' \geqslant 7$ days.

The ultimate shrinkage coefficient $\varepsilon_u^s$ is specified as follows:

$$\varepsilon_u^s = \begin{cases} 0.000800 K_H^s K_T^s K_S^s K_B^s K_F^s K_A^s & \text{for moist cured concrete,} \\ 0.000730 K_H^s K_T^s K_S^s K_B^s K_F^s K_A^s & \text{for steam cured concrete} \end{cases} \qquad (7.111)$$

where $K_H^s$, $K_T^s$, $K_S^s$, $K_B^s$, $K_F^s$, and $K_A^s$ are shrinkage correction factors. They equal 1.0 for the following standard conditions: 4 in. or less slump, 40% environmental relative humidity, and the minimum thickness of member 6 in. or less. For other than standard conditions the following shrinkage correction factors are used:

$$K_H^s = \begin{cases} 1.40 - 0.010 h_e & 40\% \leqslant h_e \leqslant 80\% \\ 3.00 - 0.030 h_e & 80\% \leqslant h_e \leqslant 100\% \end{cases}$$

$$K_T^s = \begin{cases} 1.23 - 0.038 T_m & \text{for } \leqslant 1 \text{ year loading} \\ 1.17 - 0.029 T_m & \text{for ultimate value} \end{cases}$$

$$K_S^s = 0.89 + 0.041 S_c \qquad K_B^s = 0.75 + 0.034 B_s \qquad (7.112)$$

$$K_F^s = \begin{cases} 0.30 + 0.0140 F_a & \text{for } F_a \leqslant 50\% \\ 0.90 + 0.0020 F_a & \text{for } F_a \geqslant 50\% \end{cases}$$

$$K_A^s = 0.95 + 0.0080 A_c$$

where $B_s$ is the number of 94-lb sacks of cement per cubic yard of concrete. For $f_c$ and $t_0$, the following values are recommended: $f_c = 35$ days; $t_0 = 7$ days for moist cured concrete; and $f_c = 55$ days; $t_0 = 1$ to 3 days for steam-cured concrete.

As $T_m \to 0$ the factor $K_T^c$ should approach 0.6 because an infinitely thick specimen is equivalent to concrete at pore humidity nearly 100%. Since Equation (7.109) for $K_T^c$ does not satisfy this condition, the ACI Model cannot be applicable for very thick specimens.

## 7.7.2 CEB-FIP model

The functions and coefficients in Equations (7.14) and (7.15)[68,162] are specified as

$$\varepsilon_{s_0} = \varepsilon_{s_1}\varepsilon_{s_2} \qquad \phi_d = 0.4 \qquad \phi_f = \phi_{f_1}\phi_{f_2}$$

$$F_i(t') = \frac{1}{E_c(t')} + \frac{\beta_s(t')}{E_{c28}} \qquad \beta_s(t') = 0.8\left(1 - \frac{f'_c(t')}{f'_{c_\infty}}\right) \tag{7.113}$$

$$E_c(t') = 1.25E_{c_m}(t') \qquad E_{c_m}(t') = 9500[f'_{c_m}(t')]^{1/3} \qquad E_{c28} = 9500f'^{1/3}_c \tag{7.114}$$

Here strain $\varepsilon_{s_1}$ is defined by Table 0.1, column 4, of CEB-FIP 'Model Code for Concrete Structures'[68] as a function of humidity $h_e$; $\varepsilon_{s_2}$ is defined by a graph in Figure e.5 as a function of the effective thickness defined as $H_0 = \lambda(2A_c/U)$, $A_c/U$ is the ratio of cross-sectional area to the exposed surface; $\lambda$ is a function of $h_e$ defined by Table e.1; $\beta_s(t')$ is a function of age $t$ defined by six graphs in Figure e.6 for various values of effective thickness $H_0$. Furthermore, age $t$ is corrected for temperature in Section e.5 of CEB-FIP, 'Model Code for Concrete Structures',[68] but the acceleration of creep due to temperature rise is not considered. Quantities $f'_{c_m}$, $E_{c_m}$, and $E_c$ must all be given in MPa. The strength $f'_c$ is given by a graph in Figure e.1 of CEB-FIP Model Code[68] as a function of $t'$; $\phi_{f_1}$ is given in Table e.1 of CEB-FIP Model Code[68] as a function of humidity $h_e$; $\phi_{f_2}$ is given by a graph in Figure e.2 of CEB-FIP Model Code[68] as a function of effective thickness $H_0$; $\beta_d$ is defined by a graph in Figure e.3 of CEB-FIP Model Code[68] as a function of stress duration $t - t'$, $\beta_f$ is given by six graphs in Figure e.4 for various effective thicknesses $H_0$ (Table 2.3 of CEB-FIP Model Code[68]) as a function of age $t$ (corrected for temperature).

Note that in contrast to ACI and BP Models, the CEB-FIP Model is not defined completely by formulae. Graphs consisting of sixteen curves are used to define the functions.

## 7.7.3 BP model

The complete definition of this model[42-44] is as follows.
The shrinkage is described by:

$$\varepsilon_{sh}(\hat{t}, t_0) = \varepsilon_{sh_\infty}k_hS(\hat{t}) \qquad \hat{t} = t - t_0$$

$$S(\hat{t}) = \left(\frac{\hat{t}}{\tau_{sh} + \hat{t}}\right)^{1/2} \qquad \tau_{sh} = 600\left(\frac{k_s}{150}D\right)^2\frac{10}{C_1(t_0)} \qquad D = 2\frac{v}{s} \tag{7.115}$$

For

$$h \leq 0.98: k_h = 1 - h^3; \quad \text{for } h = 1.00: k_h = -0.2$$

$$C_1(t) = C_7 k_T'[0.05 + \sqrt{(6.3/t)}] \qquad \varepsilon_{sh_\infty} = \varepsilon_{s_\infty} \frac{E(7 + 600)}{E(t_0 + \tau_{sh})} \tag{7.115}$$

$$k_T' = \frac{T}{T_0} \exp \left( \frac{5000}{T_0} - \frac{5000}{T} \right)$$

in which $D$ is an effective thickness of cross section (in mm), $v/s$ is the volume-to-surface ratio, $E(t') = 1/J(t' + 0.1, t')$ is the conventional elastic modulus, $T_0 = 23\,°C =$ reference temperature, $C_1(t)$ is drying diffusivity at age $t$ (in mm$^2$/day), $C_7$ is a given or assumed value $C_1$ at age 7 days, $k_S$ is the shape factor ($= 1.0$ for an infiite slab, 1.15 for an infinite cylinder, 1.25 for an infinite square prism, 1.30 for a sphere and 1.55 for a cube). Equations (9)–(10) of Bažant and Panula[43] give the coefficients $\varepsilon_{s_\infty}$ and $C_7$ as functions of strength $f_c'$, water/cement ratio w/c, cement content $c$, aggregate/cement ratio a/c, and sand/cement ratio s/c. If, however, at least one measured value of shrinkage on a small specimen is available, either $\varepsilon_{s_\infty}$ or $C_7$ may better be evaluated from this value, which improves the accuracy of prediction.

To take moisture effects into account, the BP Model distinguishes three long-time components of the creep function:

$$J(t, t') = \frac{1}{E_0} + C_0(t, t') + \bar{C}_d(t, t', t_0) - C_p(t, t', t_0) \tag{7.116}$$

in which $C_0(t, t')$ gives the basic creep, i.e. the creep in the absence of moisture exchange; $\bar{C}_d(t, t', t_0)$ gives the increases of creep due to simultaneous moisture exchange, in particular drying that proceeds simultaneously with creep; and $C_p(t, t', t_0)$ gives the decrease of creep due to pre-drying; this decrease occurs long after the drying process reaches the final, stable state. Time $t_0$ is the age at the time the exposure to a drying environment begins. Term $C_p(t, t', t_0)$ is negligible and may be omitted except when the cross section of concrete is very thin ($\leq 10\,cm$) or the temperature is elevated. $E_0$ represents the asymptotic modulus which gives the asymptotic value of the deformation extrapolated to extremely short load durations (less than a microsecond, beyond the range of interest).

The basic creep is given by the double power law:

$$C_0(t, t') = \frac{\phi_T}{E_0} (t_e'^{-m} + \alpha)(t - t')^{n_T} \tag{7.117}$$

in which $t_e' = \int \beta_T(t')\, dt'$, $\phi_T = \phi_1 C_T$, $n_T = n\beta_T$. Here $C_T$ and $\beta_T$ introduce the effect of temperature $T$ and may be taken as 1.0 when $T = T_0 \simeq 23\,°C =$ reference temperature; then $t_e' = t'$, $\phi_T = \phi_1$, $n_T = n$. Coefficients $\phi_1$, $n$, $m$, and $\alpha$ characterize the basic creep at reference temperature from load durations of $t-t'=10^{-7}$ day (dynamic range) through the short-time static load range (about 0.1 day) until at least 30 years. These coefficients as well as $E_0$ may be evaluated from Equations (15)–(19) of Bažant and Panula[43] as functions of standard cylinder strength $f_c'$, water/cement ratio w/c, aggregate/cement ratio a/c, aggregate/gravel ratio s/g, unit mass of concrete $\rho$, and the type of cement. Coefficients $\beta_T$, $C_T$, and $n_T$ are defined by Equations (36)–(39) from Bažant and Panula[43] as functions of temperature $T$, of age $t_T'$ at which this temperature begins, and of w/c, a/c, and the cement type.

The creep increase during drying and the creep decrease after drying are given in the BP Model as:

$$\bar{C}_d(t, t', t_0) = \frac{\phi_d'}{E_0} t_e'^{-m/2} k_h' \varepsilon_{sh_\infty} S_d(t, t')$$

$$C_p(t, t', t_0) = \varepsilon_p k_h'' S_p(t, t_0) C_0(t, t')$$

(7.118)

where

$$\phi_d' = \left(1 + \frac{\Delta\tau'}{10}\right)^{-1/2} \phi_d \qquad k_h' = 1 - h^{1.5} \qquad k_h'' = 1 - h^2$$

$$S_d(t, t') = \left(1 + \frac{10\tau_{sh}(k_T')^{1/4}}{t-t'}\right)^{-n'} \qquad S_p(t, t_0) = \left(1 + \frac{100}{\Delta\tau}\right)^{-n}$$

$$\Delta\tau' = \int_{t_0}^{t'} \frac{(k_T)^{5/4}}{\tau_{sh}} dt \qquad \Delta\tau = \int_{t_0}^{t} \frac{dt}{\tau_{sh}}$$

(7.119)

$$n' = \frac{c_d^n}{K_T^2} \qquad k_T = 0.42 + 17.6\left[1 + \left(\frac{100}{\hat{T}}\right)^4\right]^{-1}$$

$$K_T = 1 + 0.4\left[1 + \left(\frac{93.5}{\hat{T}}\right)^4\right]^{-1}$$

Here $h$ is the relative humidity of the environment. In the integrals, $\tau_{sh}$ must be evaluated for the given temperature as a function of time. When $T = T_0$, we have $k_T' = K_T = 1.0$. The material parameters $C_p$, $C_d$, and $\phi_d$ are functions of $n$, $\varepsilon_{s_\infty}$, $f_c'$ and of mix ratios s/a, g/s, and w/c as indicated by Equations (30)–(32) from Bažant and Panula.[43]

A relatively simple refinement allows one to obtain cyclic creep, i.e. creep when a cyclic load is superimposed on a static load.[43]

The composition effects in shrinkage are given by:

$$C_7 = \frac{c}{8}\frac{w}{c} - 12; \quad \text{for } C_7 < 7 \text{ set } C_7 = 7, \text{ for } C_7 > 21 \text{ set } C_7 = 21$$

$$\varepsilon_{s_\infty} = (1.21 - 0.88y)10^{-3} \qquad y = (390z^{-4} + 1)^{-1}$$

(7.120)

$$z = \left[1.25\left(\frac{a}{c}\right)^{1/2} + \frac{1}{2}\left(\frac{g}{s}\right)^2\right]\left(\frac{1 + s/c}{w/c}\right)^{1/3} (f_c')^{1/2} - 12 \quad \text{if } z \geqslant 0; \text{ else } z = 0$$

in which $f'_c$ is the 28-day cylindrical strength in ksi ($= 1000\,\text{psi} = 6.89\,\text{MN/m}^2$); $w, c, a =$ contents (masses) of water, cement and aggregate, $\text{kg/m}^3$ of concrete; $a = g + s$ where $g, s$ are the masses of gravel and sand. The composition effects in basic creep at reference temperature are:

$$\phi_1 = \frac{10^{3n}}{2(28^{-m} + \alpha)} \qquad \alpha = \frac{0.025}{w/c} \qquad m = 0.28 + f'^{-2}_c$$

$$n = \begin{cases} 0.12 + 0.07(1 + 5130x^{-6})^{-1} & \text{for } x > 4 \\ 0.12 & \text{for } x \leqslant 4 \end{cases} \qquad (7.121)$$

$$x = [2.1(a/c)(s/c)^{-1.4} + 0.1(f'_c)^{1.5}(w/c)^{1/3}(a/g)^{2.2}]a_1$$

$$a_1 = \begin{cases} 1 & \text{for cement types I and II} \\ 0.93 & \text{for type III} \\ 1.05 & \text{for type IV} \end{cases}$$

When a measured value $E$ of conventional elastic modulus is known, one substitutes $1/E = J(t' + 0.1, t')$ into Equation (7.116) for $T = T_0$ and solves for $E_0$. The same is done when any value of $J(t, t')$ is known. When there is no drying and $E$ pertains to age 28 days, one simply has $E_0 = 1.5E$. When no measured value is known, one may use:

$$\frac{1}{E_0} = 0.09 + \frac{1}{1.7z_1^2} \qquad z_1 = 0.00005\rho^2 f'_c \qquad (7.122)$$

The coefficients for the temperature effect in basic creep are:

$$\beta_T = \exp\left(\frac{4000}{T_0} - \frac{4000}{T}\right) \qquad C_T = c_T \tau_T c_0 \qquad c_0 = \frac{1}{8}\left(\frac{w}{c}\right)^2\left(\frac{a}{c}\right)a_1$$

$$c_T = \frac{19.4}{1 + (100/\hat{T})^{3.5}} - 1 \qquad \tau_T = \frac{1}{1 + 60(t'_T)^{-0.69}} + 0.78 \qquad (7.123)$$

$$n_T = B_T n \qquad B_T = \frac{0.25}{1 + (74/\hat{T})^7} + 1 \qquad \hat{T} = T - 253.2.$$

Here $f'$ and $E$ must be in ksi, $T$ in degrees Kelvin, $\hat{T}$ in degrees Celsius. The composition effects for drying creep are estimated as follows:

$$c_p = 0.83 \qquad c_d = 2.8 - 7.5n$$

$$\text{For } r > 0: \quad \phi_d = 0.008 + 0.027u \qquad u = \frac{1}{1 + 0.7r^{-1.4}}$$

$$r = 56000\left(\frac{s}{a}f'_c\right)^{0.3}\left(\frac{g}{s}\right)^{1.3}\left(\frac{w/c}{\varepsilon_{s\infty}}\right)^{1.5} - 0.85; \qquad (7.124)$$

for $r \leqslant 0: \quad \phi_d = 0.008$.

A simplified version of the BP Model can be found in Bažant and Panula.[45]

### 7.7.4 Material characterization for a general purpose program

Different characterizations of creep and shrinkage may be appropriate in various situations. For the input of material properties, the following scheme, used in one recent finite element program[46] and listed in full in Ref. 199, may be provided for the input of material properties.

The data input subroutine, MATPAR,[199] has the following options:

(1) $J(t, t')$ is specified as an array of values. No drying.
(2) $\bar{J}(t, t')$ and $\varepsilon_S(t, t_0)$ are specified as an array. Drying.
(3) $J(t, t')$ is given by double power law, for which all parameters are given, no drying.
(4) Same as (3) but all double power law parameters except $E_{c_{28}}$ are generated from the given strength and composition parameters.
(5) Same as (4) except that $E_{c_{28}}$ is also predicted from the strength and composition parameters.
(6) $J(t, t')$ is defined by the double power law plus drying term $C_d(t, t')$, and shrinkage is given by a formula. All parameters are given.
(7) Same as (6) but all parameters except $E_{c_{28}}$ and $\varepsilon_{sh_\infty}$ are predicted from the strength and composition.
(8) Same as (6) but all parameters except $E_{c_{28}}$ are predicted from the strength and composition.
(9) Same as (6) but all parameters are predicted from the strength and composition.
(10) The double power law parameters $E_0$ and $\phi_1$ are determined by the best fit of the given, array of values $J(t, t')$ which may be of limited range; $m, n, \alpha$ are given. No drying. Coefficient of variation for the deviations from given $J(t, t')$ is computed and printed.
(11) Same as (10) but $m, n, \alpha$ are predicted from given strength and composition.
(12) Same as (10) but drying is included.
(13) Same as (11) but drying is included.

The subroutine for evaluating the compliance function, COMPLF, has the following options:[199]

(1) $J(t, t')$ is evaluated by interpolation or extrapolation from a given array of values.
(2) $J(t, t')$ is evaluated from a formula without the drying term.
(3) $J(t, t')$ is evaluated from a formula with the drying term.

*Subroutine for Dirichlet series expansion,*[199] *DIREX*

The coefficients $\hat{E}_\mu(t')$ of Dirichlet series expansion of $J(t, t')$ or $R(t, t')$ at various discrete times are automatically generated from $J(t, t')$. Then, as a check, the values of $J(t, t')$ are calculated from the Dirichlet series expansion of $J(t, t')$ or $R(t, t')$, and the coefficient of variation of their deviations from the originally given $J(t, t')$ is computed and printed.

In case that Dirichlet series expansion of $R(t, t')$ is used, this subroutine[199] consists of subroutine RELAX that computes the discrete values of $R(t, t')$ from $J(t, t')$, subroutine MAXW that computes the discrete values of the moduli $E_\mu(t')$ of the Maxwell chain, and subroutine CRCURV that computes for a check the discrete values of the creep curves back from the discrete values of $E_\mu(t')$ and evaluates the coefficient of variation of deviations.

The subroutine for shrinkage function, SHRF, has the options:

(1)  $\bar{\varepsilon}_S(t, t_0)$ is evaluated by interpolation or extrapolation from given array of values.
(2)  $\bar{\varepsilon}_S(t, t_0)$ is evaluated from a formula.

*Subroutine for Dirichlet series coefficients or Maxwell chain moduli $E_\mu(t')$*

These coefficients at any time are evaluated by interpolation from the values of $E_\mu$ at certain discrete times.

### 7.7.5  Proof of age-adjusted effective modulus method[23]

Assume that the strain in excess of the shrinkage strain $\varepsilon^0(t)$ varies linearly with $J(t, t_0)$. This means that it also varies linearly with $\phi(t, t_0)$, i.e.

$$\varepsilon(t) - \varepsilon^0(t) = \varepsilon_0 + c\phi(t, t_0) \quad \text{(for } t > t_0\text{)} \tag{7.125}$$

and $\varepsilon(t) - \varepsilon^0(t) = 0$ for $t < t_0$. Substituting $\phi(t, t_0) = E(t_0)J(t, t_0) - 1$, and noting that, by definition, $J(t, t_0) = \mathbf{E}^{-1}H(t - t_0)$ where $\mathbf{E}^{-1} = $ creep operator such that Equation (7.16) has the form $\varepsilon(t) = \mathbf{E}^{-1}\sigma(t) + \varepsilon^0(t)$, and $H(t - t_0) = $ Heaviside step function ($= 1$ for $t > t_0$, $0$ for $t < t_0$), Equation (7.125) may be rewritten as

$$\varepsilon(t) - \varepsilon^0(t) = (\varepsilon_0 - c)H(t - t_0) + cE(t_0)\mathbf{E}^{-1}H(t - t_0) \tag{7.126}$$

Observing that

$$\mathbf{E}H(t - t_0) = R(t, t_0), \qquad \mathbf{E}\mathbf{E}^{-1}H(t - t_0) = H(t - t_0) \tag{7.127}$$

where $\mathbf{E} = $ relaxation operator such that Equation (7.19) has the form $\sigma(t) = \mathbf{E}[\varepsilon(t) - \varepsilon^0(t)]$, we may apply operator $\mathbf{E}$ to both sides of Equation

(7.126). This yields

$$\sigma(t) = (\varepsilon_0 - c)R(t, t_0) + cE(t_0) \qquad (7.128)$$

We conclude that if the strain varies linearly with $J(t, t_0)$ or $\phi(t, t_0)$, then the stress varies linearly with $R(t, t_0)$.

Denote now

$$\Delta\sigma(t) = \sigma(t) - \sigma(t_0), \qquad \Delta\varepsilon(t) = \varepsilon(t) - \varepsilon(t_0), \qquad \Delta\varepsilon^0(t) = \varepsilon^0(t) - \varepsilon^0(t_0) \qquad (7.129)$$

Substituting this and the relations

$$\varepsilon_0 = \frac{\sigma(t_0)}{E(t_0)} + \varepsilon^0(t_0), \qquad c = \frac{\Delta\varepsilon(t) - \Delta\varepsilon^0(t)}{\phi(t, t_0)} \qquad (7.130)$$

into Equation (7.128), we obtain

$$\sigma(t_0) + \Delta\sigma(t) = \left[\frac{\sigma(t_0)}{E(t_0)} + \varepsilon^0(t_0)\right]R(t, t_0) + [E(t_0) - R(t, t_0)]\frac{\Delta\varepsilon(t) - \Delta\varepsilon^0(t)}{\phi(t, t_0)} \qquad (7.131)$$

or

$$\Delta\sigma = \frac{E(t_0) - R(t, t_0)}{\phi(t, t_0)}(\Delta\varepsilon - \Delta\varepsilon^0 - \Delta\varepsilon'') \quad \text{with} \quad \Delta\varepsilon'' = \frac{\sigma(t_0)}{E(t_0)}\phi(t, t_0) \qquad (7.132)$$

This is identical to Equations (7.28)–(7.29), which completes the proof.[23]

### 7.7.6  Sign of $\partial^2 J/\partial t \, \partial t'$ for Maxwell chain

The fact that for Maxwell chain model the sign of this mixed derivative is not restricted to be positive (Section 7.5.1) was proven by Bažant and Kim.[33] A shorter proof may be given as follows. We consider a strain history $\varepsilon(t)$ that starts with a jump at $t'$ and is smooth afterwards. Equation (7.19) may then be written as

$$R(t, t')\varepsilon(t') + \int_{t'+}^{t} R(t, \tau)\frac{d\varepsilon(\tau)}{d\tau}\,d\tau = \sigma(t) \qquad (7.133)$$

In particular we consider that $\varepsilon(t) = J(t, t')$, in which case $\sigma = 1$ for $t > t'$. Equation (7.133) then becomes

$$\frac{R(t, t')}{E(t')} + \int_{t'+}^{t} R(t, \tau)\frac{\partial J(\tau, t')}{\partial\tau}\,d\tau = 1 \qquad (7.134)$$

where $E(t') = R(t', t')$. Now we substitute Equation (7.40) for $R(t, t')$ according to the Maxwell chain model, and we get

$$\frac{1}{E(t')} \sum_\mu E_\mu(t') \exp\left[y_\mu(t') - y_\mu(t)\right] + \int_{t'+}^{t} \sum_\mu E_\mu(\tau)$$

$$\times \exp\left[y_\mu(\tau) - y_\mu(t)\right] \frac{\partial J(\tau, t')}{\partial \tau} d\tau = 1 \quad (7.135)$$

where $E(t') = \sum_\mu E_\mu(t')$. Differentiating this first with respect to $t$ we get rid of the integral, and differentiating again with respect to $t'$ we obtain

$$\frac{\partial^2 J(t, t')}{\partial t\, \partial t'} = \frac{1}{\sum_\mu E_\mu(t)} \sum_\mu \dot{y}_\mu(t) \exp\left[-y_\mu(t)\right] \frac{d}{dt'} \left(\frac{E_\mu(t')}{\sum_\mu E_\mu(t')} \exp\left[y_\mu(t')\right]\right)$$

$$(7.136)$$

Here $E_\mu(t')$ and $y_\mu(t')$ are increasing functions, but because of the increasing sum $\sum_\mu E_\mu(t')$ in the denominator, there is no reason for this expression to be always non-negative.

## 7.8 SUMMARY AND CONCLUSIONS

The greatest uncertainty in the analysis of creep and shrinkage effects in concrete structures stems from material characterization, both the numerical values of material parameters defining creep and shrinkage and the form of the constitutive relations. The determination of creep at constant stress is described first and various practical prediction models are given. For the case of exposed concrete subjected to drying, the existing models do not specify material properties but the mean compliance function and the mean shrinkage of the entire drying cross section. Such material characteristics allow determining the mean forces and deformations in the cross sections of beams, frames, plates or shells but not the stresses and strains at various points of the cross section.

As for the constitutive relations applicable at arbitrarily variable stress, only the linear theory based on the principle of superposition is well developed. Various forms of the integral-type constitutive relations, based on the compliance function, the impulse memory function, the relaxation function, and their multiaxial forms are outlined. Subsequently, step-by-step-algorithms based on approximating the history integral by a finite sum are presented. Finally, a simple method called the age-adjusted effective modulus method, which allows an easy approximate determination of creep and shrinkage effects by means of a single elastic finite element analysis, is indicated.

For the analysis of large structural systems it is necessary to avoid history integrals and use an equivalent rate-type formulation. Such a formulation is generally obtained by expanding the compliance function or the relaxation function into a series of real exponentials, called a Dirichlet series. This representation is equivalent to assuming an age-dependent Kelvin or Maxwell chain model. The rate-type formulation also allows a simple extension to variable temperature, in which the activation energy concept is used to model both the acceleration of creep rate due to heating and the acceleration of aging (hydration) which offsets the increase in creep rate. Special step-by-step algorithms, called exponential algorithms, are required to allow an unrestricted increase of the time step as the rate of change of stresses and strains declines with the passage of time.

Most of the discrepancies between measurements and linear theory applications can be traced back to various non-linear effects. First, the phenomenon of aging in the context of linear constitutive relations leads to certain violations of thermodynamic restrictions relative to the dissipation of energy in the chemical hydration process. It seems that such violations cannot be avoided without passing to a non-linear theory. The main non-linear phenomena in creep are the flow, which consists in an increase of creep well beyond proportionality as the stress approaches the strength limit, and the adaptation or stiffening non-linearity, which describes the stiffening of the material due to previous sustained compression. A proper model for cracking and tensile non-linear behaviour is also an important ingredient of a finite element program if realistic results should be obtained.

The most complicated aspect of concrete creep is the moisture effect. The pore humidity as well as temperature affect the rate of aging (hydration). Shrinkage, when considered as a constitutive (material) property rather than a cross-section mean property, is not a function of time but a function of pore humidity or specific water content. Regarding the constitutive relations for creep at the presence of drying, it is not clear at present whether the acceleration of creep observed at drying is due mainly to microcracking and tensile non-linear behaviour, or whether some intrinsic mechanism on the microscale, such as, for example, the diffusion of moisture (water) between gel micropores and capillary macropores, causes a significant increase of creep rate. Calculation of creep and shrinkage effects requires, of course, numerical determination of pore humidity distributions at various times. For this purpose a non-linear diffusion model, which agrees with experiments relatively well, is available. When both water content and temperature vary in time and space, a coupled moisture and heat transfer must be considered.

Overall, it may be concluded that the theory of creep and shrinkage has seen a tremendous progress during the last decade. However, a number of important questions are still open and much further research, which is likely to lead to many revisions in the foregoing presentation, will have to be carried out.

246 *Creep and Shrinkage in Concrete Structures*

## ACKNOWLEDGEMENT

Financial support by the US National Science Foundation provided through Grant No. CME-8009050 to Northwestern University is gratefully acknowledged.

## REFERENCES

1. Aleksandrovskii, S. V. (1966). *Analysis of Plain and Reinforced Concrete Structures for Temperature and Moisture Effects with Account of Creep*, (in Russian), Stroyizdat, Moscow, 443 pp.
2. Aleksandrovskii, S. V., and Kolensnikov, N. S. (1971), 'Non-linear creep of concrete at step-wise varying stress' (in Russian), *Beton i Zhelezobeton*, **16**, 24–7.
3. Aleksandrovskii, S. V., and Popkova, O. M. (1970), 'Nonlinear creep strains of concrete at complex load histories' (in Russian), *Beton i Zhelezobeton*, **16**, 27–32.
4. Ali, J., and Kesler, C. E. (1963), 'Creep in concrete with and without exchange of moisture with the environment', *Univ. Illinois, Urbana, Dept Theor. Appl. Mech., Rep. No.* 641.
5. American Concrete Institute Comm. 209, Subcomm. II (1971), 'Prediction of creep, shrinkage and temperature effects in concrete structures', in *Designing for Effects of Creep, Shrinkage and Temperature, Am. Concr. Inst. Spec. Publ. No. 27.*
6. Anderson, C. A. (1980), 'Numerical creep analysis of structures', *Los Alamos Scientific Laboratory Rep.* LA-UR-80-2585, 1980, also in Z. P. Bažant and F. H. Wittman (Eds) *Creep and Shrinkage in Concrete Structures*, Wiley, London, pp. 259–303.
7. Argyris, J. H., Pister, K. S., Szimmat, J., and Willam, K. J. (1977), 'Unified concepts of constitutive modelling and numerical solution methods for concrete creep problems', *Comput. Meth. Appl. Mech. Eng.*, **10**, 199–246.
8. Argyris, J. H., Vaz, L. E., and Willam, K. J. (1978), 'Improved solution methods for inelastic rate problems', *Comput. Meth. Appl. Mech. Eng.*, **16**, 231–277.
9. Arthanari, S., and Yu, C. W. (1967), 'Creep of concrete under uniaxial and biaxial stresses at elevated temperatures', *Mag. Concr. Res.*, **19**, 149–56.
10. Arutyunian, N. Kh. (1952), *Some Problems in the Theory of Creep* (in Russian), Techteorizdat, Moscow; Engl. Transl. Pergamon Press (1966).
11. Bažant, Z. P., (1962), *Theory of Creep and Shrinkage of Concrete in Nonhomogeneous Structures and Cross Sections* (in Czech), Stavebnícky Časopis SAV **10**, pp. 552–76.
12. Bažant, Z. P. (1964), 'Die Berechnung des Kriechens und Schwindens Nicht-omogener Betonkonstruktionen', *Proc. 7th Congr., Int. Assoc. for Bridge and Struct. Eng., Rio de Janeiro*, pp. 887–96.
13. Bažant, Z. P. (1966a), *Creep of Concrete in Structural Analysis* (in Czech), State Publishers of Technical Literature SNTL, Prague.
14. Bažant, Z. P. (1966b), 'Phenomenological theories for creep of concrete based on rheological models', *Acta Technica ČSAV, Prague*, **11**, 82–109.
15. Bažant, Z. P. (1968), 'Langzeitige Durchbiegungen von Spannbetonbrücken infolge des Schwingkriechens unter Verkehrslasten', *Beton und Stahlbetonbau* **63**, 282–5.

16. Bažant, Z. P. (1970a), 'Constitutive equation for concrete creep and shrinkage based on thermodynamics of multiphase systems', *Materials and Structures* (RILEM, Paris), **3**, 3–36; see also Rep. 68/1, Dept Civ. Eng., University of Toronto (1968).
17. Bažant, Z. P. (1970b), 'Delayed thermal dilatations of cement paste and concrete due to mass transport', *Nucl Eng. Des.*, **14**, 308–18.
18. Bažant, Z. P. (1970c), 'Numerical analysis of creep of an indeterminate composite beam', *J. Appl. Mech.*, *ASME*, **37**, 1161–4.
19. Bažant, Z. P. (1971a), 'Numerical analysis of creep of reinforced plates', *Acta Technica Hung.*, **70**, 415–8.
20. Bažant, Z. P. (1971b), 'Numerical solution of non-linear creep problems with application to plates', *Int. J. Solids Struct.*, **7**, 83–97.
21. Bažant, Z. P. (1971c), 'Numerically stable algorithm with increasing time steps for integral-type aging creep', *Proc. 1st Int. Conf. on Structural Mechanics in Reactor Technology*, *West Berlin*, Vol. 3, Paper H2/3.
22. Bažant, Z. P. (1972a), 'Numerical determination of long-range stress history from strain history in concrete', *Materials and Structures* RILEM, Paris, **5**, 135–41.
23. Bažant, Z. P. (1972b), 'Prediction of concrete creep effects using age-adjusted effective modulus method', *Am. Concr. Inst. J.*, **19**, 212–7.
24. Bažant, Z. P. (1972c), 'Thermodynamics of interacting continua with surfaces and creep analysis of concrete structures', *Nucl. Eng. Des.*, **20**, 477–505; see also *Cem. Concr. Res.*, **2**, 1–16.
25. Bažant, Z. P. (1975), 'Theory of creep and shrinkage in concrete structures: a précis of recent developments', *Mech. Today*, **2**, 1–93.
26. Bažant, Z. P. (1977), 'Viscoelasticity of porous solidifying material—concrete', *J. Eng. Mech. Div. ASCE*, **102**, 1049–67.
27. Bažant, Z. P. (1979), 'Thermodynamics of solidifying or melting viscoelastic material', *J. Eng. Mech. Div. ASCE*, **105**, 933–52.
28. Bažant, Z. P., and Asghari, A. (1974), 'Computation of age-dependent relaxation spectra', *Cem. Concr. Res.*, **4**, 567–79; see also 'Computation of Kelvin-chain retardation spectra', *Cem. Concr. Res.*, **4**, 797–806.
29. Bažant, Z. P., and Asghari, A. A. (1977), 'Constitutive law for non-linear creep of concrete', *J. Eng. Mech. Div. ASCE*, **103**, No. EM1, Proc. Paper 12729, 113–24.
30. Bažant, Z. P., Carreira, D. J., and Walser, A. (1975), 'Creep and shrinkage in reactor containment shells', *J. Struct. Div. ASCE*, **202** No. ST10, Proc. Paper 11632, 2117–31.
31. Bažant, Z. P., and Chern, J. C. (1982), 'Comment on the use of Ross' hyperbola and recent comparisons of various practical creep prediction models', *Cem. Concr. Res.*, **12**, 527–532.
32. Bažant, Z. P., Chern, J. C., and Thonguthai, W. (1981), 'Finite element program for moisture and heat transfer in heated concrete', *Nucl. Eng. Des.*, **68,** 61–70.
33. Bažant, Z. P., and Kim, S. S. (1978), 'Can the creep curves for different ages at loading diverge?', *Cem. Concr. Res.*, **8**, No. 5, 601–12.
34. Bažant, Z. P., and Kim, S. S. (1979a), 'Approximate relaxation function for concrete', *J. Struct. Div. ASCE*, **105**, ST12, 2695–705.
35. Bažant, Z. P., and Kim, S. S. (1979b), 'Nonlinear creep of concrete—adaptation and flow', *J. Eng. Mech. Div.*, **105**, EM3, 419–46.
36. Bažant, Z. P., Kim. S. S., and Meiri, S. (1979), 'Triaxial moisture-controlled

creep tests of hardened cement paste at high temperature', *Materials and Structures* (RILEM, Paris), **12,** 447–56.

37. Bažant, Z. P., and Najjar, L. J. (1972), 'Nonlinear water diffusion in nonsaturated concrete', *Materials and Structures*, (RILEM, Paris) **5,** 3–20.
38. Bažant, Z. P., and Najjar, L. J. (1973), 'Comparison of approximate linear methods for concrete creep', *J. Struct. Div. ASCE*, **99,** 1851–74.
39. Bažant, Z. P., and Oh, B. (1980), 'Strain-rate effect in rapid nonlinear triaxial deformation of concrete', *Struct. Eng. Rep. No. 80–8/640s, Northwestern University, Evanston, Ill.*; also *J. Eng. Mech. Div. ASCE*, **108,** Nov. 1982.
40. Bažant, Z. P., and Osman, E. (1975), 'On the choice of creep function for standard recommendations on practical analysis of structures', *Cem. Concr. Res.*, **5,** 631–41; 1976, **6,** 149–57; 1977, **7,** 111–30; 1978, **8,** 129–30.
41. Bažant, Z. P., and Osman, E. (1976), 'Double power law for basic creep of concrete', *Materials and Structures*, (RILEM, Paris) **9,** No. 49, 3–11.
42. Bažant, Z. P., Osman, E., and Thonguthai, W. (1976), 'Practical formulation of shrinkage and creep of concrete', *Materials and Structures*, (RILEM, Paris) **9,** No. 54, 395–406.
43. Bažant, Z. P., and Panula, L. (1978a), 'Practical prediction of creep and shrinkage of concrete', *Materials and Structures*, (RILEM, Paris), Parts I and II, No. 69, 1978, 415–34; Parts V and VI, **12,** No. 72, 1979.
44. Bažant, Z. P., and Panula, L. (1978b), 'New model for practical prediction of creep and shrinkage', Presented at *A. Pauw Symp. on Creep, ACI Convention, Houston, October* 1978, to be published as ACI Special Publication, 1982.
45. Bažant, Z. P., and Panula, L. (1980), 'Creep and shrinkage chracterization for analyzing prestressed concrete structures, *Prestr. Concr. Inst. J.*, **25,** No. 3, 86–122.
46. Bažant, Z. P., Rossow, E. C., and Horrigmoe, G. (1981), 'Finite element program for creep analysis of concrete structures', *Proc. 6th Int. Conf. on Structural Mechanics in Reactor Technology, Paris,* Paper H2/1.
47. Bažant, Z. P., and Thonguthai, W. (1976), 'Optimization check of certain recent practical formulations for concrete creep', *Materials and Structures,* (RILEM, Paris), **9,** 91–6.
48. Bažant, Z. P., and Thonguthai, W. (1978), 'Pore pressure and drying of concrete at high temperature', *J. Eng. Mech. Div. ASCE,* **104,** EM5, 1059–79.
49. Bažant, Z. P. and Thonguthai, W. (1979), 'Pore pressure in heated concrete walls: theoretical prediction', *Mag. Concr. Res.*, **32,** No. 107, 67–76.
50. Bažant, Z. P., and Tsubaki, T. (1980), 'Weakly singular integral for creep rate of concrete', *Mech. Res. Commun.*, **7,** 335–40.
51. Bažant, Z. P., and Wu, S. T. (1973), 'Dirichlet series creep function for aging concrete', *J. Eng. Mech. Div. ASCE*, **99,** No. EM2, Proc. Paper 9645.
52. Bažant, Z. P., and Wu, S. T. (1974a), 'Creep and shrinkage law for concrete at variable humidity', *J. Eng. Mech. Div. ASCE*, **100,** EM6, 1183–209.
53. Bažant, Z. P., and Wu, S. T. (1974b), 'Rate-type creep law of aging concrete based on Maxwell chain', *Materials and Structures*, (RILEM, Paris), **7,** 45–60.
54. Bažant, Z. P., and Wu, S. T. (1974c), 'Thermoviscoelasticity of aging concrete', *J. Eng. Mech. Div. ASCE*, **100,** 575–97; also (1973), *ASCE Preprint* 2110.
55. Berwanger, C. (1971), 'The modulus of concrete and the coefficient of thermal expansion below normal temperatures', *Am. Concr. Inst. Spec. Publ. No. 25, Temperature and Concrete, Detroit,* pp. 191–234.
56. Biot, M. A. (1955), 'Variational principles of irreversible thermodynamics with application to viscoelasticity', *Phys. Rev.*, **97,** 1463–69.

57. Boltzmann, Z. (1876), 'Zur Theorie der Elastischen Nachwirkung', *Sitzber. Akad. Wiss.*, *Wiener Bericht* 70, *Wiss. Abh.* **1** (1874), 275–306; see also *Pogg. Ann. Phys.*, **7**, 624.
58. Branson, D. E., Meyers, B. L., and Kripanarayanan, K. M. (1970a), 'Time-dependent deformation of non-composite and composite prestressed concrete structures', *Highway Res. Rec. No.* 324, pp. 15–33.
59. Branson, D. E., Meyers, B. L., and Kripanarayanan, K. M. (1970b), 'Loss of prestress, camber, and deflection of noncomposite and composite structures using different weight concretes', *Final Rep. No.* 70–6 *Iowa Highway Commision, Aug.* 1970, pp. 1–229. Also, condensed papers presented at the *49th Annual Meeting, Highway Research Board, Washington, DC, Jan. 1970*, pp. 1–42, and at the *Sixth Congress, Fédération Internationale de la Précontrainte, Prague*, pp. 1–28.
60. Branson, D. E., and Christianson, M. L. (1971), 'Time-dependent concrete properties related to design strength and elastic properties, creep and shrinkage', *Am. Concr. Inst. Spec. Publ.* SP-27, *Designing for Creep, Shrinkage and Temperature, Detroit*, pp. 257–77.
61. Branson, D. E. (1977), *Deformations of Concrete Structures*, McGraw-Hill, New York.
62. Bresler, B., and Selna, L. (1964), 'Analysis of time-dependent behavior of reinforced concrete structures', *ACI Symp. on Creep of Concrete, ACI Spec. Publ.* SP-9.
63. Brettle, H. J., (1958), 'Increase in concrete modulus of elasticity due to prestress and its effect on beam deflections', *Constructional Rev. (Sydney)*, **31**, 32–5.
64. Browne, R. D. (1967), 'Properties of concrete in reactor vessels', *Proc. Conf. on Prestressed Concrete Reactor Pressure Vessels, Inst. Civ. Eng., London*, Paper 13, pp. 11–13.
65. Browne, R. D., and Blundell, R. (1969), 'The influence of loading age and temperature on the long term creep behavior of concrete in a sealed, moisture stable state', *Materials and Structures*, (RILEM, Paris) **2**, 133–44.
66. Browne, R. D., and Bamforth, P. P. (1975), 'The long term creep of the Wylfa P. V. concrete for loading ages up to $12\frac{1}{2}$ years', *Proc. 3rd Int. Conf. on Structural Mechanics in Reactor Technology, London*, Paper H1/8.
67. Carlson, R. W. (1937), 'Drying shrinkage of large concrete members', *J. Am. Concr. Inst.*, **33**, 327–36.
68. CEB-FIP (1978), *Model Code for Concrete Structures*, Comité Euro-International du Béton, Paris, Vol. 2, Appendix e.
69. Cederberg, H. and Davis, M. (1969), 'Computation of creep effects in prestressed concrete pressure vessels using dynamic relaxation', *Nucl. Eng. Des.*, **9**, 439–48.
70. Chiorino, M. A., and Levi, R. (1967), 'Influence de l'élasticité différée sur le régime des contraintes des constructions en béton', *Cah. Rech. No.* 24, Institut Technique du Bâtiment et des Travaux Publics, Eyrolles, Paris, France (see also Giornale del Genio Civile, 1967; and Academia Nazionale dei Lincei, Fasc. 5, Seri. 8, **28**, May, 1965).
71. Çinlar, E., Bažant, Z. P., and Osman, E. (1977), 'Stochastic process for extrapolating concrete creep', *J. Eng. Mech. Div., ASCE*, **103**, No. EM6, Proc. Paper 13447, 1069–88.
72. Copeland, L. E., Kantro, D. L. and Verbeck, G. (1960), *Chemistry of Hydration of Portland Cement, Natl. Bur. Stand. Monogr.* 43, 4th Int. Symp. on the Chemistry of Cement, Washington, DC, pp. 429–65.

73. Cost, T. M., (1964), 'Approximate Laplace transform inversions in viscoelastic stress analysis', *AIAA J.* **2**, 2157–66.
74. Cottrell, A. H. (1965), *The Mechanical Properties of Matter*, Wiley, New York.
75. Cruz, C. R. (1967), 'Elastic properties of concrete at high temperatures', *J. Portland Cem. Assoc. Res. Dev. Lab.*, **9**, 37–45.
76. Davies, R. D., (1957), 'Some experiments on the applicability of the principle of superposition to the strain of concrete subjected to changes of stress, with particular reference to prestressed concrete', *Mag. Concr. Res.*, **9**, 161–72.
77. Dischinger, F. (1937), 'Untersuchungen über die Knicksicherheit, die Elastische Verformung und das Kriechen des Betons bei Bogenbrücken', *Der Bauingenieur*, **18**, 487–520, 539–52, 595–621.
78. Dischinger, F., (1939), 'Elastische und plastische Verformungen bei Eisenbetontragwerke', *Der Bauingenieur*, **20**, 53–63, 286–94, 426–37, 563–72.
79. England, G. L., and Illston, J. M. (1965), 'Methods of computing stress in concrete from a history of measured strain', *Civ. Eng. Publ. Works Rev.*, 513–7, 692–4, 845–7.
80. Faber, H. (1927–28), 'Plastic yield, shrinkage and other problems of concrete and their effect on design', *Minutes Proc. Inst. Civ. Eng.*, **225**, London, England, pp. 27–76; disc. pp. 75–130.
81. Fahmi, H. M., Polivka, M., and Bresler, B. (1972), 'Effect of sustained and cyclic elevated temperature on creep of concrete', *Cem. Concr. Res.*, **2**, 591–606.
82. Ferry, J. D., (1970), *Viscoelastic Properties of Polymers*, 2nd Edn, Wiley, New York.
83. Freudenthal, A. M., and Roll, F. (1958), 'Creep and creep recovery of concrete under high compressive stress', *J. Am. Concr. Inst.*, **54**, 1111–42.
84. Gaede, K. (1962), 'Versuche über die Festigkeit und die Verformungen von Beton bei Druck-Schwellbeanspruchung', *Deutscher Ausschuss für Stahlbeton, Schriftenr. Heft* 144.
85. Glanville, W. H., (1930), 'Studies in reinforced concrete III—creep or flow of concrete under load', *Building Res. Tech. Paper No.* 12, *Dept. Sci. Ind. Res.*, London; also (1933), *Struct. Eng.*, **II**, 54.
86. Glasston, S., Laidler, K. J., and Eyring, H. (1941), *The Theory of Rate Processes*, McGraw-Hill, New York.
87. Glucklich, J., and Ishai, O. (1962), 'Creep mechanism in cement mortar', *J. Am. Concr. Inst.*, **59**, 923–48.
88. Hannant, D. J. (1967), 'Strain behavior of concrete up to 95 °C under compressive stresses', *Proc. Conf. on Prestressed Concrete Pressure Vessels*, Group C, Paper 17, Institution of Civil Engineers, London, pp. 57–71.
89. Hannant, D. (1968), 'The mechanism of creep of concrete', *Mater. Struct.*, (RILEM, Paris) **1**, 403–10.
90. Hansen, T. C. (1960), 'Creep and stress relaxation in concrete', *Proc., Swed. Cem. Concr. Res. Inst.* (CBI), *Royal Inst. Tech.*, Stockholm, No. 31.
91. Hansen, T. C. (1964), 'Estimating stress relaxation from creep data', *Mater. Res. Stand. (ASTM)*, **4**, 12–14.
92. Hansen, T. C., and Mattock, A. H. (1966), 'Influence of size and shape of member on the shrinkage and creep of concrete', *J. Am. Concr. Inst.*, **63**, 267–90.
93. Harboe, E. M., et al. (1958), 'A comparison of the instantaneous and the sustained modulus of elasticity of concrete', *Concr. Lab. Rep. No.* C-854, Division of Engineering Laboratories, US Department of the Interior, Bureau of Reclamation, Denver, Colo.

94. Hanson, J. A. (1953), 'A 10-year study of creep properties of concrete', *Concr. Lab. Rep. No.* Sp-38, US Department of the Interior, Bureau of Reclamation, Denver, Colorado.
95. Hanson, J. A. (1968), 'Effects of curing and drying environments on splitting tensile strength', *J. Am. Concr. Inst.*, **65**, 535–43 (also *PCA Bull.* D141).
96. Hardy, G. M., and Riesz, M. (1915), *The General Theory of Dirichlet Series, Cambridge Tracts in Mathematics and Mathematical Physics, No.* 18, Cambridge Univ. Press, Cambridge.
97. Hatt, W. K. (1907), 'Notes on the effect of time element in loading reinforced concrete beams', *Proc. ASTM*, **7**, 421–33.
98. Hellesland, J., and Green, R. (1972), 'A stress and time-dependent strength law for concrete', *Cem. Concr. Res.*, **2**, 261–75.
99. Hickey, K. B. (1967), 'Creep, strength, and elasticity of concrete at elevated temperatures', *Rep. No.* C-1257, *Concr. Struct. Branch*, Division of Research, United States Department of the Interior, Bureau of Reclamation, Denver, Colorado.
100. Hirst, G. A., and Neville, A. M. (1977), 'Activation energy of creep of concrete under short-term static and cyclic stresses', *Mag. Concr. Res.*, **29**, No. 98, 13–18.
101. Honk, I. E., Orville, E. B., and Houghton, D. L. (1969), 'Studies of autogeneous volume change in concrete for Dworshak Dam', *J. Am. Concr. Inst.*, **66**, 560–8.
102. Huet, C. (1980), 'Adaptation of Bazant's algorithm to the analysis of aging viscoelastic composite structures', *Materials and Structures*, (RILEM, Paris), **13**, No. 75, 91–98, (in French).
103. Illston, J. M., and Jordaan, I. J. (1972), 'Creep prediction for concrete under multiaxial stress', *J. Am. Concr. Inst.*, **69**, 158–64.
104. Illston, J. M., and Sanders, P. D. (1973), 'The effect of temperature upon the creep of mortar under torsional loading', *Mag. Concr. Res.*, **25**, No. 84, 136–66.
105. Ishai, O. (1964), 'Elastic and inelastic behavior of cement mortar in torsion', *Am. Concr. Inst. Spec. Publ.* SP-9, *Symp. on Creep, Detroit*, pp. 65–94, 115–28.
106. Jonasson, J. E. (1978), 'Analysis of creep and shrinkage in concrete and its application to concrete top layers', *Cem. Concr. Res.*, **8**, 397–518.
107. Kabir, A. F. (1976), 'Nonlinear analysis of reinforced concrete panels, slabs and shells for time-dependent effects', *PhD Dissertation*, Division of Structural Engineering and Structural Mechanics, University of California, Berkeley, *Rep. No.* UC-SESM 76-6.
108. Kabir, A. F., and Scordelis, A. C. (1979), 'Analysis of RC shells for time dependent effects', *IASS Bull.* **XXI**, No. 69.
109. Kang, Y. J. (1977), 'Nonlinear geometric, material and time-dependent analysis of reinforced and prestressed concrete frames', *PhD Dissertation*, Division of Structural Engineering and Structural Mechanics, University of California, Berkeley, *Rep. No.* UC-SESM 77-1.
110. Kang, Y. J., and Scordelis, A. C. (1980), 'Nonlinear analysis of prestressed concrete frames', *J. Struct. Div. ASCE*, **106**, No. 1.
111. Keeton, J. R. (1965), 'Study of creep in concrete', *Tech. Reps.* R333-I, R333-II, R333-III, *US Naval Civil Engineering Laboratory, Port Hueneme, California*.
112. Kesler, C. E., and Kung, S. H. (1964), 'A study of free and restrained shrinkage of mortar', *T. & A.M. Rep. No.* 647, *Department of Theoretical and Applied Mechanics, University of Illinois, Urbana, Illinois*.

252     *Creep and Shrinkage in Concrete Structures*

113. Kesler, C. E., Wallo, E. M., Yuan, R. L., and Lott, J. L. (1965), 'Prediction of creep in structural concrete from short-time tests', *6th Prog. Rep., Department of Theoretical and Applied Mechanics, University of Illinois, Urbana, Illinois.*
114. Kimishima, H., and Kitahara, H. (1964), 'Creep and creep recovery of mass concrete', *Tech. Rep.* C-64001, *Central Research Institute of Electric Power Industry, Tokyo, Japan.*
115. Klug, P., and Wittmann, T. (1970), 'The correlation between creep deformation and stress relaxation in concrete', *Materials and Structures* (RILEM, Paris) **3,** 75–80.
116. Komendant, G. J., Polivka, M., and Pirtz, D. (1976), 'Study of concrete properties for prestressed concrete reactor vessels', *Final Rep. Part II,* 'Creep and strength characteristics of concrete at elevated temperatures', *Rep. No.* UCSESM76-3, *Structures and Materials Research, Department of Civil Engineering, Report to General Atomic Company, San Diego, California, Berkeley, California.*
117. Lambotte, H., and Mommens, A. (1976), 'L'évolution du fluage du béton en fonction de sa composition, du taux de contrainte et de l'age', *Groupe de travail GT 22, Centre national de recherches scientifiques et techniques pour' l'industrie cimentière, Bruxelles.*
118. Lanczos, C. (1964), *Applied Analysis,* Prentice-Hall, Englewood Cliffs, pp. 272–80.
119. LeCamus, B. (1947), 'Recherches expérimentales sur la déformation du béton et du béton armé, Part II', *Annales Inst. Techn, du Bâtiment et des Travaux Publics, Paris.*
120. Levi, F., and Pizzetti, G. (1951), *Fluage, Plasticité, Précontrainte,* Dunod, Paris.
121. L'Hermite, R. (1970), *Volume Changes of Concrete, US Natl Bur. Stand. Monogr.* 43, *4th Int. Symp. on Chemistry of Cements, Washington,* Vol. 3, pp. 659–94.
122. L'Hermite, R., and Mamillan, M. (1968), 'Retrait et fluage des bétons', *Annales, Inst. Techn. du Bâtiment et des Travaus Publics (Suppl.),* **21,** 1334 (No. 249); and (1969), 'Nouveaux résultats et récentes études sur le fluage du béton', *Materials and Structures,* **2,** 35–41; and Mamillan, M., and Bouineau, A. (1970), 'Influence de la dimension des éprouvettes sur le retrait', *Ann. Inst. Tech. Bat. Trav. Publics (Suppl.),* **23,** 5–6 (No. 270).
123. L'Hermite, R., Mamillan, M., and Lefèvre, C. (1965), 'Nouveaux résultats de recherches sur la déformation et la rupture du béton', *Annales de l'Institut Technique du Bâtiment et des Travaux Publics* **18,** pp. 325–360; see also (1968), *Int. Conf. on the Structure of Concrete,* Cement and Concrete Assoc., pp. 423–33.
124. McDonald, J. E. (1972), 'An experimental study of multiaxial creep in concrete', *Amr. Concr. Inst. Spec. Publ. No. 34, Concrete for Nuclear Reactors,* Detroit, pp. 732–68.
125. McHenry, D. (1943), 'A new aspect of creep in concrete and its application to design', *Proc. ASTM,* **43,** 1069–86.
126. McMillan, F. R. (1916), 'Method of designing reinforced concrete slabs', discussion by A. C. Janni, *Trans. ASCE,* **80,** 1738.
127. Mamillan, M. (1969), 'Evolution du fluage et des propriétés de béton', *Ann. Inst. Tech. Bat. Trav. Publics* **21,** 1033; and (1960), **13,** 1017–52.
128. Mamillan, M., and Lelan, M. (1970), 'Le Fluage de béton', *Ann. Inst. Tech. Bât. Trav. Publics (Suppl.)* **23,** 7–13, (No. 270), and (1968), **21,** 847–50 (No. 246).

129. Mandel, J. (1958), 'Sur les corps viscoélastiques linéaire dont les propriétés dépendent de l'age', *C.R. Séances Acad. Sci.*, **247**, 175–8.
130. Maréchal, J. C. (1969), 'Fluage du béton en fonction de la température', *Materials and Structures* (RILEM, Paris) **2**, 111–15; see also (1970), *Mater. Struct.*, **3**, 395–406.
131. Máréchal, J. C. (1970a), 'Contribution a l'étude des propriétés thermiques et mécaniques du béton en fonction de la température', *Annales de l'Institut Technique du Bâtiment et des Travaux Publics*, **23**, No. 274, 123–145.
132. Maréchal, J. C. (1970b), 'Fluage du béton en fonction de la température', *Ann. Inst. Tech. Bât. Trav. Publics*, **23**, No. 266, 13–24.
133. Maslov, G. N. (1940), 'Thermal stress states in concrete masses, with consideration of concrete creep', (in Russian), *Izvestia Nauchno-Issledovatel'skogo Instituta Gidrotekhniki, Gosenergoizdat*, **28**, 175–88.
134. Mehmel, A., and Kern, E. (1962), 'Elastische und Plastische Stauchungen von Beton infolge Druckschwell-und Standbelastung', *Dtsch. Ausschuss Stahlbeton Schriftenr. Heft* 153.
135. Meyer, H. G. (1969), 'On the influence of water content and of drying conditions on lateral creep of plain concrete', *Mater. Struct.* (RILEM, Paris) **2**, 125–31.
136. Meyers, B. L., and Slate, F. O. (1970), 'Creep and creep recovery of plain concrete as influenced by moisture conditions and associated variables', *Mag. Concr. Res.*, **22**, 37–8.
137. Mörsch, E. (1947), *Statik der Gewölbe und Rahmen*, Teil, A., Wittwer, Stuttgart.
138. Mukaddam, M. (1974), 'Creep analysis of concrete at elevated temperatures', *ACI J.*, **71**, No. 2.
139. Mukaddam, M. A., and Bresler, B. (1972), 'Behavior of concrete under variable temperature and loading', *Am. Concr. Inst. Spec. Publ. No. 34, Concrete for Nuclear Reactors*, Detroit, pp. 771–97.
140. Mullen, W. G., and Dolch, W. L. (1966), 'Creep of Portland cement paste', *Proc. ASTM*, **64**, 1146–70.
141. Mullick, A. K. (1972), 'Effect of stress-history on the microstructure and creep properties of maturing concrete', Ph.D. Thesis, *University of Calgary*, Alberta, Canada.
142. Nasser, K. W., and Neville, A. M. (1965), 'Creep of concrete at elevated temperatures', *J. Am. Concr. Inst.*, **62**, 1567–79.
143. Nasser, J. W., and Neville, A. M. (1967), 'Creep of old concrete at normal and elevated temperatures', *J. Am. Concr. Inst.*, **64**, 97–103.
144. Neville, A. M. (1973), *Properties of Concrete*, 2nd Edn., Wiley, New York.
145. Neville, A. M., and Dilger, W. (1970), *Creep of Concrete, Plain, Reinforced, Prestressed*, North-Holland, Amsterdam.
146. Nielsen, L. F. (1970), 'Kriechen und relaxation des betons', *Beton-und Stahlbetonbau*, **65**, 272–5.
147. Nielsen, L. F. (1977), 'On the applicability of modified Dischinger equations', *Cem. Concr. Res.*, **7**, 149; (discussion, **8**, 117).
148. Pickett, G. (1946), 'The effect of change in moisture content on the creep of concrete under a sustained load', *J. Am. Concr. Inst.*, **36**, 333–5; see also 'Shrinkage stresses in concrete', *J. Am. Concr. Inst.*, **42**, 165–204, 361–400.
149. Pirtz, D. (1968), 'Creep characteristics of mass concrete for Dworshak Dam', *Rep. No. 65-2, Structural Enginerring Laboratory, University of California*, Berkeley.

254 *Creep and Shrinkage in Concrete Structures*

150. Pister, K. S., Argyris, J. H., and Willam, K. J. (1976), 'Creep and shrinkage of aging concrete, *ACI Symp. on Concrete and Concrete Structures*, 24–29 Oct. 1976, *Mexico City*, ACI SP55–1.
151. Powers, T. C. (1965), 'Mechanism of shrinkage and reversible creep of hardened cement paste', *Proc. Int. Conf. on the Sstructure of Concrete, London*, Cem. Concr. Assoc., London, pp. 319–44.
152. Powers, T. C. (1966), 'Some observations on the interpretation of creep data', *RILEM Bull.* (Paris) No. 33, 381–91.
153. Powers, T. C. and Brownyard, T. C. (1946), 'Studies of the physical properties of hardened Portland cement paste', *J. Am. Concr. Inst.* **42**, 101–32, 249–366, 469–504; **43** (1947) 549–602, 669–712, 854–80, 933–92.
154. Rashid, Y. R. (1972), 'Nonlinear analysis of two-dimensional problems in concrete creep', *J. Appl. Mech.*, **39**, 475–82.
155. Rice, J. (1971), 'Inelastic constitutive relations for solids: an internal variable theory and its application to metal plasticity', *J. Mech. Phys. Solids*, **19**, 433–55.
156. Roll, F. (1964), 'Long-time creep-recovery of highly stressed concrete cylinders', *Am. Concr. Inst. Spec. Publ.* SP-9, *Symp. on Creep, Detroit*, pp. 115–28.
157. Roscoe, R. (1950), 'Mechanical models for the representation of viscoelastic properties', *Br. J. Appl. Phys.*, **1**, 171–3.
158. Ross, A. D. (1958), 'Creep of concrete under variable stress', *J. Am. Concr. Inst.*, **54**, 739–58.
159. Rostasy, F. S., Teichen, K. Th., and Engleke, H. (1972), 'Beitrag zur Klärung der Zusammenhanges von Kriechen und Relaxation bei Normal-beton', *Amtliche Forschungs-und Materialprufüngsanstalt für das Bauwesen, Otto-Graf-Institut, Universität, Stuttgart, Strassenbau und Strassenverkehrstechnik, Heft* 139.
160. Ruetz, W. (1968), 'A hypothesis for the creep of hardened cement paste and the influence of simultaneous shrinkage', *Proc. Int. Conf. on the Structure of Concrete, London, 1965*, Cem. Concr. Assoc., London, pp. 365–87; see also (1966), *Deutscher Ausschuss für Stahlbeton, Schriftenr. Heft* 183.
161. Rüsch, H., *et al.* (1968), 'Festigkeit und Verformung von unbewehrten Beton unter konstanter Dauerlast', *Deutscher Ausschuss für Stahlbeton, Schriftenr. Heft* 198; see also (1968), *J. Am. Concr. Inst.* **57**, 1–58.
162. Rüsch, H., Jungwirth, D., and Hilsdorf, H. (1973), 'Kritische Sichtung der Verfahren zur Berücksightigung der Einfüsse von Kriechen", *Beton-und Stahlbetonbau*, **68**, 49–60, 76–86, 152–8; (discussion, **69**, (1974), 150–1).
163. Sackman, J. L. (1963), 'Creep in concrete and concrete structures', *Proc. Princeton Univ. Conf. on Solid Mech.*, pp. 15–48.
164. Saugy, B., Zimmermann, Th., and Hussain, J. (1976), 'Three-dimensional rupture analysis of prestressed concrete pressure vessel including creep effects', *Nucl. Eng. Des.*, **28**, 97–102.
165. Schade, D., and Haas, W. (1975), 'Elektronische Berechnung der Auswirkungen von Kriechen und Schwinden bei abschnittsweise *hergestellten verbundtragwerken, Deutscher Ausschuss Stahlbeton, Schriftenr. Heft* 244.
166. Schapery, R. A. (1962), 'Approximate methods of transform inversion for viscoelastic stress analysis', *Proc. 4th US Natl. Congr. on Appl. Mech., Berkeley, Calif.*, Vol. 2, ASME, pp. 1075–85.
167. Seki, S., and Kawasumi, M. (1972), 'Creep of concrete at elevated temperature', *Am. Concr. Inst. Spec. Publ.* SP-34, *Symp. on Concrete for Nuclear Reactors*, Vol. 1, pp. 591–638.

168. Selna, L. G. (1967), 'Time-dependent behavior of reinforced concrete structures', *UC-SESM Rep. No.* 67–19, *University of California, Berkeley*.
169. Selna, L. G. (1969), 'A concrete creep, shrinkage and cracking law for frame structures', *J. Am. Concr. Inst.*, **66**, 847–8; with *Suppl. No.* 66–76.
170. Shank, J. R. (1935), 'The plastic flow of concrete, *Ohio State Univ. Eng. Exp. Stn., Bull. No.* 91.
171. Smith, P. D., Cook, W. A., and Anderson, C. A. (1977), 'Finite element analysis of prestressed concrete reactor vessels', *Proc. 4th Int. Conf. on Structural Mechanics in Reactor Technology, San Francisco, Ca.*
172. Smith, P. D., and Anderson, C. A. (1978), 'NONSAP-C: a nonlinear stress analysis program for concrete containments under static, dynamic, and long-term loadings', *Los Alamos Sci. Lab. Rep.* LA-7496-MS.
173. Straub, H. (1930), 'Plastic flow in concrete arches', *Proc. ASCE*, **56**, 49–114.
174. Suter, G. T., and Mickelborough, N. C. (1975), 'Creep of concrete under cyclically varying dynamic loads', *Cem. Concr. Res.*, **5**, No. 6, 565–76.
175. Taylor, R. L., Pister, K. S., and Goudreau, G. L. (1970), 'Thermomechanical analysis of viscoelastic solids', *Int. J. Num. Meth. Eng.*, **2**, 45–60.
176. Trost, H. (1967). 'Auswirkungen des Superpositionsprinzips auf Kriech- und Relaxationsprobleme bei Beton und Spannbeton', *Beton-und Stahlbetonbau*, **61**, 230–8; see also (1970) **65**, 155–79.
177. Troxell, G. E., Raphael, J. M., and Davis, R. W. (1958), 'Long-time creep and shrinkage tests of plain and reinforced concrete', *Proc. ASTM*, **58**, 1101–20.
178. VanGreunen, J. (1979), 'Nonlinear geometric, material and time dependent analysis of reinforced and prestressed concrete slabs and panels', *PhD Dissertation*, Division of Structural Engineering and Structural Mechanics, University of California, Berkeley, *Rep. No.* UC-SESM 79–3.
179. Van Zyl, S. F. (1978), 'Analysis of curved segmentally erected prestressed concrete box girder bridges', *PhD Dissertation*, Division of Structural Engineering and Structural Mechanics, University of California, Berkeley, *Rep. No.* UC-SESM 78–2.
180. Van Zyl, S. F., and Scordelis, A. C. (1979), 'Analysis of curved prestresssed segmental bridges', *J. Struct. Div. ASCE*, **105**, No. 11.
181. Volterra, V. (1913), *Leçons sur les Fonctions de Ligne*, Gauthier-Villars, Paris; and (1959), *Theory of Functionals and of Integral and Integro-differential Equations*, Dover, New York.
182. Wagner, O. (1958), 'Das Kriechen unbewehrten Betons', *Deutscher Ausschuss für Stahlbeton, Schriftenr. Heft* 13.
183. Weil, G. (1959), 'Influence des dimensions et des tensions sur le retrait et le fluage de béton', *RILEM Bull.*, No. 3, 4–14.
184. Whaley, C. P., and Neville, A. M. (1973), 'Non-elastic deformation of concrete under cyclic compression', *Mag. Concr. Res.*, **25**, No. 84, 145–54.
185. Whitney, G. S. (1932), 'Plain and reinforced concrete arches', *J. Am. Concr. Inst.*, **28**, No. 7, 479–519; Disc., **29**, 87–100.
186. Willam, K. J. (1978), 'Numerical solution of inelastic rate processes', *Comput. Struct.* **8**, 511–31.
187. Williams, M. L. (1964), 'The structural analysis of viscoelastic materials', *AIAA J.*, **2**, 785–808.
188. Williamson, R. B. (1972), 'Solidification of Portland cement', in B. Chalmers *et al.* (Eds), *Progress in Materials Science*, Vol. 15, No. 3, Pergamon Press, New York, London.

189. Wittmann, F. (1966), 'Kriechen bei Gleichzeitigem Schwinden des Zementsteins', *Rheol. Acta*, **5**, 198–204.
190. Wittmann, F. (1968), 'Surface tension, shrinkage and strength of hardened cement paste', *Materials and Structures* (RILEM, Paris) **1**, 547–52.
191. Wittmann, F. (1970), 'Einfluss des Feuchtigkeitsgelhaltes auf das Kriechen des Zementsteines', *Rheol. Acta*, **9**, 282–7.
192. Wittmann, F. (1971a), 'Kriechverformung des Betons unter Statischer und unter Dynamischer Belastung', *Rheol. Acta*, **20**, 422–8.
193. Wittman, F. H. (1971b), 'Vergleich einiger Kriechfunktionen mit Versuchsergebnissen', *Cem. Concr. Res.*, **1**, 679–90.
194. Wittmann, F. (1974), 'Bestimmung Physikalischer Eigenschaften des Zementsteins', *Deutscher Ausschuss für Stahlbeton, Schriftenr. Heft 232*.
195. Wittmann, F. H., and Roelfstra, P. (1980), 'Total deformation of loaded drying concrete', *Cem. Concr. Res.*, **10**, 601–10.
196. York, G. P., Kennedy, T. W., and Perry, E. S. (1970) 'Experimental investigation of creep in concrete subjected to multiaxial compressive stresses and elevated temperatures', *Res. Rep.* 2864–2, *University of Texas, Austin*; see also (1971), *Am. Concr. Inst. Spec. Publ. No.* 34, 641–700.
197. Zienkiewicz, O. C., Watson, M., and King, I. P. (1968), 'A numerical method of viscoelastic stress analysis', *Int. J. Mech. Sci.*, **10**, 807–27.
198. Zienkiewicz, O. C., and Watson, M. (1966), 'Some creep effects in stress analysis with particular reference to concrete pressure vessels', *Nucl. Eng. Des.*, No. 4.
199. Bažant, Z. P. (1982), 'Input of creep and shrinkage characteristics for a structural analysis program', *Materials and Structures* (RILEM; Paris), Vol. 15.
200. Bažant, Z. P., Ashgari, A. A., and Schmidt, J. (1976), 'Experimental study of creep of hardened cement paste at variable water content', *Materials and structures* (RILEM, Paris) **9**, 279–290.
201. Bažant, Z. P., and Ong. J. S. (1981), 'Creep in continuous beams built span-by-span', *Report No.* 81–12/665c, Center for Concrete and Geomaterials, Northwestern University, Evanston, IL. (submitted to ASCE Struct. Div. J.)
202. Bažant, Z. P., and Raftshol, W. J. (1982), 'Effect of cracking in drying and shrinkage specimens', *Cement and Concrete Research*, **12**, 209–226.
203. Madsen, H. O., and Bažant, Z. P. (1982) 'Uncertainty analysis of creep and shrinkage effects in concrete structures', *Report No.* 82–2/665u, Center for Concrete and Geomaterials, Northwestern University, Evanston, IL.
204. Bažant, Z. P., and Zebich, S. (1982), 'Statistical regression analysis of creep prediction models', *Report No.* 82–4/665s, Center for Concrete and Geomaterials, Northwestern University, Evanston, IL.
205. Tsubaki, T., and Bazant. Z. P. (1982), 'Random shrinkage stresses in aging viscoelastic vessel, *J. Engng. Mech. Div.*, ASCE **108**, EM3, June, 527–45.
206. Bažant, Z. P., Tsubaki, T., and Celep, Z., 'Singular History Integral for Creep Rate of Concrete, *J. Engng. Mech. Div.*, ASCE, in press.
207. Anderson, C. A., Smith, P. D., and Carruthers, L. M., "NONSAP-C: A Nonlinear Strain Analysis Program for Concrete Containments under Static, Dynamic and Long-Term Loadings, Report NUREG/CR0416, LA-7496-MS, Rev. 1, R7 & R8, Los Alamos National Laboratory, New Mexico, Jan. 1982 (available from NTIS, Springfield, Va.).

# PART III
# STRUCTURAL ANALYSIS
# AND BEHAVIOUR

Creep and Shrinkage in Concrete Structures
Edited by Z. P. Bažant and F. H. Wittmann
© 1982 John Wiley & Sons Ltd

*Chapter 8*

# Numerical Creep Analysis of Structures†

*C. A. Anderson*

## 8.1 INTRODUCTION

Concurrent with the development of computing machines with expanded memory and faster execution times has been the development of numerical methods for solving non-linear physical problems in two and three dimensions. In the area of structural mechanics, the analysis of reinforced concrete structures has in the past been treated in an *ad hoc* fashion with the use of numerous approximations (such as elastic, perfectly plastic, or viscoelastic behaviour) to represent the behaviour of the reinforced concrete. The development of fast computers with large memories, together with inelastic finite element methods and codes, has rendered such approaches obsolete. It is the purpose of this chapter to describe the finite element method and its implementation into a computer code for the creep analysis of structures and, particularly, concrete pressure vessels and containments.

Analytical procedures that are accurate for prediction of the stress and deformation fields of homogeneous linear continua encounter numerous difficulties when applied to concrete structures. Among these difficulties are the incorporation of the non-homogeneous behaviour caused by the presence of the reinforcement, concrete cracking, moisture and temperature effects, the effect of triaxial stress, and non-linear effects over an extended range of stress. Consequently, although early attempts at modelling the behaviour of reinforced concrete structures relied on an elastic finite element analysis,[1,2] the preponderance of research in this field in the last five years has been concerned with modeling the inelastic behaviour of reinforced concrete structures by use of the finite element method.[3-5]

Representation of the composite nature of reinforced concrete proceeded historically with the use of separate finite elements (initially, constant strain triangles) for the reinforcement and concrete.[2] More recently, with the widespread adoption of higher-order finite elements to represent complicated geometrical shapes with a small number of elements, the emphasis has

---

† The work described was supported by the US Nuclear Regulatory Commission's Office of Nuclear Regulatory Research.

changed to a 'smeared' element stiffness representation.[6] In this method, the mechanical and geometrical properties of the plain concrete and the reinforcement are integrated over the finite element to provide a stress state-dependent element stiffness, which can include yielding of the reinforcement and cracking of the concrete. The method, though often used for elastic and strength calculations, is not suitable for concrete creep calculations.

Undoubtedly, the most difficult non-linearity in the analysis of concrete structures is crack formation and propagation. Concrete cracking is accounted for by introduction of a crack oriented perpendicular to the maximum principal stress direction whenever the stress or strain state at the point in question satisfies the cracking criterion. The constitutive matrix is subsequently modified to prevent transmission of normal tensile stress across the crack plane. The cracking process is carried out pointwise (at the integration points for isoparametric elements), and an integrated element stiffness, reflecting the reduction in stiffness caused by the presence of the crack, is determined. Unfortunately, the extreme material 'softening' of crack formation can sometimes cause numerical difficulties. Finally, since incremental equations are usually employed, progressive crack growth can be followed using this method. Dodge *et al.*[7] give an up-to-date survey of analysis methods for concrete structures, including a discussion of the numerical treatment of cracking in concrete.

For predicting time-dependent behaviour of concrete under long-term loads, viscoelastic models have been used and continue to be applied in finite element codes. Bazant[8] has reviewed the recent development of physically based viscoelastic theories to represent concrete creep including attempts to account for temperature, shrinkage, and moisture changes. The most commonly applied theory for concrete structures is the linear aging viscoelastic theory for which the superposition theorem holds; here, the only numerical complication is the problem of storing the past stress history. Expansion of the creep function in a finite Dirichlet series, however, simplifies the storage problem and renders possible the numerical solution of complex three-dimensional concrete structures. Numerical difficulties notwithstanding, practical creep analysis of real structures is difficult even presuming that the concrete follows the linear aging viscoelastic theory. The main difficulties are, again, the treatment of cracking as creep progresses and the accommodation of reinforcing and prestressing tendons for prestressed or post-tensioned structures. These structural elements, for which creep is insignificant compared to the creep of concrete, must be modelled individually if changes in prestressing forces are to be evaluated over the life of the structure.

Section 8.2 briefly discusses the basic ingredients of the finite element displacement theory and the process of structural discretization when the structural materials obey a general first-order rate constitutive law. Explicit and implicit numerical schemes for solving the discretized structural equations for inelastic rate processes have been described previously[9,10] and are briefly discussed in Section 8.2. Also in Section 8.2, the stiffness matrices for tendon, membrane, and three-dimensional continuum elements, used in the construction of conventional and nuclear concrete containment structures, are developed.

Section 8.3 is devoted to the constitutive theory appropriate to concrete structural elements described in Section 8.2. The linear aging theory for concrete elements, its Dirichlet series representation, and the incorporation of the Dirichlet series into finite element codes are described. The accommodation of moisture and temperature effects into a finite element analysis is also discussed. Finally, the treatment of cracking of concrete elements during the creep process is described. Now, the concrete creep problem becomes non-linear regardless of the material model and non-linear methods long associated with finite element analysis, such as stiffness reformulation and equilibrium iteration, must be employed. These non-linear methods are discussed and an algorithm based on cracking at a limiting tensile strain is presented.

Section 8.4 discusses the mechanics of incorporating the kinematical and constitutive representations in Sections 8.2 and 8.3 in a finite element code. Emphasis here is on the problems arising in computer application and how they show up in storage and running time requirements for creep analysis of large concrete structures.

Section 8.5 is devoted to a presentation of the results of creep analyses of simple engineering structures—where analytical solutions are available for comparison—and large scale concrete nuclear structures where structural design for reliability and safety is an important factor. The analyses were carried out using the finite element code described in Smith *et al.*[11] and Smith and Anderson.[12] Of commensurate importance to the accuracy of solutions to finite element creep problems is the cost of the concrete creep analysis. Data are given on the cost of large scale creep analyses using CDC-7600 machines and the CRAY-1 machine, which is the largest and fastest machine currently available for scientific computing.

Section 8.6 is concerned with how the stochastic nature of concrete properties can be taken into account in the process of analysing concrete structures for creep behaviour. It is well known[13] that the strength of concrete is variable and the same causes for the variability in strength are likely to be reflected in variability in creep properties. (This variability shows up as a variation in strength of samples of the same size and a decreased mean strength as the sample volume increases.) How these physical observations can be used to obtain statistics on the creep of concrete structures—as determined by numerical methods—is discussed. Here, the Latin hypercube sampling of randomly distributed creep and load parameters is proposed as a method for obtaining statistics of the creep of concrete structures. This method has been used previously for assessing the reliability of concrete reactor vessels.[14]

## 8.2 FORMULATION OF FINITE ELEMENT EQUATIONS FOR CREEP ANALYSIS OF STRUCTURES

The foundations of the finite element method are now firmly established and can be found in Zienkiewicz[15] and Martin and Carey[16] for instance. In this section we will review those aspects of the finite element method that are peculiar to the analysis of complex concrete structures where the concrete obeys a general first-order inelastic rate constitutive law. As is true of reinforced concrete structural behaviour in general, we will assume that small displacement theory applies and, thus, we will not distinguish between structural equilibrium in the deformed or undeformed configuration. We will also briefly discuss the finite element formulation for a multinode frictionless tendon, a 4- to 8-node membrane element that does not support bending moments, and a fully three-dimensional continuum element with from 8 to 21 nodes. The three finite elements are characterized by one-, two-, and three-dimensional strain, respectively, and have been promoted in the past for their efficiency in analysing concrete reactor vessels and containments.[17] All formulations are in a three-dimensional Cartesian coordinate system $(x, y, z)$; consistent with the notation of Zienkiewicz,[15] we will designate matrices by $[M]$ and vectors by $\{v\}$.

The basic idea of the finite element displacement method is to represent the displacements $(u, v, w)$ within the element by continuous interpolation (or shape) functions of the form

$$u(x, y, z) = \sum_{j=1}^{L} N_j(x, y, z)u_j \qquad (8.1)$$

where $N_j$, $j = 1, 2, \ldots, L$ are the shape functions associated with the $L$ nodes and $u_j$, $j = 1, 2, 3, \ldots, L$ are the nodal values of the displacement $u(x, y, z)$. Similar expressions apply for $v$ and $w$. For given shape functions and nodal velocities, a strain rate field within the element can be defined

$$\{\dot{\varepsilon}(x, y, z)\} = [B(x, y, z)]\{\dot{\delta}\} \qquad (8.2)$$

where $\{\dot{\varepsilon}\}$ is a vector of strain rate components, $[B]$ is a matrix of gradients of the shape functions $N_j$, $j = 1, 2, \ldots, L$, and $\{\dot{\delta}\}$ is the element nodal velocity vector,

$$\{\dot{\delta}\}^T = (\dot{u}_1, \dot{v}_1, \dot{w}_1, \dot{u}_2, \dot{v}_2, \dot{w}_2, \ldots, \dot{u}_L, \dot{v}_L, \dot{w}_L)$$

Once the gradient matrix $[B]$ has been formed, equilibrium of the stress rates $\{\dot{\sigma}\}$ is given by

$$\sum_e \int_{V^e} [B]^{\mathrm{T}}\{\dot{\sigma}\}\, \mathrm{d}V = \{\dot{R}\}^a \tag{8.3}$$

where the summation is taken over all elements composing the finite element mesh and $\{\dot{R}\}^a$ is the applied nodal load rate vector. For materials obeying inelastic rate processes, the stress rate is given by

$$\{\dot{\sigma}\} = [D](\{\dot{\varepsilon}\} - \{\dot{\eta}\}) \tag{8.4}$$

where $[D]$ is the elasticity matrix, $\{\dot{\varepsilon}\}$ is the total strain rate vector given in terms of element nodal velocities by Equation (8.2), and $\{\dot{\eta}\}$ are inelastic (creep) strain rates. Inelastic strain rates are usually given by a vector function of the stress, the temperature, and either the accumulated inelastic strain (strain hardening) or the time (time hardening). In the case of time hardening behaviour, for instance,

$$\{\dot{\eta}\} = \{f(\{\sigma\}, t, T)\} \tag{8.5}$$

Forms of the inelastic material law, Equation (8.5), for representing concrete behaviour, are discussed in Section 8.3.

Substituting Equation (8.4) for the stress rates into Equation (8.3) for equilibrium then gives a system of equations for the unknown nodal velocities $\{\dot{\delta}\}$

$$[K]\{\dot{\delta}\} = \{\dot{R}\}^a + \{\dot{R}\}^n \tag{8.6}$$

where $[K]$ is the assembled elastic stiffness matrix

$$[K] = \sum_e \int_{V^e} [B]^{\mathrm{T}}[D][B]\, \mathrm{d}V \tag{8.7}$$

and $\{\dot{R}\}^n$ is the initial creep rate vector for the creep process

$$\{\dot{R}\}^n = \sum_e \int_{V^e} [B]^{\mathrm{T}}[D]\{\dot{\eta}\}\, \mathrm{d}V \tag{8.8}$$

Equations (8.4) through (8.8) can be written in incremental form over a time interval for the incremental displacements $\{\Delta\delta\}$ and incremental stresses $\{\Delta\sigma\}$. Thus, Equation (8.6) becomes

$$[K]\{\Delta\delta\} = \{\Delta R\}^a + \{\Delta R\}^n \tag{8.6'}$$

while Equation (8.8) takes the form

$$\{\Delta R\}^n = \sum_e \int_{V^e} [B]^{\mathrm{T}}[D]\{\Delta\eta\}\, \mathrm{d}V \tag{8.8'}$$

For non-aging materials $[K]$ remains constant, and the main question is how to approximate $\{\Delta R\}^n$, which depends on varying stresses over the time interval of interest.

Numerical schemes for time advancing solutions to Equations (8.4)–(8.8) are discussed in Willam[9] and Argyris *et al.*[10] Traditional explicit methods characterized by strict time step size limitations for numerical stability use an initial creep rate vector $\{\dot{R}^n\}$ or creep increment $\{\Delta R\}^n$ evaluated for conditions at the beginning of the time interval. In order to use time increments of unlimited size, implicit methods (which do not assume constant stress over the time interval) must be resorted to. For each time step these methods involve much more calculation in that iteration must be carried out or a gradient matrix must be evaluated at each integration point. Predictor–corrector and tangent stiffness methods are often used in implicit methods. Willam[9] advocates the use of constant stiffness matrices while keeping the local gradient matrices on the right-hand side of Equation (8.6) as a pseudo-load vector.

A stable and accurate method of representing the incremental creep vector of Equation (8.8)′ was proposed in Anderson,[18] which enabled the use of very large time steps to reach steady-state creep states. However, at each time step it was necessary to formulate a new elastic-creep stiffness matrix. In Anderson and Bridwell[19] large geometry changes were incorporated in the creep numerical algorithm and long-term non-linear time-dependent geomechanical problems were studied. A stable numerical algorithm for solving Equations (8.4)–(8.8) will be further discussed in Section 8.3 with references to aging viscoelastic materials. (For aging viscoelastic materials, $[K]$ is no longer a time-independent matrix.)

We conclude this section with a discussion of the development of the stiffness matrix $[K]$ for tendon, membrane, and three-dimensional continuum elements.

### 8.2.1 Tendon element

The strain-displacement and stiffness matrices for the frictionless tendon shown in Figure 8.1 can be obtained in a straightforward manner. At time $t = 0$ the length of the tendon $S(0)$ is given by

$$S(0) = \sum_{n=1}^{L} [(x_{i_{n+1}} - x_{i_n})^2 + (y_{i_{n+1}} - y_{i_n})^2 + (z_{i_{n+1}} - z_{i_n})^2]^{1/2}$$

while the length $S(t)$ at any later time $t$ is

$$S(t) = \sum_{n=1}^{L} [(x_{i_{n+1}} - x_{i_n} + u_{i_{n+1}} - u_{i_n})^2 + (y_{i_{n+1}} - y_{i_n} + v_{i_{n+1}} - v_{i_n})^2 + (z_{i_{n+1}} - z_{i_n} + w_{i_{n+1}} - w_{i_n})^2]^{1/2}$$

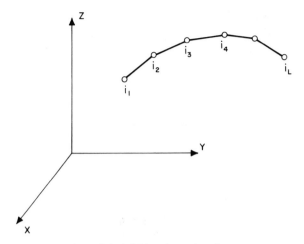

**Figure 8.1** Multi-node tendon element

where $(u_{i_n}, v_{i_n}, w_{i_n})$, $n = 1, 2, \ldots, L$, are the nodal displacements. Calculating the tendon strain as $\varepsilon(t) = [S(t) - S(0)]/S(0)$ and using the binomial series expansion and throwing away quadratic and higher terms gives

$$\varepsilon(t) = -\frac{1}{S(0)} \left[ \frac{1}{l_1} (x_{i_2} - x_{i_1}), \frac{1}{l_1} (y_{i_2} - y_{i_1}), \right.$$

$$\left. \frac{1}{l_1} (z_{i_2} - z_{i_1}), \frac{1}{l_2} (x_{i_3} - x_{i_2}), \ldots \right] \begin{Bmatrix} u_{i_1} \\ v_{i_1} \\ w_{i_1} \\ u_{i_2} \\ \vdots \\ w_{i_L} \end{Bmatrix} \quad (8.9)$$

where $l_1, l_2, \ldots, l_{L-1}$ are the distances between the nodes of the tendon. The terms of the $(1 \times 3L)$ strain-displacement matrix $[B]$ are now obvious and the stiffness matrix can be formed directly from Equation (8.7) with $[D]$ having a single term equal to the modulus of elasticity $E_s$ of the material of the tendon.

If the tendon is grouted into the concrete and there is no slip between tendon and concrete, then the stiffness matrix for the tendon can be formed as the sum of $L - 1$ stiffness matrices for its individual links. The stiffness of each link is the stiffness of the well-known rod element (see Zienkiewicz[15] and Martin and Carey,[16] for example). For intermediate values of friction the problem becomes non-linear and is not easily solvable. A simplified formulation, which indicates the effect of a uniform constraint between tendon and concrete and which reduces to the frictionless and fully bonded situations, respectively, can be developed, however. Let $\mu$ be a constraint parameter with $0 \leqslant \mu \leqslant 1$. We write the strain-displacement relation as

$$\{\varepsilon\} = ((1-\mu)[B]_1 + \mu[B]_2) \begin{Bmatrix} u_{i_1} \\ v_{i_1} \\ w_{i_1} \\ u_{i_2} \\ \vdots \\ w_{i_1} \end{Bmatrix} \tag{8.10}$$

where $[B]_1$ is the strain-displacement matrix for the frictionless case ($\mu = 0$) given by Equation (8.9) and $[B]_2$ is the strain-displacement matrix for the fully bonded case ($\mu = 1$). The stiffness matrix is again formed according to Equation (8.7) with $[D]$ a $1 \times 1$ matrix of magnitude $E_s$.

### 8.2.2 Membrane element

A synopsis of the mathematical formulation for the stiffness of a 4- to 8-node membrane finite element is given. Here, it is convenient to switch to indicial notation with Latin subscripts referring to a three-dimensional Cartesian space $x_i$, $i = 1, 2, 3$ and Greek subscripts referring to a two-dimensional membrane space. The material follows Green and Zerna[20] with modifications for finite element representations.

Let $\{g_\alpha\}$, $\alpha = 1, 2$ be non-unit covariant base vectors tangent to the non-orthogonal curvilinear (isoparametric) coordinates of the membrane as shown in Figure 8.2. For small strains, the strain tensor of the membrane, neglecting bending strain, is given by

$$\gamma_{\alpha\beta} = \frac{1}{2} \left( \{g_\alpha\}^{\mathrm{T}} \left\{ \frac{\partial u}{\partial \beta} \right\} + \{g_\beta\}^{\mathrm{T}} \left\{ \frac{\partial u}{\partial \alpha} \right\} \right) \tag{8.11}$$

where

$$\{g_\alpha\} = \sum_{i=1}^{3} \frac{\partial x_i}{\partial \alpha} \{e\}_i \qquad \left\{ \frac{\partial u}{\partial \beta} \right\} = \sum_{i=1}^{3} \frac{\partial u_i}{\partial \beta} \{e\}_i \tag{8.12}$$

and $u_i$, $i = 1, 2, 3$ are displacement components and $\{e\}_i$, $i = 1, 2, 3$ are Cartesian unit vectors. The partial derivatives in Equations (8.11) and (8.12) can be expressed in terms of element nodal coordinates $x_i^l$, $i = 1, 2, 3$, $l = 1, 2, \ldots, L$ and displacements $u_i^l$, $i = 1, 2, 3$, $l = 1, 2, \ldots, L$ by use of the shape functions (Equation (8.1)). This then becomes an isoparametric rep-

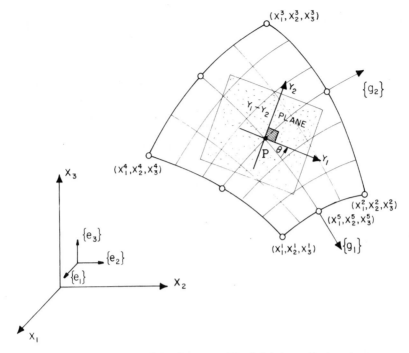

**Figure 8.2** Membrane finite element and local global coordinate systems

resentation. Thus,

$$\frac{\partial x_i}{\partial \alpha} = \sum_{l=1}^{L} \frac{\partial N_l}{\partial \alpha} x_i^l, \quad i = 1, 2, 3 \tag{8.13}$$

$$\frac{\partial u_i}{\partial \beta} = \sum_{l=1}^{L} \frac{\partial N_l}{\partial \beta} u_i^l \quad i = 1, 2, 3. \tag{8.14}$$

Metric coefficients $g_{\alpha\beta}$, $\alpha = 1, 2$, and $\beta = 1, 2$ used in coordinate transformations are expressed as

$$g_{\alpha\beta} = \{g_\alpha\}^T \{g_\beta\} \tag{8.15}$$

and a unit membrane surface area is then transformed from Cartesian to membrane coordinates by

$$dA = (g_{11}g_{22} - g_{12}g_{21})^{1/2} \, d\xi \, d\eta = g^{1/2} \, d\xi \, d\eta \tag{8.16}$$

Constitutive laws are formulated in a local Cartesian system oriented with respect to the membrane coordinates as shown in Figure 8.2. The $y_1$–$y_2$ plane is coplanar with the $\{g_1\}$–$\{g_2\}$ plane and at a point P are rotated with respect to the other by angle $\theta$. Transformations of stresses and strains between the local Cartesian plane and the membrane coordinates are given by

$$\{\varepsilon\}_c = [A]\{\varepsilon\}_m \qquad \{\sigma\}_m = [A]\{\sigma\}_c \qquad (8.17)$$

where the subscripts c and m refer to the local Cartesian and membrane coordinates, respectively. The constitutive matrix $[C]$ transforms as

$$[C]_m = [A]^T[C][A] \qquad (8.18)$$

In Equations (8.17) and (8.18) the transformation matrix $[A]$ has components

$$[A] = \begin{bmatrix} (d_1^1 d_1^1) & (d_1^2 d_1^2) & (d_1^1 d_1^2) \\ (d_2^1 d_2^1) & (d_2^2 d_2^2) & (d_2^1 d_2^2) \\ 2(d_1^1 d_2^1) & 2(d_1^2 d_2^2) & (d_1^1 d_2^2 + d_2^1 d_1^2) \end{bmatrix} \qquad (8.19)$$

where

$$\begin{bmatrix} d_1^1 d_1^2 \\ d_2^1 d_2^2 \end{bmatrix} = \begin{bmatrix} \dfrac{\cos \theta}{(g_{11})^{1/2}} - \dfrac{g_{12} \sin \theta}{(gg_{11})^{1/2}} & \left(\dfrac{g_{11}}{g}\right)^{1/2} \sin \theta \\ -\left(\dfrac{g_{12} \cos \theta}{(gg_{11})^{1/2}} + \dfrac{\sin \theta}{(g_{11})^{1/2}}\right) & \left(\dfrac{g_{11}}{g}\right)^{1/2} \cos \theta \end{bmatrix} \qquad (8.20)$$

Finally, the stiffness of an individual membrane element, from Equation (8.7), becomes

$$[K]^e = \int_{-1}^{+1} \int_{-1}^{+1} ([A][b])^T [C] ([A][b]) g^{1/2} \, d\xi \, d\eta \qquad (8.21)$$

where $[b]$ is the strain-displacement matrix and relates the membrane strains $(\gamma_{11}, \gamma_{22}, \gamma_{12})$ to the nodal displacements. The matrix $[b]$ is obtained by combining Equations (8.11) through (8.14).

### 8.2.3 Three-dimensional solid or thick-shell element

A general three-dimensional isoparametric element with a variable number of nodes is the standard continuum element for representing the displacement field in concrete structures. As shown in Figure 8.3 the element possesses eight corner nodes and a varible number of mid-side nodes plus a central node. By restricting the number of nodes and the order of integration through the thickness of a shell structure, one can enforce a shell displacement field producing membrane and bending strain.

Formation of the element stiffness matrix according to Equation (8.7) by using the isoparametric transformations is straightforward and is covered in many textbooks on finite elements (see Zienkiewicz[15] and Martin and Carey,[16] for instance). Methods of improving the stiffness and stress re-

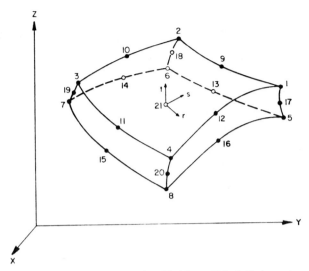

**Figure 8.3** Three-dimensional brick or thick shell element

sponse of higher-order finite elements are found in Argyris and Willam[21] and some comparative results are given.

## 8.3 CONSTITUTIVE LAWS FOR CONCRETE CREEP

The physical mechanisms underlying the creep of concrete, just now being fully understood, are very complex,[8] and their complete incorporation into numerical codes for the prediction of the creep of complex concrete structures is still impossible. Useful numerical results can often be obtained though, by using simplified creep constitutive models that do not incorporate the complicating effects of moisture change, temperature change, and an extended working range of stress and strain. In this section we will first present the simplest such constitutive model, the so-called linear aging viscoelastic model, that has been proposed and often used for the numerical creep analysis of massive concrete structures. Later in the section we will discuss the issues of moisture and temperature change and cracking, and how these effects can be accommodated in the constitutive law.

### 8.3.1 The linear aging model

The characteristic of concrete that distinguishes it from the traditional viscoelastic material is the aging effect. Thus, as function of time the constitutive law is changing through the chemical action of hydration, and it is extremely important to its creep response as to when in its lifetime an

aging viscoelastic structure is loaded. Fortunately, experiments indicate that
the response due to an increment in load is independent of all other past
load increments, and the superposition principle applies. Thus, for small
increments in stress vector $\{d\sigma\}$ occurring at $t'$ measured from the time of
casting, we have

$$\{\varepsilon(t)\} = \int_0^t [J(t, t')]\{d\sigma\} \qquad (8.22)$$

If the material is isotropic with the creep and elastic Poisson's ratios being
equal, then for three-dimensional stress states the matrix $[J]$ can be written
as

$$[J(t, t')] = J(t, t')[\bar{D}]$$

$$= J(t, t') \begin{bmatrix} 1 & -\nu & -\nu & 0 & 0 & 0 \\ -\nu & 1 & -\nu & 0 & 0 & 0 \\ -\nu & -\nu & 1 & 0 & 0 & 0 \\ 0 & 0 & 0 & \frac{1}{2}(1+\nu) & 0 & 0 \\ 0 & 0 & 0 & 0 & \frac{1}{2}(1+\nu) & 0 \\ 0 & 0 & 0 & 0 & 0 & \frac{1}{2}(1+\nu) \end{bmatrix} \qquad (8.23)$$

where $J(t, t')$ is a scalar function.

Numerical creep analysis based on the stress–strain law of Equations
(8.22) and (8.23) may be performed by subdividing the total time interval of
interest into time steps $\Delta t$ and discrete times $t_r$ $(r = 1, 2, \ldots)$. The integral in
Equation (8.22) can then be approximated by finite sums involving incre-
mental stress changes over the time steps. Details of this method, which is
generally applicable to any form of the creep function $J(t, t')$, are given in
Bažant.[8] Because the numerical method results in extensive storage and
computational requirements, it has been superseded by methods that involve
approximating the creep functional $J(t, t')$ by a Dirichlet series and thus
tying the constitutive model physically to Kelvin chain models (or Maxwell
chain models if the relaxation functional is approximated), which then yield
structural equations of the form (8.6)–(8.8).

The creep function $J(t, t')$ is approximated by a series of real exponentials
of the form

$$J(t, t') = \frac{1}{E(t')} + \sum_{i=1}^n \frac{1}{\hat{E}_i(t')} \{1 - \exp[-(t - t')/\tau_i]\} \qquad (8.24)$$

in which $\tau_i$ are constants called retardation times and $E_i$ are aging coefficients. When this function is introduced into the superposition integral
(Equations (8.22) and (8.23)) the integrand degenerates into the product of a
function of $t'$ and a function of $t$. The latter function does not involve the

variable of integration and can be extracted from the integral, leaving only an integration of functions that are independent of $t$. Thus, at each new time step, it is only necessary to compute the change in value of the integral from the last time step rather than from the time of initial loading, as is required in a general case.

A completely stable incremental analysis can be obtained by numerically approximating the integral at discrete times $t_0, t_1, \ldots, t_N$ and using the assumption that $\{d\sigma\}/dt$ and $E_i(t')$ are constant within each time interval $(t_{r-1} \leq t \leq t_r)$. The incremental stress–strain law analogous to Equation (8.4) becomes

$$\{\Delta\sigma\}_r = E_r''[\bar{D}](\{\Delta\varepsilon\}_r - \{\Delta\eta\}_r) \tag{8.25}$$

in which

$$\frac{1}{E_r''} = \frac{1}{E_{r-1/2}} + \sum_{i=1}^{n} \frac{1 - [1 - \exp(-\Delta t_r/\tau_i)]\tau_i/\Delta t_r}{(\hat{E}_i)_{r-1/2}} \tag{8.26}$$

and

$$\{\Delta\eta\}_r = \sum_{i=1}^{n} [1 - \exp(\Delta t_r/\tau_i)]\{\varepsilon_i^*\}_{r-1} \tag{8.27}$$

The effect of prior stress histories is contained in the set of vectors $\{\varepsilon_i^*\}$, which are defined by the recurrence formula

$$\{\varepsilon_i^*\}_r = \frac{\{\Delta\sigma\}_r[1 - \exp(-\Delta t/\tau_i)]\tau_i/\Delta t_r}{(E_i)_{r-1/2}} + \{\varepsilon_i^*\}_{r-1} \exp(-\Delta t_r/\tau_i) \tag{8.28}$$

Thus, the difficulty of computing sums involving the complete stress history at each time step is avoided.

Restricting ourselves to situations of one-dimensional stress, it can be shown[22] that the Dirichlet series is the solution to the system of differential equations

$$\dot{\varepsilon} = \dot{\sigma}/E + \sum_{i=1}^{n} \dot{\varepsilon}_i$$

$$\dot{\varepsilon}_i = \sigma_i/\eta_i \qquad i = 1, 2, \ldots, n \tag{8.29}$$

$$\dot{\sigma} = \dot{\sigma}_i + E_i\dot{\varepsilon}_i \qquad i = 1, 2, \ldots, n$$

when a unit step stress $\sigma(t)$ is applied at time $t'$. This system of equations, which corresponds to the physical system shown at the top of Figure 8.4 and with variables $E, E_i, \eta_i$ as shown, is called the Kelvin chain model. Since the formulation defined by Equation (8.29) states the relations between the rates of stress and strain, it is referred to as a rate-type formulation.

Substituting $\{\Delta\eta\}_r$ and $E_r[D]$ for $\{\Delta\eta\}$ and for $[D]$ in Equations (8.6)', (8.7), and (8.8)' then gives a set of linear equations for the incremental

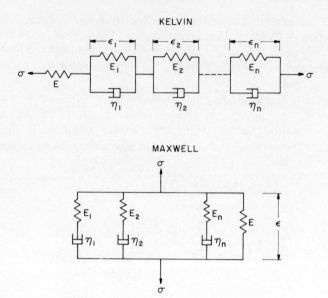

**Figure 8.4**  Kelvin and Maxwell chain models

displacement $\{\Delta\delta\}$ over the time interval. This method for creep analysis has proved to be both stable and accurate and suitable for use with arbitrary sized time steps. However, the method requires the formation of the stiffness matrix $[K]$ at each time step. In contrast numerical methods that assume constant stress conditions over the time interval[23,24] will have time step size limitations to ensure numerical stability.

Another formulation of the viscoelasticity problem is through the use of the relaxation function $[H(t, t')]$ rather than the creep compliance function $[J(t, t')]$,

$$\{\sigma(t)\} = \int_0^t [H(t, t')]\{d\varepsilon(t')\} \tag{8.30}$$

If the relaxation function is expanded in a Dirichlet series and substituted into Equation (8.30), it can be shown that the differential equations (for one dimensional stress)

$$\sigma = \sum_{i=1}^n \sigma_i$$

$$\dot{\varepsilon} = \frac{\dot{\sigma}_i + \sigma_i/\tau_i}{E_i} \qquad i = 1, 2, \ldots, n \tag{8.31}$$

are obtained.[25] This system of rate equations corresponds to the physical system shown at the bottom of Figure 8.4 and is called the Maxwell chain model. The quantity, $\tau_i = \eta_i/E_i$, is the relaxation time of the $i$th unit of the Maxwell chain. Incremental stress–strain laws for the Maxwell chain model have been formulated and structural discretized equations can then be formed as described in Section 8.2. As with the Kelvin model the stress history in the incremental law is defined by a recurrence relation, and the need to sum the complete stress history at each time step is eliminated.

Kelvin and Maxwell chain models can be used interchangeably to solve structural creep problems provided the relaxation function $H$ can be determined from $J$, or vice versa. However, advantages accrue to the Maxwell model when the effects of temperature and humidity change are included since these involve summing hidden stress variables.

## 8.3.2  Modelling moisture and temperature effects

It is not the purpose of this chapter to discuss the many physical phenomena that affect the creep of concrete as that has been described in detail previously.[8] The main purpose of the chapter is to outline the numerical procedures required to analyse concrete structures and to illustrate some typical creep analyses of practical concrete structures. However, finite element codes for creep analysis of concrete structures must have the capability to handle aging viscoelastic constitutive materials and the aging process is itself dependent on temperature $(T)$ and pore moisture $(h)$.

Accommodating the Kelvin chain model to pore moisture and temperature changes is not possible based on the underlying physics of the creep mechanism for concrete. On the other hand, as shown in Bazant,[8] incorporating the temperature effect into the creep law corresponding to the Maxwell model is rather simple.[25] Equation (8.31) is rewritten as

$$\sigma = \sum \sigma_i \qquad \dot{\varepsilon} = \frac{\dot{\sigma}_i}{E_i(t_e)} + \frac{\sigma_i}{\eta_i(t_e)} \tag{8.32}$$

where $\eta_i$ is the age-dependent viscosity associated with the $i$th Maxwell unit, which equals $\tau_i E_i$ at constant temperature $T$. Since creep is a thermally activated process, it is known from physics that $\eta_i$ should depend on temperature according to the Arrhenius equation,

$$\frac{1}{\eta_i} = \frac{1}{\eta_0} \exp\left[\frac{U_i}{R}\left(\frac{1}{T_0} - \frac{1}{T}\right)\right] \tag{8.33}$$

where $T_0$ is the reference temperature, $\eta_0$ is the value of $\eta$ at $T_0$, $R$ is the universal gas constant, and $U_i$ is the activation energy of the $i$th Maxwell unit. The effect of temperature on aging is represented by making $E_i$ and $\eta_i$ dependent on $t_e$ rather than on $t$, where $\Delta t_e$ is an equivalent hydration period that yields the same degree of hydration at temperature $T$ as occurs during a period $\Delta t$ at temperature $T_0$. This equivalent time is given by

$$\Delta t_e = \beta_T \Delta T \tag{8.34}$$

with

$$\beta_T = \exp\left[\frac{U_h}{R}\left(\frac{1}{T_0} - \frac{1}{T}\right)\right] \tag{8.35}$$

in which $U_h$ equals the activation energy for hydration.

The behaviour of concrete becomes even more complex when the moisture content varies within the pores. The movement of moisture produces an acceleration of the creep in the zones of changing moisture content. The development of constitutive equations for this condition is enhanced by utilizing the principles of irreversible thermodynamics of multiphase systems. The details are again given in Bazant,[8] where the Maxwell chain model provides the physical model in whose context the effect of the moisture motion phenomenon on concrete creep can be explained. Computationally significant is the fact that for non-constant moisture it is necessary to distinguish between two types of stresses (those carried by the water in the micropores and those carried by the solid) and by two sets of material properties. Thus, the computer data storage problem that will be described in Section 8.4 will be exacerbated.

The calculation of pore moisture and temperature for a concrete structure is based on the solution of two diffusion equations both of which are uncoupled from the structural creep problem. The partial differential equation for pore moisture is highly non-linear, however, with the diffusivity a strongly non-linear function of pore moisture. Finite difference numerical solutions for drying of simple slab, cylinder, and spherical structures are described in Bazant and Najjar;[26] it is clear that finite element solutions to the non-linear equation for pore moisture in complex geometries can also be obtained.

In conclusion, moisture and temperature effects can be incorporated into the numerical creep analysis of structures where the creep constitutive model is of the Maxwell chain type. However, there is a very significant increase in computer storage requirements and running times will be longer. At the present time there is no general purpose structural creep analysis code that properly handles these effects. There is, however, a European finite element code SMART,[27] developed at the University of Stuttgart, that incorporates temperature and moisture effects in concrete creep analysis. The code has not been widely distributed.

### 8.3.3 Modelling cracking

In certain concrete structures, such as reactor pressure vessels and containment buildings and radioactive waste isolation tanks, the occurrence of cracking during the creep process is of concern in that it may lead to loss of contained fluid (through increased permeability) even though the structure remains intact due to the reinforcing and prestressing tendons. Thus, the prediction of cracking for intact sections as well as the prediction of the rupture (ultimate) load for long-term creep is important for the reliability and safety of these structures.

The prediction of cracking of brittle materials using the finite element method has evolved using variations of the so-called 'initial stress' approach.[28] In this approach the maximum principal stress at each integration point is compared against the tensile strength. If the tensile strength is exceeded, then a crack plane, perpendicular to the principal direction, is introduced and the traction on the plane is set to zero by subjecting the structure to a set of nodal loads that are equipollent to the negative of the traction. The procedure must be carried out iteratively as new tensile stresses may be developed and may have to be removed. Furthermore, in the following load step the stiffness matrix of the cracked element can be reduced to reflect the presence of the crack, which generally improves the convergence of the iteration process. Thus, although the structure may be initially isotropic, as cracking develops the material will become more and more orthotropic. This method, without stiffness reformation, was used in Saugy *et al.*[29] to predict the creep rupture of a prestressed concrete reactor vessel (PCRV). However, a constitutive model not of the rate type was used for representing the concrete creep behaviour.

For predicting simultaneous cracking and creep of concrete structures under constant load, the use of a tensile stress failure criterion is not practical from physical considerations. For, it the structure is statically determinate, then it either cracks during the first time step or not at all; for indeterminate structures, it is well known that creep will redistribute (and ameliorate) stresses caused by the applied loads and, hence, the same conclusion applies. In addition, there is experimental evidence that failure of concrete occurs at an ultimate tensile strain rather than an ultimate tensile stress,[30] which is consistent with the observation that the long-term strength of concrete is less than the short-term strength.

The following procedure has been adapted to numerical calculation of simultaneous creep and cracking of concrete structures where the material behaviour is viscoelastic aging. For each time step the stress and strain is saved and updated at each integration point as described previously. At the end of each time step the principal strains are then computed and compared with the ultimate tensile strain. Supposing that at a certain integration point a principal tensile strain exceeds the ultimate tensile strain, then a crack plane is introduced perpendicular to the principal strain direction as shown in Figure 8.5, and a $(r, s, t)$ coordinate system is introduced. The stress components $\sigma_{ss}, \sigma_{rs}, \sigma_{ts}$ are then annihilated by resolving for equilibrium of

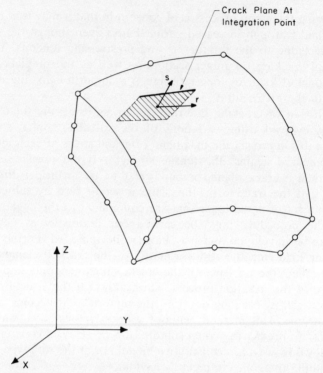

**Figure 8.5** Crack formation in numerically integrated isoparametric element

the structure with an incremental element nodal force vector (from Equation (8.3))

$$\{\Delta R\}_e^a = -\int_{V_e} [B]^{\mathrm{T}} \{\Delta\sigma\} \, \mathrm{d}V \qquad (8.36)$$

where $\{\Delta\sigma\}$ is the stress vector found by rotating the components $\sigma_{ss}$, $\sigma_{rs}$, $\sigma_{ts}$ back to the global coordinate system. All cracks are treated similarly and a structural internal nodal force vector is developed and applied to the structure. New elastic-creep incremental displacements, incremental stresses, and total stresses are then computed. The procedure is repeated until the change of incremental stress is negligible.

Upon the next load increment (or creep time increment for constant loading) the material properties at the cracked integration points are modified to reflect the presence of the crack and its effect on the element stiffness matrix. Thus, in the usual procedure, the instantaneous normal and shear elastic moduli, $E_s$ and $E_{rs}$, are reduced to one per cent of the isotropic values while the creep properties are unchanged. The procedure described here requires that total strains, stresses, and a crack closure strain be stored for each integration point. If, during the life of the structure, a crack closes up then this is detected by comparing the normal strain with the crack closure strain.

## 8.4 NUMERICAL IMPLEMENTATION

This section briefly discusses the features of a large, general purpose, finite element code that would be used for the creep analysis of complex three dimensional structures; such a code is described in detail in Smith and Anderson.[12] This code has capabilities beyond those required for long-term creep analysis. Thus, both static and dynamic problems can be analysed with material models including elastic, elastic–plastic, orthotropic concrete strength, elastic–plastic concrete strength, as well as aging viscoelastic creep. The primary goal of any such large computer code is the assemblage and solution of Equations (8.6)–(8.8); the presentation of the results in compacted form for easy use by the analyst is a secondary goal.

The computational process can be divided into four phases, which consist of:

(i) Input phase where control data and nodal point data are read and equation numbers for active degrees of freedom are established. Applied loads for each time step are formed and element data (e.g. node connectivity and material properties) are stored.

(ii) In the assembly phase the element structural stiffness matrices (these may incorporate creep properties as can be seen in Equations (8.25)–(8.28)) and the internal creep load vectors are formed. The system stiffness matrix and nodal load vector are formed and boundary conditions are imposed.

(iii) During the solution phase the velocity solution of Equation (8.6) is obtained and the element stresses, strains, and hidden variables are calculated and stored.

(iv) Finally, in the output phase results of the analysis are printed and/or displayed in compacted form graphically.

Large aging viscoelastic creep problems will tax the fast core memory capacity of all modern computers except possibly for the CRAY computer, which possesses 500,000 or more words of fast core memory. In addition to storing and solving a large assembled stiffness matrix, a large amount of storage is required in a viscoelastic creep analysis for storing data at the integration points—even using the Dirichlet series creep functional form to simplify the storage problem. Thus, if IDW is the vector stored at each integration point, NGAUS is the number of integration points, and NUME is the number of elements, then IDW * NGAUS * NUME is required for storage of the past structural history. For example, in a three-dimensional problem with 5 Dirichlet series terms (IDW = 42 for 6 stresses, 6 strains, and 5 * 6 hidden variables), 2 * 2 * 2 integration points, and 500 elements the required storage is 168,000. Hence, for anything less than a CRAY computer an out-of-core solver and disc storage of past history structural data are required. The data are then read to and from a large common block which can be sized for the computer at hand.

For the finite element code described in Smith and Anderson,[12] the finite elements of the complete assemblage are divided into element groups according to their type (e.g. tendon, membrane, or brick), the non-linear formulations, and the material models used. Each element group must consist of a single element type, non-linear formulation, and specific material model (i.e. one cannot mix purely elastic and creep brick elements in the same grouping even though they occur in the same problem). The use of element groups greatly reduces time consuming input–output transfers during the solution process, since the element data are read into and from the common block array during the solution of Equations (8.6)–(8.8) and during stress calculations. This common block array resides in the fast core memory. Tables 8.1, 8.2, and 8.3 show the storage allocation within this array for three element types—tendon, membrane, and three-dimensional continuum element—that are used in the code described in Smith and Anderson.[12] In Tables 8.1–8.3, MXNODS is the maximum number of nodes of any one element in an element group, NUME is the number of elements in the group, NUMMAT is the number of different materials, NGAUS is the number of Gauss (integration) points, NCON is the number of material property constants, etc. The number of storage locations assigned in the main program to the common block array determines the number of elements that can be placed in a single element group.

Table 8.1　Tendon element data storage

| Starting address | Length | Array | Description |
|---|---|---|---|
| N101 = 1 | 3 * MXNODS * NUME | LM | element connectivity |
| N102 | 3 * MXNODS * NUME | XYZ | nodal coordinates |
| N103 | NUME | IELT | nodes per element |
| N104 | NUME | IPST | stress output table number |
| N105 | NUME | MATP | material properties set numbers |
| N106 | NUMMAT | E | Young's modulus |
| N107 | NUMMAT | AREA | cross-sectional area |
| N108 | NUMMAT | STRAI | initial tendon strain |
| N109 | NCON * NUMMAT | PROP | material properties |
| N110 | NDM2 | S | element stiffness matrix |

Table 8.2   Membrane element data storage

| Starting address | Length | Array | Description |
|---|---|---|---|
| N101 = 1 | 3 * MXNODS * NUME | LM | element connectivity |
| N102 | 3 * MXNODS * NUME | XYZ | nodal coordinates |
| N103 | NUME | IELT | nodes per element |
| N104 | NUME | IPST | stress output table number |
| N105 | NUME | BETA | orientation of orthotropic material axes |
| N106 | NUME | THICK | element thickness |
| N107 | NUME | MATP | material properties set numbers |
| N108 | NUMMAT | DEN | material density |
| N109 | NCON * NUMMAT | PROP | material properties |
| N110 | IDW * NGAUS * NUME | WA | deformation history stored at each time step |
| N111 | (MXNODS-4) * NUME | NOD5 | mid-side node numbers |
| N112 | 9 * NTABLE | ITABLE | stress output locations |

The general purpose creep code described in Smith and Anderson[12] is an overlaid program, which further improves its high speed storage capacity. Then, the maximum common block array length is limited by the amount of fast core memory that is available after the lengthiest chain of overlays is loaded into the compiler. Currently, the lengthiest chain of overlays is associated with the three-dimensional continuum creep element.

Table 8.3   Three-dimensional continuum element data storage

| Starting address | Length | Array | Description |
|---|---|---|---|
| N101 = 1 | 3 * MXNODS * NUME | LM | element connectivity |
| N102 | 3 * MXNODS * NUME | XYZ | nodal coordinates |
| N103 | NUME | IELTD | nodes per element for displacements |
| N104 | NUME | IELTX | nodes per element for geometry |
| N105 | NUME | IPSI | stress output table number |
| N107 | NUME | MATP | material properties set number |
| N108 | (MXNODS-8) * NUME | NOD9 | mid-side node numbers |
| N109 | NUME | IREUSE | marker for reuse of preceding element stiffness |
| N110 | NUMMAT | DEN | material density |
| N111 | NCON * NUMMAT | PROP | material properties |
| N112 | IDW * NGAUS * NUME | WA | deformation history stored at each load step |
| N113 | 16 * NTABLE | ITABLE | stress output locations |
| N114 | MXNODS * NUME | NODGL | map of local and global node numbers |
| N115 | NSTE + 1 | T | non-uniform spaced solution times (for viscoelastic material model) |
| N120 | NDM2 | S | element stiffness |
| N121 | 2 * NDM | XM | lumped mass and unit nodal body loads |
| N122 | NDM | B | strain-displacement coefficients |
| N123 | NDM | RE | internal loads |
| N124 | NDM | EDIS | nodal displacements |

In the code described in Smith and Anderson,[12] the assembled stiffness matrix is stored by columns in blocks of 50,000 words each. Only the data contained below the skyline of the global matrix are processed. Details of the out-of-core solution scheme are found in Mondkar and Powell.[31] The global matrix is assembled by element group. The upper triangle of each element stiffness matrix is computed in compacted vector form, and the element stiffness and element connectivity array are written on a sequential disk file for each element in the group. Upon completion of all of the element stiffness computations within an element group, blocks 1 and 2 of the global matrix are read from disk into core. Element data are read from disk file NFROM and assembled to the extent they fit into blocks 1 and 2. Data for elements that do not fit entirely in blocks 1 and 2 are written to disk file NTO. When all elements have been processed, blocks 1 and 2 are sent to disk and blocks 3 and 4 are brought into core. Element data are now read from disk file NTO, assembled, and written to disk file NFROM. Assembly continues until all global matrix blocks have been processed.

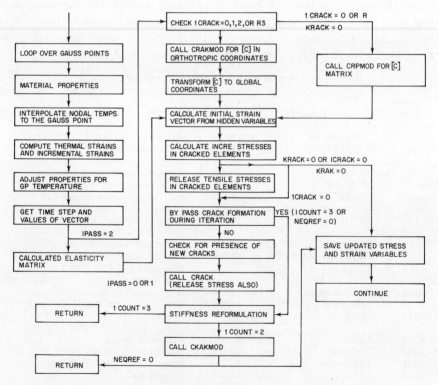

**Figure 8.6**   Flow chart for combined cracking and creep in isoparametric elements

General purpose codes allow for great flexibility in handling non-linearities such as might be caused by concrete cracking. Thus, in the code described in Smith and Anderson,[12] various combinations of stiffness reformulation and equilibrium iteration can be used during a step-by-step creep analysis. Figure 8.6 is a flow chart of the three-dimensional continuum element stiffness formation for viscoelastic creep and cracking where can be seen the various options for stiffness reformulation and iteration. The procedure has been described previously in Section 8.3.3.

General purpose codes for structural creep analysis have many other features that make the task of the engineering analyst less time consuming. For example, a cylindrical coordinates option where element stiffness matrices and internal loads calculated in rectangular coordinates are transformed to cylindrical coordinates prior to assembly of the global stiffness matrix is very important since cylindrical coordinates is the natural coordinate system for applying boundary conditions for many three-dimensional nuclear structures (e.g. PCRVs and concrete containment buildings). Transformation of pressure and gravity loads to nodal loads consistent with the shape functions representative of the displacement field is an important option. Finally, a means of numerically generating finite element meshes[32] is an important capability for a creep analysis code. All of these capabilities will be demonstrated in Section 8.5.

## 8.5 APPLICATIONS

This section deals with the numerical solutions of some viscoelastic aging creep problems. The problems that are treated include three relatively simple problems for which analytical solutions or other numerical data are available for comparison with the finite element solutions, and two problems involving relatively complex geometry that show the power of a general purpose creep code in analysing complex structures. For the latter problems, a system of auxiliary graphics developed by Brown[33] is used to display compactly the results of a three-dimensional creep analysis.

The computer program that was used to generate the numerical solutions is the one described in Smith and Anderson.[12] This computer program accepts viscoelastic aging functionals of the Dirichlet series form as given in Equation (8.24) where the coefficients $E_i(t')$ are restricted to the form

$$\frac{1}{E_i(t')} = a_i + b_i(t')^{n_i} \tag{8.37}$$

or

$$\frac{1}{E_i(T)} = a_i + b_i T \tag{8.38}$$

where $a_i$, $b_i$, and $n_i$ are material constants. The former does not handle temperature effects and the latter takes into account temperature effects in an empirical way but does not handle aging effects.

### 8.5.1   Creep of Fort St Vrain cruciform

Figure 8.7 illustrates a three-dimensional finite element mesh that has been suggested for the cruciform[34] for which experimental creep data are available. Note that this is not a compatible mesh (i.e. displacements are not continuous across all element faces). The modulus of elasticity was experimentally measured (see Table E 20-6 of Reference 34); a value of 31,000 MPa ($4.5 \times 10^6$ psi) at the age of test (33 days) and constant thereafter appears to be justified by the data and was used in the calculation. Utilizing the non-aging creep functional

$$J(t) = 4.0T \times 10^{-9}[(1 - e^{-1.5t}) + (1 - e^{-0.035t})] \qquad (8.39)$$

where $T$ is temperature in °C, $t$ is time in days, and $\sigma$ is stress, calculation of the creep behaviour of the cruciform was made for constant axial and side pressure loading of 10.8 MPa (1585 psi) over 63 days. The cruciform was assumed to be equilibrated at 64.5 °C. A creep Poisson's ratio of 0.20 was used. Eight integration points per element were used for this mesh; the mesh itself consists of 46 nodes. Eighteen time steps were employed in the

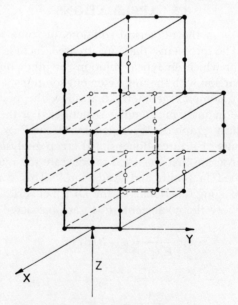

**Figure 8.7**   Cruciform finite element mesh

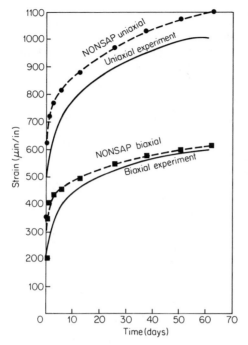

**Figure 8.8** Cruciform experimental strain-time
data and results of finite element calculation

calculation. Figure 8.8 illustrates the uniaxial strain ($\varepsilon_z$) measured at the centre of the top element and the biaxial strain ($\frac{1}{2}\varepsilon_y + \frac{1}{2}\varepsilon_z$) measured at the junction of the middle two elements, together with the experimentally measured values. Good agreement between the numerical and experimental values is seen.

From Equation (8.39) one can evaluate the creep strain at large times to be equal to $8 \times 10^{-9}T$. The limiting creep strain was found to be 815 $\mu$strain which added to an initial elastic strain of 350 $\mu$strain, gives a total limiting strain of 1165 $\mu$strain. By taking large time steps (25, 50, 100, 200, and 400 days), the limiting state of strain in the top element of the cruciform mesh, which is in a state of uniaxial stress, can be simulated numerically. The calculated value of limiting strain was 1140 $\mu$strain which is within 2% of the analytical value.

On the CDC-7600 computer this small problem requires 50 s of running time with the major fraction of time (44 s) spent in I/O and only 6 s in central processing time. On the CRAY computer a total of 13 s was required for executing this problem.

### 8.5.2　Post-tensioned concrete cube

Figure 8.9 illustrates a very simple composite steel and concrete test problem in which a concrete cube is prestressed by four tendons along its edges and then allowed to deform as a function of time through concrete creep. The corresponding Kelvin chain analogy is also shown in Figure 8.9. Results of calculating the creep response, using the code described in Smith and Anderson,[12] are shown in Figure 8.10; the results are in excellent agreement with analytical predictions.

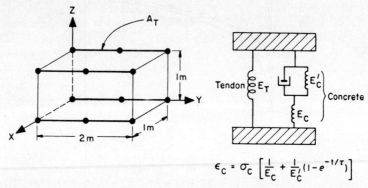

$$\epsilon_c = \sigma_c \left[ \frac{1}{E_c} + \frac{1}{E_c'} (1 - e^{-t/\tau}) \right]$$

**Figure 8.9**　Concrete cube with four tendons and one-dimensional viscoelastic analogy

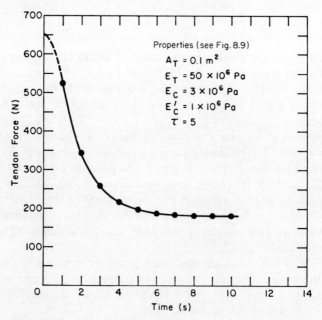

**Figure 8.10**　Decay of tendon preload because of concrete creep

In this problem there is no error caused by spatial discretization since the problem involves only a time varying axial stress. The error caused by the numerical approximation of the creep functional (Equation (8.22)) is small for the constant time step used in this problem. The final tendon strain differed from the exact value at long time by 0.25%.

### 8.5.3 Post-tensioned concrete ring

Figure 8.11 illustrates a concrete ring that is post-tensioned by two cables on the exterior surface as shown. This problem illustrates use of the cylindrical coordinates option that allows the tendon element stiffness matrix and internal load vector to be formed in cylindrical coordinates, which is the natural coordinate system for applying the boundary conditions of no shear traction and no normal displacement on the ends of the ring.

In this problem the steel cables are elastic whereas the concrete creeps according to the viscoelastic aging relation of the Dirichlet series form given in Equation (8.24). For the problem, the coefficients $E_i(t')$, $i = 1, 2$, are of the form given in Equation (8.37), where $a_i = 7.5 \times 10^{-9}$, $b_i = 0.233 \times 10^{-6}$ and $n_i = -0.333$ for $i = 1, 2$. The relaxation times were taken to be $\tau_1 = 5.6$ d and $\tau_2 = 56$ d, respectively.

The concrete ring was modeled using 12 eight-node bricks. As mentioned above no transverse shear and no normal displacement boundary conditions were enforced (in cylindrical coordinates) on the ends of the ring. The ring was also restrained by smooth rigid plattens on the top and bottom surfaces. Finally, a tendon cross-sectional area of 10 was taken and a constant elastic modulus of $4.0 \times 10^6$ and Poisson's ratio of 0.2 were used in the calculation.

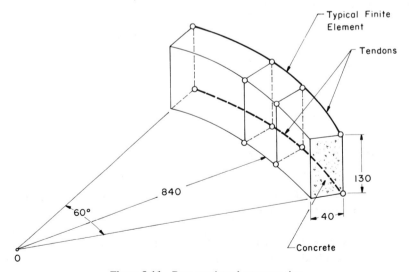

**Figure 8.11**  Post-tensioned concrete ring

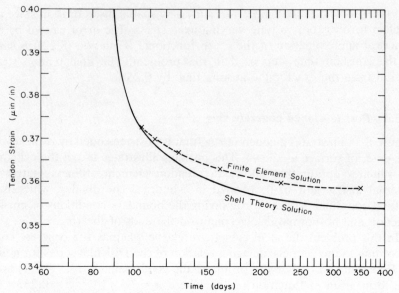

**Figure 8.12**    Decay of post-tensioning strain for concrete ring problem

Figure 8.12 illustrates the variation of the post-tensioning strain in the cable as a function of time out of 400 d starting from an initial strain in the cable of 0.0005 applied at 90 d from concrete casting. Also shown for comparison is a shell solution to this problem for a complete composite concrete steel shell with the same geometry, material quantity, and material properties as in the finite element model. At 350 d there is a 1.3% difference in the finite element cable strain and the cable strain predicted by the shell theory solution.

Fifteen time steps were used ind the calculation beginning with a 0.5 d time step initially when the post-tensioning loads were applied. Fifty-two equations were solved with a half-bandwidth of 41 and 1100 words in the global stiffness matrix. Twenty-five stress, strain and hidden variables were stored at each of the 96 integration points. The problem was executed in 105 s on the CDC-7600.

### 8.5.4 Oak Ridge thermal cylinder

A concrete cylinder with axial and circumferential prestressing was subjected to a radial temperature gradient corresponding to a uniform internal heat flux and a time-varying internal pressure at the Oak Ridge National Laboratory. Dimensions and the prestress configuration are shown in Figure 8.13. The vessel was prestressed 90 days after casting. The temperature gradient was applied over the 40th to 55th days after the prestress was

Selected typical section

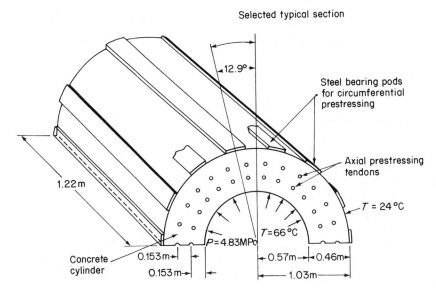

**Figure 8.13** Thermal cylinder

applied. From the 135th day to the 330th day after prestressing, an internal pressure of 4.83 MPa was applied. Details of the experiment are found in Callahan.[35]

A 12.8° sector of the cylinder was modelled with 20-node isoparametric elements as shown in Figure 8.14. The circumferential prestress was applied

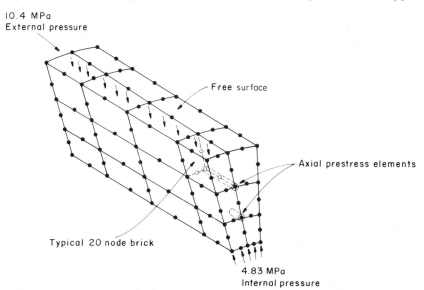

**Figure 8.14** Finite element mesh for the thermal cylinder

Table 8.4 Values of coefficients in Equation (8.38)

| Series term | Retardation time $\tau$ (days) | $a$ ($\times 10^6$) | $b$ ($\times 10^6$) |
|---|---|---|---|
| 1 | 1 | 0.0169 | 0.00036 |
| 2 | 10 | 0.0206 | 0.00047 |
| 3 | 100 | 0.0308 | 0.00118 |
| 4 | 1000 | 0.0797 | 0.00077 |
| 5 | 10000 | 0.0741 | −0.00070 |

to only one row of elements to simulate the bearing pads shown in Figure 8.13. The Young's modulus of 38,300 MPa and the Poisson's ratio of 0.2 were taken from Callahan.[35] The coefficient of thermal expansion for limestone concrete was taken from Figure 51 of Nanstand[36] as 5.4 $\mu$strain/°C.

For concrete creep properties we referred to the eight-year study of multiaxial creep behaviour of concrete summarized in Kennedy.[37] During this investigation, strains were measured in cylindrical specimens subjected to a variety of multiaxial loading conditions, three curing times, two curing

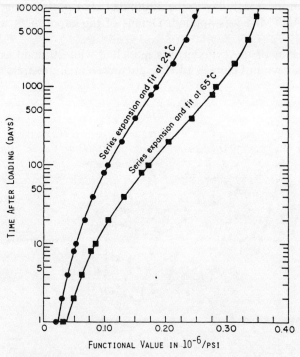

**Figure 8.15** Creep functionals for as-cast PCRV concrete at 24 °C and 65 °C with Dirichlet series approximations

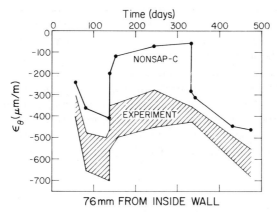

**Figure 8.16** Thermal cylinder hoop strain-time results
at the inner surface

histories, and two curing temperatures (24 °C and 65 °C). Experimental data
were taken for as long as five years. This experimental work is particularly
important because the various test conditions roughly approximate the
conditions for concrete in a prestressed concrete reactor vessel (PCRV). The
results obtained in this study were fit to a non-aging five-term Dirichlet
series of the form shown in Equation (8.24) with constants $E_i$ of the form of
Equation (8.38). Retardation times differing by decades were arbitrarily
selected in advance. The values obtained for the Dirichlet series parameters
are shown in Table 8.4. The experimental data were for concrete loaded 90
days after casting. Figure 8.15 illustrates the closeness of the fit between the
Dirichlet series representation of Equation (8.24) and Equation (8.38), using
parameters given in Table 8.4 and the experimental data of Kennedy.[37]

The creep behaviour of the thermal cylinder was calculated using the
temperature-dependent creep coefficients and retardation times given in
Table 8.4, with the mesh shown in Figure 8.14, and the temperature and
traction boundary conditions shown in Figure 8.13.

Figures 8.16 and 8.17 are plots of circumferential strain histories obtained

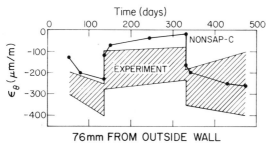

**Figure 8.17** Thermal cylinder hoop strain-time results at the outer surface

with the finite element code. Also shown are banded plots of strains measured during the experiment. More extensive calculational results are presented in Smith and Anderson,[38] and explanations are advanced for discrepancies between the calculated and experimental data.

We used 24 quadratic isoparametric brick elements and 193 nodes to describe the geometry of this problem. Eight integration points per element stored the 42 words representing stress, strain, and hidden variable data which was stored in-core in a single block of length 14,222. There were 476 active equations, and there were 44,086 words in the global stiffness matrix whose half-bandwidth was 129. The 39 time steps required 15 minutes of running time on a CDC-7600, of which approximately 30% was CPU time and 70% was I/O time.

### 8.5.5  Post tensioned reactor containment building

Figure 8.18 represents a cross-sectional view of a model of a pressurized water reactor containment building that uses a post-tensioning system to

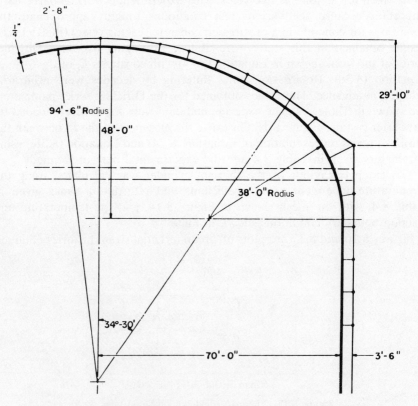

**Figure 8.18**  Containment building schematic

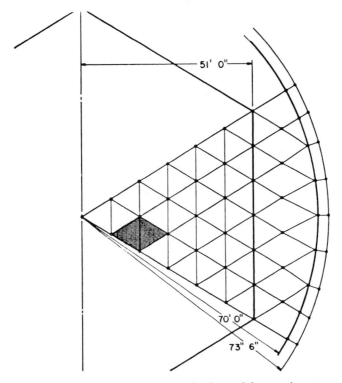

**Figure 8.19**   Tendon arrangement on the dome of the containment
building

keep the concrete in compression during operational and design basis
accident loadings. This problem was chosen for creep analysis since it
possesses more stringent geometrical, data storage, and data output requirements than in the previous problems.

The creep analysis of the problem is rendered three-dimensional because
of the post-tensioning system on the dome of the containment building.
Typically, the tendons are arranged on a triangular grid inside of a hexagon
with centre on the same axis of symmetry, as shown in Figure 8.19. This
provides a 60° symmetry in the stress state in the dome. The dome tendons
then attach to the vertical sidewall tendons as shown, and circumferential
tendons are used to prestress the cylindrical portion of the containment
building. The containment building is lined on the inside with a relatively
thin steel membrane, which serves to contain fission product gases in the
event of an accident.

Eight-node bricks, four-node membranes, and multinode tendons were
chosen to represent the concrete, steel liner, and post-tensioning cables,

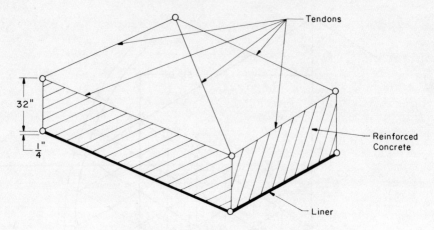

**Figure 8.20**  Typical finite element in the dome

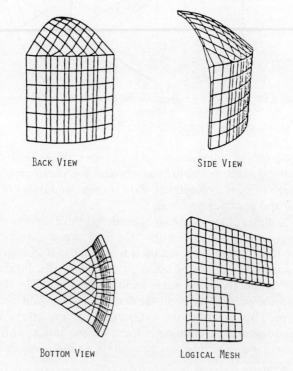

**Figure 8.21**  Views of the finite element mesh for the
containment building

respectively. A typical finite element is shown in Figure 8.20. To generate the nodes and elements for the finite element model of a 60° section of the containment building, the mesh generator code described in Cook[32] was used. Figure 8.21 illustrates three views of the resulting finite element mesh as well as the logical mesh used to describe the brick element connectivity (the tendon connectivity is not shown). Finally, Figure 8.22 illustrates a view of the finite element mesh after it has been smoothed and colour toned by the post processor code MOVIE.[33]

Although the number of nodes (276) and elements (117 each concrete and membrane, 41 tendon) were modest, the connectivity of the tendons produces a fairly large stiffness matrix with 118,994 entries, 744 equations, and half-bandwidth of 716. Using a three-term Dirichlet series for representing the creep compliance function, three separate blocks of concrete elements were required for storage of the stress, strain, and hidden variable arrays (see Section 8.4) for the CDC-7600 computer. Thus, the problem

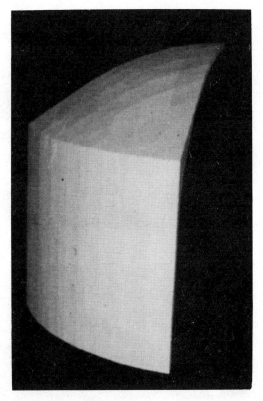

**Figure 8.22** View reconstructed from the finite element mesh

294    *Creep and Shrinkage in Concrete Structures*

provides a good test of a computer code and computing facility to handle realistic structural creep problems.

For this problem a three-term viscoelastic aging function with coefficients of the form of Equation (8.37) was used and with relaxation times of 5.6 d, 56 d, and 560 d corresponding to the Browne–Wylfa data of Bažant and Wu.[22] The constants $a_i$, $b_i$, and $n_i$ in Equation (8.37) are $7.5 \times 10^{-9}$, $0.233 \times 10^{-6}$, and $-0.333$, respectively, for $i = 1, 2, 3$. A constant elastic modulus of $4.0 \times 10^6$ and a Poisson's ratio of 0.2 were used for concrete in the calculations. Elastic behaviour of the membrane and tendons was assumed with an elastic modulus of $3.0 \times 10^7$.

The tendons were post-tensioned to 0.0005 strain at 90 d from concrete casting. The ensuing creep was calculated using 15 time steps with an initial time step of 1 d. Figure 8.23 illustrates a colour toned plot of the deformed shape of the dome model at 135 d from load application. (The deflections have been scaled up by 500.) The dome vertical deflection (measured at the apex) as a function of time is shown in Figure 8.24. To

**Figure 8.23**   View of the deformed shape, $t = 225$ d

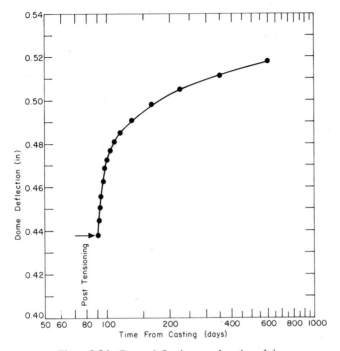

**Figure 8.24** Dome deflection as a function of time

execute this problem required 52 min on the CDC-7600. On the CRAY computer, where only one block of concrete elements is required instead of three and where the equations are solved in-core, the corresponding running time was 21 min.

## 8.6 NUMERICAL TREATMENT OF STATISTICAL VARIABILITY

The previous five sections have dealt with deterministic numerical methods for predicting the creep behaviour of complex structures. However, loadings on these structures are often not well known, both as to their magnitude and temporal and spatial description, and the creep properties are often unknown, ill-defined, or subject to large statistical variations as has been observed for concrete strength.[13,39] Thus, probabilistic methods, which can take into account the random nature of loadings and material properties, would seem to be of great importance in the creep analysis of structures. This is especially true in the case of nuclear plant structures where safety and reliability are of paramount importance.

In the design of structures for strength, many questions have been raised concerning the deterministic nature of design loads, the oversimplified assumptions concerning the load resistance of structural material, and the notion of safety factors that are used in various design codes. Since Freudenthal[40] presented a rational approach to structural safety problems some 30 years ago, an ever increasing effort has been directed toward the application of probabilistic methods to structural engineering. The state-of-the-art in this regard is summarized in Reference 41. In 1974, the ASCE Task Committee on Structural Safety presented a reliability-based design code format for the civil engineering profession.[42] Probabilistic methods have also been incorporated in the comprehensive manual[43] for structural analysis and design of nuclear plant facilities prepared under the direction of the ASCE's Committee of Nuclear Structures. Meanwhile, applications of structural reliability theory are being made in earthquake engineering, wind engineering, and for ocean and aerospace structures.

Reliability methods have been applied to the safety analysis of simple structures—both statically determinate and indeterminate—such as trusses and frames, and the results have been published in the literature.[44] For more complex structures such as concrete reactor vessels, the random nature of the strength of the structure is not so easily determined since it can itself be a non-linear function of the many material properties (which are in turn random variables) and the prestressing loads. Although Shinozuka and Shao[45] attempted a reliability analysis of a prestressed concrete reactor vessel (PCRV), the difficulties of carrying out the assessment of reliability of such a vessel are formidable even though resort is made to approximate techniques and idealized material behaviour.

As far as creep behaviour of structures is concerned, there does not yet appear to be a recommended method for incorporating the random nature of creep properties into a structural reliability theory. Here, as in the conventional reliability theory, the incorporation of randomness of load magnitudes and structural creep resistance are basic ingredients of the reliability calculation; however, the situation is more complex—at least for aging viscoelastic materials—in that the time of load application must be accounted for as a random variable. Below is outlined a general procedure for establishing the probability density function $f_\varepsilon(x)$ for the creep strain of an aging viscoelastic structure whenever $\varepsilon$ depends on a number of randomly distributed creep resistance parameters and random loads. If the failure criterion is that the strain not exceed a specified critical value $\varepsilon_u$ (usually tensile) then the structural reliability $L(t)$ is given by

$$L(t) = P(\varepsilon(\tau) < \varepsilon_u, 0 < \tau < t) \qquad (8.40)$$

Since the creep strain will be, in general, a monotonic increasing function of time, then the reliability function will be a non-increasing function of time. Thus, at the end of life, $T$, of the structure we can evaluate the reliability as

$$L(T) = \int_{-\infty}^{\varepsilon_u} f_\varepsilon(x)\,dx \qquad (8.41)$$

### 8.6.1 Latin hypercube sampling

The procedure that we employ uses Latin hypercube sampling[46] of the creep and load parameters to establish the density function $f_\varepsilon(x)$ with a small number of samples relative to random sampling such as that used in the Monte Carlo method. The method is particularly efficient (in terms of a decrease in number of necessary computer runs) for establishing trends from large finite element analyses for complex structures. In this chapter, however, the method is illustrated on a relatively simple structure where several parameters contribute to the overall creep resistance of the structure.

Details of the Latin hypercube sampling procedure are found in McKay *et al.*[46] Each of $n$ computer runs is to be made with a set of $m$ input parameters the values of which are to be selected from $m$ random variables, $x_i$ of known probability distribution $f_i(x)$. The special form of Latin hypercube sampling used in this study consists of dividing the range of each random variable into $n$ intervals of equal probability. When each computer run is made, the set of input parameters is composed of a sample from a selected interval of each of the $m$ random variables. The intervals from which the samples are taken are selected at random and without replacement. In this fashion, each interval is sampled only once during $n$ computer runs. The random assignment of intervals to be sampled for a particular computer run is accomplished by associating with each random variable $x_i$ a sequence of uniform distribution numbers $u_1, u_2, \ldots, u_n$. The rank $r_k$ of $u_k$ is defined such that $r_k - 1$ is the number of $u_j$'s less than $u_k$. The interval from which the $i$th random variable is sampled for the $k$th computer run is then $r_k$. Figure 8.25 illustrates the Latin hypercube sampling method for three random variables and five computer runs.

It has been shown in McKay *et al.*[46] that an increase in the precision of estimators of the mean and cumulative distribution function can be expected over the corresponding estimators from a completely random (ordinary Monte Carlo) sample when the output of the computer code is a monotonic function of the input variables.

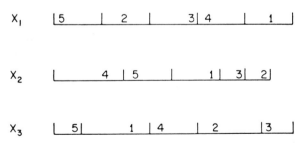

**Figure 8.25**   Latin hypercube sampling with three random variables and five computer runs

### 8.6.2 Reliability analysis of a prestressed concrete beam

Figure 8.26 illustrates a uniform, simply-supported concrete beam of length $2l$ and height $h$ under the action of a constant transverse load $P_0$ applied at the centre of the beam at time $t_0$. The beam is axially prestressed with a tendon initial strain of $\varepsilon_0$ applied at time $t^*$. Time is measured from the time of concrete casting. Under the action of the transverse load $P_0$ and the prestress load $P^*(t)$ the beam will creep. For simplicity, the creep compliance function will be taken to be of the form

$$J(t, t') = \frac{1}{E_c} + [a + b(t')^n]\{1 - \exp[-(t - t')/\tau]\} \qquad (8.42)$$

where $E_c$, $a$, $b$, $n$, and $\tau$ are material constants. The beam will be assumed to fail if the fibre strain on the under side of the beam changes from compressive to tensile.

From equilibrium of forces and strain compatibility it can be shown that the prestressing force is given by

$$P^*(t) = \varepsilon_0 A_s \left[\frac{1}{E_s} + \left(\frac{A_s}{A_c}\right)J(t, t^*)\right]^{-1} \qquad (8.43)$$

where $E_s$ is the steel modulus and $A_s$ and $A_c$ are the cross-sectional steel and concrete areas respectively. The fibre strain at the centre of the beam on the underside is then

$$\varepsilon(-h/2, t) = \frac{P_0 h l}{4I} J(t, t_0) - \frac{\varepsilon_0 J(t, t^*)}{(A_c/A_s E_s) + J(t, t^*)} \qquad (8.44)$$

where $I$ is the cross-sectional moment of inertia.

The probability density function $f_\varepsilon(x)$ of the fibre strain at the base of the beam was then estimated by using the Latin hypercube sampling of the assumed random load variables $P_0$, $t_0$ and random material property variables $a, b, n$. For illustrative purposes $P_0$ was taken to be a normally distributed unit load with a standard deviation of 0.2 and $t_0$ to be normally distributed variable with a mean of 90 d and a standard deviation of 10 d.

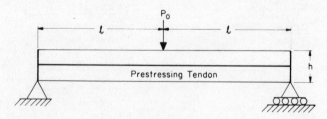

**Figure 8.26** Prestressed simply supported beam under transverse loading

Table 8.5  Latin hypercube samples and calculated fibre strain

| $P_0$ | $t_0$ | $a \ (\times 10^6)$ | $b \ (\times 10^6)$ | $n$ | $\varepsilon \ (\times 10^6)$ |
|---|---|---|---|---|---|
| 1.178 | 65.0 | 0.0758 | 0.6359 | −0.3904 | −1.838 |
| 0.985 | 80.4 | 0.0575 | 0.3670 | −0.1713 | −6.327 |
| 1.137 | 101.8 | 0.1064 | 0.7967 | −0.3426 | −4.328 |
| 0.818 | 85.3 | 0.1384 | 0.6936 | −0.3145 | −12.105 |
| 1.362 | 104.9 | 0.0893 | 0.5378 | −0.2355 | 2.850 |
| 0.895 | 88.7 | 0.1245 | 0.4454 | −0.2193 | −9.133 |
| 0.703 | 97.0 | 0.0603 | 0.5830 | −0.4808 | −14.333 |
| 1.019 | 81.7 | 0.1468 | 0.2562 | −0.2914 | −5.248 |
| 1.062 | 94.0 | 0.1108 | 0.4156 | −0.4021 | −5.584 |
| 0.916 | 90.6 | 0.0990 | 0.2913 | −0.4625 | −8.432 |

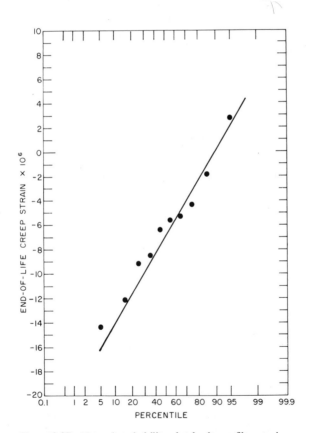

**Figure 8.27**  Normal probability plot for beam fibre strain—
sample size 10

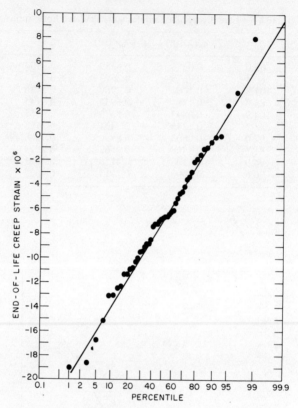

**Figure 8.28** Normal probability plot for the beam fibre strain—sample size 50

The material property variables $a, b, n$ were assumed to be uniformly distributed with $a$ defined between $0.05 \times 10^{-6}$ and $0.15 \times 10^{-6}$, $b$ defined between $0.2 \times 10^{-6}$ and $0.8 \times 10^{-6}$, and $n$ between $-1/2$ and $-1/6$.

For deterministic variables we used

$$
\begin{aligned}
E_s &= 30 \times 10^6 & h &= 1 \\
E_c &= 4 \times 10^6 & \varepsilon_0 &= 0.0001 \\
\tau &= 50 & t^* &= 30 \\
l &= 20 & A_c/A_s &= 30
\end{aligned}
$$

Ten Latin hypercube samples were initially used to estimate the probability density function for the fibre strain at the end of 200 d. This is only twice the number of assumed random variables. The sample values and the resulting fibre strain calculated by Equation (8.44) are given in Table 8.5 and Figure 8.27 illustrates the cumulative frequency distribution of the results plotted on normal probability paper. A linear relationship (implying normality of the results) is indicated, and a straight line fit to the data is shown in Figure 8.27 also. We can then read off the mean and standard deviation for the distribution as 7.0 $\mu$strain and 5.5 $\mu$strain, respectively. Figure 8.28 illustrates the results for a sample size of 50 and here a strong case for normality of the data can be made.

One can now evaluate the reliability (Equation (8.41) with $\varepsilon_u = 0$) directly from Figures 8.27 and 8.28 as approximately 0.90 in both cases.

The procedure of Latin hypercube sampling, illustrated above on a simple structure, is directly applicable to large and complex structures, which can be analysed by numerical means. Efficient sampling of uncertain material property data and undertain load data can provide trends in the creep response of these structures—a procedure which, though expensive, is not prohibitively so.

## REFERENCES

1. Rashid, Y. R., and Rockenhauser, W. (1968), 'Pressure vessel analysis by finite element techniques', *Proc. Conf. on Prestressed Concrete Pressure Vessels*, Inst. Civ. Eng., London.
2. Ngo, D., and Scordelis, A. C. (1967), 'Finite element analysis of reinforced concrete beams', *J. Am. Concr. Inst.*, **64**, No. 3, 152–63.
3. Phillips, D. V., and Zienkiewicz, O. C. (1976), 'Finite element non-linear analysis of concrete structures', *Proc. Inst. Civ. Eng.*, **61**, Part 2, 59–88.
4. Argyris, J. H., Buck, K. E., Scharpf, D. W., and Willam, K. J. (1972), 'Non-linear methods of structural analysis', *Nucl. Eng. Des.*, **19**, 169–97.
5. Argyris, J. H., Faust, G., and Willam, K. J. (1976), 'Limit load analysis of thick-walled concrete structures—a finite element approach to fracture', *Comput. Meth. Appl. Mech. Eng.*, **8**, 215–43.
6. Isenberg, J., and Adham, S. (1970), 'Analysis of orthotropic reinforced concrete structures', *J. Struct. Div. ASCE*, ST12, 2607–24.
7. Dodge, W. G., Bazant, Z. P., and Gallagher, R. H. (1977), 'A review of analysis methods for prestressed concrete reactor vessels', *Oak Ridge National Laboratory Rep.* ORNL-5173.
8. Bazant, Z. P. (1975) 'Theory of creep and shrinkage in concrete structures: a precis of recent developments', *Mech. Today*, **2**, 1–93.
9. Willam, K. J. (1978), 'Numerical solution of inelastic rate processes', *Comput. Struct.* **8**, 511–31.
10. Argyris, J. H., St. Doltsinis, J., and Willam, K. J. (1979), 'New developments in the inelastic analysis of quasistatic and dynamic problems', *Int. J. Num. Meth. Eng.*, **14**, No. 12, 1813–50.
11. Smith, P. D., Cook, W. A., and Anderson, C. A. (1977), 'Finite element analysis of prestressed concrete reactor vessels', *Proc. 4th Structural Mechanics in Reactor Tech. Conf.*, San Francisco, Ca, 15–19 August, 1977.
12. Smith, P. D., and Anderson, C. A. (1978), 'NONSAP-C: a nonlinear stress analysis program for concrete containments under static, dynamic, and long-term loadings', *Los Alamos Scientific Laboratory Rep.* LA-7496-MS.

302 Creep and Shrinkage in Concrete Structures

13. Mirza, S. A., Hatzinikolas, M., and MacGregor, J. G. (1979), 'Statistical descriptions of strength of concrete', *J. Struct. Div. ASCE*, **105**, No. ST6.
14. Anderson, C. A., Whiteman, D. E., Smith, P. D., and Yao, J. T. P. (1978), 'A method for reliability analysis of concrete reactor vessels', paper 78-PUP-100, presented at the *Joint ASME/CSME Pressure VEssels and Piping Conf., Montreal, Canada.*
15. Zienkiewicz, O. C. (1977), *The Finite Element Method*, McGraw-Hill, New York.
16. Martin, H. C., and Carey, G. F. (1973), *Introduction to Finite Element Analysis*, McGraw-Hill, New York.
17. Zienkiewicz, O. C., Owen, D. R. J., Phillips, D. V., and Nayak, G. C. (1972), 'Finite element methods in the analysis of reactor vessels', *Nucl. Eng. Des.*, **20**, 507–41.
18. Anderson, C. A. (1976), 'An investigation of the steady creep of a spherical cavity in a half space', *J. Appl. Mech.*, **98**, No. 2, 254–8.
19. Anderson, C. A., and Bridwell, R. J. (1980), 'A finite element method for studying the transient nonlinear thermal creep of geological structures', *Int. J. Num. Anal. Meth. Geomech.* **4**, No. 3, 255–76.
20. Green, A. E., and Zerna, W. (1954), *Theoretical Elasticity*, Clarendon Press, Oxford, Chaps 4, 5.
21. Argyris, J. H., and Willam, K. J. (1974), 'Some considerations for the evaluation of finite element models', *Nucl. Eng. Des.*, **28**, 76–96.
22. Bazant, Z. P., and Wu, S. T. (1973), 'Dirichlet series creep function for aging concrete', *J. Eng. Mech. Div. ASCE*, **99**, EM2, 367–87.
23. Zienkiewicz, O. C., Watson, M., and King, I. P. (1968), 'A numerical method of visco-elastic stress analysis', *Int. J. Mech. Sci.*, **10**, 807–27.
24. Greenbaum, G. A., and Rubinstein, M. F. (1968), 'Creep analysis of axisymmetric solids', *Nucl. Eng. Des.*, **7**, 379–97.
25. Bazant, Z. P., and Wu, S. T. (1974), 'Rate-type creep law of aging concrete based on Maxwell chain', *Mater, Constr. (Paris)/Mater. Struct.*, **7**, 45–60.
26. Bazant, Z. P., and Najjar, L. J. (1972), 'Nonlinear water diffusion in nonsaturated concrete', *Mater. Const. (Paris)*, **5**, No. 25, 3–20.
27. Szimmat, J. (1979), *SMART I, 2 Langzeitverhalten—Benutzerhandbuch, ISD-Bericht No.* 190, Universitat Stuttgart 1980; sowie *SMART II, 2 Instationare Diffusion—Benutzerhandbuch, ISD-Bericht No.* 192, Universitat Stuttgart.
28. Zienkiewicz, O. C., Villiappan, S., and King, I. P. (1968), 'Stress analysis of rock as a 'no-tension' material', *Geotechnique*, **18**, 56–66.
29. Saugy, B., Zimmermann, Th., and Hussain, M. (1974), 'Three-dimensional rupture analysis of prestressed concrete pressure vessel including creep effects', *Nucl. Eng. Des.*, **28**, 97–120.
30. Carino, N. J., and Slate, F. O. (1976), 'Limiting tensile strain criterion for failure of concrete', *J. Am. Concr. Inst.*, **73**, 160–5.
31. Mondkar, D. P., and Powell, G. H. (1974), 'Large capacity equation solver for structural analysis', *Comput. Struct.*, **4**, 699–728.
32. Cook, W. A. (1978), 'INGEN: a general-purpose mesh generator for finite element codes', *Los Alamos Scientific Laboratory Rep.* LA-NUREG-7135-MS.
33. Brown, B. E. (1976), 'MOVIE: LASL version 1.0 user's manual', *Los Alamos Scientific Laboratory Rep.* LA-NUREG-6532-MS.
34. *Fort St Vrain Nuclear Generating Station—Final Safety Analysis Report*, Appendix E, PCRV Data, Public Service Co. of Colorado report (no date).
35. Callahan, J. P. (1977), 'Prestressed concrete reactor vessel thermal cylinder model study', *Oak Ridge National Laboratory Rep.* ORNL/TM-5613.

36. Nanstand, R. K. (1976), 'A review of concrete properties for prestressed concrete pressure vessels', *Oak Ridge National Laboratory Rep.* ORNL/TM-5497.
37. Kennedy, T. W. (1975), 'An evaluation and summary of a study of the long-term multiaxial creep behavior of concrete', *Oak Ridge National Laboratory Rep.* ORNL/TM-5300.
38. Smith, P. D., and Anderson, C. A. (1977), 'Constitutive models for concrete and finite element analysis of prestressed concrete reactor vessels', *Brookhaven National Laboratory Rep.* BNL-NUREG-50689, Vol. I, pp. 283–97.
39. Gerstle, K. H. (1976), 'Strength of concrete under multiaxial stress states', *Proc. McHenry Symp., sponsored by ACI, Mexico City, Mexico.*
40. Freudenthal, A. M. (1947), 'Safety of structures', *Trans. ASCE,* **112,** 125–80.
41. Task Committee on Structural Safety (1972), 'Structural safety—a literature review', *J. Struct. Div. ASCE,* **98,** ST4, 845–84.
42. Task Committee on Structural Safety (1974), *J. Struct. Div. ASCE,* **100,** ST9, 1753–836.
43. J. D. Stevenson, Chairman of the Editing Board (1976), *Manual of Structural Analysis and Design of Nuclear Power Plant Facilities,* ASCE, rough draft.
44. Stevenson, J. D., and Moses, T. (1970), 'Reliability analysis of frame structures', *J. Struct. Div. ASCE,* **96,** ST11, 2409–27.
45. Shinozuka, M., and Shao, L. C. (1973), 'Basic probabilistic considerations on the safety of prestressed concrete reactor vessels', *Proc. 2nd Int. Conf. on Structural Mechanics in Reactor Technology,* #H1/4.
46. McKay, M. D., Canover, W. J., and Whiteman, D. E. (1976), 'Report on the applications of statistical techniques to the analysis of computer codes', *Los Alamos Scientific Laboratory Rep.* LA-NUREG-6526-MS.

Creep and Shrinkage in Concrete Structures
Edited by Z. P. Bažant and F. H. Wittmann
© 1982 John Wiley & Sons Ltd

*Chapter 9*

# Methods of Structural Creep Analysis

*W. H. Dilger*

## 9.1 INTRODUCTION

The enhanced interest in time-dependent effects in concrete structures is reflected by the rapidly increasing number of papers and books published on this subject during the last decade or so. The aim of the research going on in this field is twofold: first, to provide information for the prediction of the time-dependent phenomena of creep and shrinkage; and second, to develop tools for the engineer to apply this information in the design of concrete structures.

The main concern of this chapter is to discuss methods of analysis of the time-dependent effects in concrete structures and in this context the material properties and their constitutive equations are of interest only insofar as the method of analysis is affected by them. Particular emphasis will be put on recent developments concerning the Trost–Bazant method of creep analysis.

Before discussing the analytical procedures, let us find out what the effect of creep is and for what type of structure and under what conditions we have to investigate the time-dependent effects. Creep in a structure may have the following results:

(a)   redistribution of internal stresses;
(b)   redistribution of the stress resultants caused by external loads;
(c)   reduction of eigenstresses;
(d)   reduction of strength due to deformations (second-order effects).

In addition, creep is always associated with an increase of deformation of a concrete structure.

The effect of creep on the stress redistribution and the magnitude of the stress resultants are most pronounced in members composed of materials with significantly different creep properties, in structures where the boundary or support conditions are changed during the lifetime of the structure or where forces develop due to imposed deformations.

Examples for significant redistribution of stresses are steel–concrete and concrete–concrete composite members. In reinforced and prestressed members creep also results in redistribution of stress but this is normally of interest only in the latter case because of the loss of prestress associated with it.

Significant redistribution of stresses also occurs in structures in which the creep properties vary throughout the thickness of the members. Good examples are nuclear reactors and pressure vessels where differences in the thermal and hygral conditions may generate vastly different creep properties within the thickness of a member.

The time-dependent redistribution of stress resultants (moments, reactions, etc.) are most prominent in beams made continuous by cast-*in-situ* concrete joints, and in continuous structures composed of creeping and non-creeping (or low-creeping) structural elements.

All the forces generated by imposed deformations such as shrinkage, temperature or differential settlement are significantly reduced by creep. For the causes just mentioned this is most desirable, but for intentionally generated forces or stresses (e.g. preflex girders) this relaxation of stresses is, of course, undesirable. Eigenstresses within a section, as caused by a non-linear distribution of free shrinkage or temperature strains are also significantly reduced by creep.

Time-dependent effects do not normally affect the strength of structural concrete members, except when the creep deformations result in an increase of the stress resultants, as in the case of long columns where the creep deflection leads to a higher moment, or as in the case of the sagging floor where ponding may result in additional load.

Finally, the question of long-term deformations due to bending, shear, and torsion may be of interest to the designer.

Although this chapter deals primarily with methods of analysis of creep effects, it should be pointed out that there is a very intimate interrelationship between the creep function used for predicting creep and creep analysis.

## 9.2 CREEP FUNCTIONS

There exist two schools of thought regarding the formulation of the creep function. The first one attempts to represent the experimentally obtained creep surfaces of virgin concrete as a *product* of age and duration functions. The resulting creep function is of the general form:

$$\Phi(t, t') = \frac{1}{E(t')} [1 + K_0 f_1(t') g_1(t - t')] \qquad (9.1)$$

This type of creep function was used by CEB in 1964[1] and 1970,[2] by ACI in 1971[3] and by Bazant and Osman[4] in his 'double power law' formulation of the creep function.

The other school of thought presents creep as the *sum* of two (or more) components, namely a delayed elastic (recoverable) component and an irrecoverable flow component. Special features of this formulation are that the delayed elastic component is assumed to be independent of the age at loading and that the flow is represented by a set of parallel curves. The creep function corresponding to these assumptions is of the form:

$$\Phi(t, t') = \frac{1}{E(t')} + K_1 f_2(t - t') + K_2[g_2(t) - g_2(t')] \tag{9.2}$$

This type of creep function is used in the 1978 CEB-FIP Model Code.[5]

## 9.3 METHODS OF CREEP ANALYSIS

There are a number of important features which are common to all methods of creep analysis. All methods are based on the assumption that creep varies linearly with stress, and they all obey Boltzmann's principle of superposition which was first applied to concrete by McHenry[6] and Maslov.[7] The principle of superposition means that the total strain at time $t$ of a concrete subjected to varying stress is obtained by summing the strain caused by each stress increment (or decrement), $\Delta\sigma(t')$ applied at age $t'$. The total strain is thus obtained by the summation term

$$\varepsilon(t) = \sum_{t_0}^{t} \Delta\sigma(t')\Phi(t, t') \tag{9.3}$$

If the stress varies continuously

$$\varepsilon(t) = \int_{t_0}^{t} \Phi(t, t') \frac{\partial\sigma}{\partial t'} \, dt' \tag{9.4}$$

It is the particular formulation of the creep function which leads to different results when evaluating the above equations.

### 9.3.1 Effective modulus method (EM method)

For this model, creep of concrete is accounted for by reducing the modulus of elasticity of concrete by a factor $[1 + \phi(t, t_0)]$ so that

$$E_{\text{eff}} = E(t_0)/[1 + \phi(t, t_0)] \tag{9.5}$$

The EM method gives good results only when the concrete stress does not vary significantly during the period under investigation and when the aging of the concrete is negligible as in old concrete. Under decreasing stress, the strains are underestimated and under increasing stress the strains are overestimated. The EM method is still used extensively for predicting creep deflections of reinforced concrete beams, and to calculate the effect of creep in slender columns.

### 9.3.2 Rate of creep method (RC method)

The basis for the RC method was established by Glanville,[8] its mathematical formulation was developed by Whitney,[9] and its application to complex structural problems was introduced by Dischinger.[10]

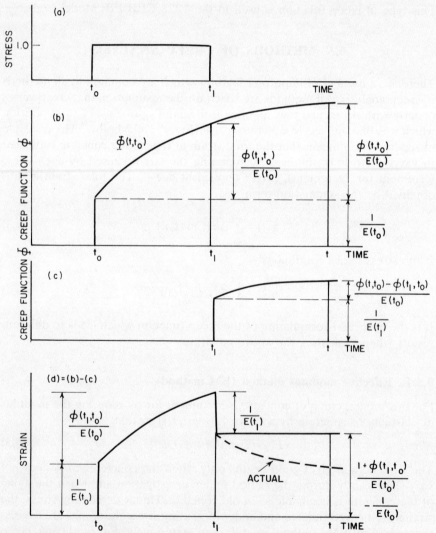

**Figure 9.1** A plot of strain versus time for a unit stress applied during the period $t_0$ to $t_1$ according to the RC method

The RC method is based on the assumption that the rate of creep is independent of the age at loading (Figure 9.1). This leads to the differential equation

$$\frac{d\varepsilon}{dt} = \frac{\sigma(t)}{E(t_0)}\frac{d\phi}{dt} + \frac{1}{E(t)}\frac{d\sigma}{dt} + \frac{d\varepsilon_{sh}}{dt} \qquad (9.6a)$$

or, using the creep coefficient as an independent variable

$$\frac{d\varepsilon}{d\phi} = \frac{\sigma(t)}{E(t_0)} + \frac{1}{E(t)}\frac{d\sigma}{d\phi} + \frac{d\varepsilon_{sh}}{d\phi} \qquad (9.6b)$$

The assumption that the rate of creep is independent of the age at loading means that the creep curves are parallel and implies that creep decreases rapidly with increasing age and becomes zero for old concrete. This is contrary to experimental observations and leads to an overestimation of creep under a decreasing stress (Figure 9.1) and of relaxation under a constant strain. Because of these shortcomings the RC method can be considered obsolete in its original form. However, recent developments to overcome these shortcomings have revived interest in this method.

### 9.3.3 Rate of flow method (RF method)

In order to overcome the deficiencies of the RC method, England and Illston[11] proposed to represent the creep function as the sum of three components: the elastic strain, $\varepsilon_{el}$, the recoverable delayed elastic strain $\varepsilon_d$, and the irrecoverable flow, $\varepsilon_f$ (Figure 9.2). England and Illston concluded from experiments of their own[11] and of others[12] that the delayed elastic strain is independent of the age at loading and reaches a final value much faster than the flow. The flow is assumed to represent the irrecoverable component of creep in the same way as in the RC method. The authors[11] used a step-by-step numerical procedure for the solution of practical creep problems.

### 9.3.4 Improved Dischinger method (ID method)

In the discussion of the RF method it was mentioned that the delayed elastic strain develops much faster than the flow component. In order to allow a simple analytical solution, Nielsen[13] proposed to add the delayed elastic strain, $\varepsilon_d$, to the instantaneous elastic strain, $\varepsilon_{el}$, and to treat the flow component in the same way as the total creep strain in the RC method (Figure 9.3). With this simplification, the creep function can be written in the form

$$\Phi(t, t') = \frac{1}{E_d} + \frac{\phi_f(t) - \phi_f(t')}{E(t')} \qquad (9.7)$$

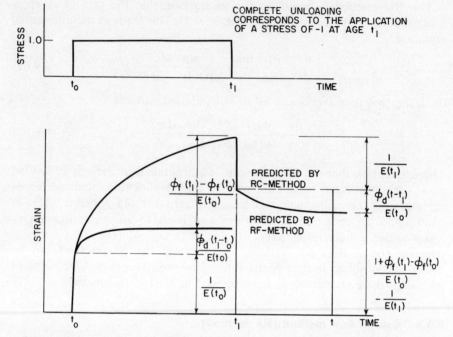

**Figure 9.2**   Strains due to a unit stress acting betwen $t_0$ and $t_1$ according to the RF method

where $E_d$ is a fictitious modulus of elasticity defined by the relation

$$E_d = E(t_0)/(1 + \phi_d) \qquad (9.8)$$

Nielsen[13] recommends $\phi_d = 1/3$ so that $E_d = 0.75 E(t_0)$. Later Rüsch et al.[14] proposed a value of $\phi_d = 0.4$ for $t - t' > 90$ days which was adopted in the 1978 CEB-FIP Recommendations.[5] Combining the instantaneous and delayed elastic strains into one term makes the ID method a hybrid of the EM and RC methods.[15] The ID method allows relatively simple analytical treatment, and gives good accuracy, for some practical problems in which the time since loading exceeds about 3 months. However, for old concrete, creep is underestimated.[18]

Assuming $E(t_0) = E(t) = E_{28}$, the differential equation according to the ID method is of the form (cf. Equation (9.6))

$$\frac{d\varepsilon}{dt} = \frac{\sigma(t)}{E_{28}}\frac{d\phi_f}{dt} + \frac{1+\phi_d}{E_{28}}\frac{d\sigma}{dt} + \frac{d\varepsilon_{sh}}{dt} \qquad (9.9a)$$

or

$$\frac{d\varepsilon}{d\phi_f} = \frac{\sigma(t)}{E_{28}} + \frac{1+\phi_d}{E_{28}}\frac{d\sigma}{d\phi_f} + \frac{d\varepsilon_{sh}}{d\phi_f} \qquad (9.9b)$$

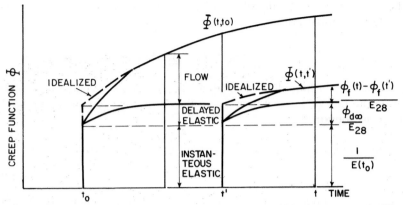

**Figure 9.3** A comparison between actual and idealized creep functions for the ID method

### 9.3.5 Rheological models (RM method)

The use of rheological models is advocated by a number of researchers.[13,16,17] According to Poulsen concrete can best be represented by the Burgers model of Figure 9.4. According to Jordaan *et al.*[17], the creep function is represented by the equation

$$\Phi(t, t') = \frac{1}{E(t')} + [f_3(t) - f_3(t')] + A(t')\left[1 - \exp\left(-\frac{t - t'}{h(t')}\right)\right] \qquad (9.10)$$

in which the function, $f_3(t)$, represents the irreversible creep of the dashpot element at time $t$; and the parameters $A(t')$ and $h(t')$ apply to the Kelvin unit of Figure 9.4. The effect of temperature or humidity on the creep function is included in Equation (9.10) by introducing creep normalizing functions $\psi(T)$ or $\rho(H)$ to the irreversible component of creep, the delayed

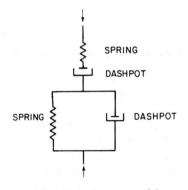

**Figure 9.4** Burgers model

elastic component not being affected by a change in temperature or humidity. Including temperature,

$$\Phi(t, t', T) = \frac{1}{E(t')} + \psi(T)[f(t) - f(t')] + A\left[1 - \exp\left(-\frac{t - t'}{h(t')}\right)\right] \quad (9.11)$$

If the variation in humidity is to be considered $\psi(T)$ is replaced by $\rho(H)$. Relating the rate of reversible creep to that of the irreversible creep by the factor $\lambda$, we obtain

$$\Phi(t, t', T) = \frac{1}{E(t)} + \psi(T)[f(t) - f(t')] + A\{1 - \exp[-\lambda(f(t) - f(t'))]\} \quad (9.12)$$

The use of a pseudo-time, $\tau = f(t)$, $\tau' = f(t')$, allows further simplification through the use of viscoelastic methods of solution.[16] Equation (9.12) then becomes Equation (9.13) and corresponds to a Burgers model with non-aging coefficients when $E(t')$ is taken as a constant, with pseudo-time $\tau$ as an independent variable:

$$\Phi(\tau - \tau', T) = \frac{1}{E} + \psi(T)(\tau - \tau') + A\{1 - \exp[-\lambda(\tau - \tau')]\} \quad (9.13)$$

The method of analysis is based on the linear viscoelasticity theory using correspondence principles for the general non-homogeneous case in which temperature and humidity variation are taken into account. Using the creep function of Equation (9.13) the total strain may be written as

$$\varepsilon(x, \tau) = \int_0^\tau \frac{\partial\sigma(x, \tau')}{\partial\tau'} \left(\frac{1}{E} + \psi(T)(\tau - \tau') + A\{1 - \exp[\lambda(\tau - \tau')]\}\right) \quad (9.14)$$

This equation is solved using the Laplace transformation.

A creep function similar to Equation (9.10) has recently been used by Nielsen[25] to formulate the updated Dischinger method (UD method).

### 9.3.6 Superposition of virgin creep curves: step-by-step method

The principle of superposition is generally accepted as applicable to concrete. Correctly interpreted, this means that strain produced at any time $t$ by a stress increment applied at age $t' < t$ is independent of the effects of any stress applied earlier or later. In other words, virgin creep curves are to be superimposed (Figure 9.5). For increasing stress and slightly decreasing stress, superposition of virgin creep curves gives good agreement with experimental data, but for complete unloading, the recovery is somewhat overestimated, as indicated in Figure 9.5. However, this is not considered a

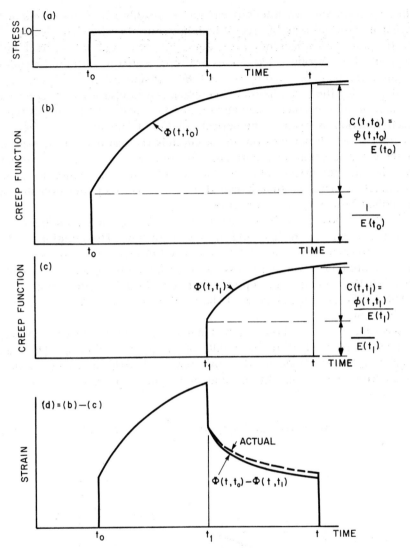

**Figure 9.5** Superposition of virgin creep curves

serious deficiency because the case of sudden, complete unloading is not a common practical case.[15]

The total strain at time $t$, due to a varying stress starting from and initial value $\sigma_0$,

$$\varepsilon(t) = \frac{\sigma_0}{E(t_0)}[1+\phi(t,t_0)] + \int_{t_0}^{t} \Phi(t,t')\,d\sigma(t') \tag{9.15}$$

314 *Creep and Shrinkage in Concrete Structures*

This integral (Stieltjes integral) cannot be readily solved in closed form, except if a special type of creep function is introduced which specifies that $d\phi/dt =$ constant, which however, does, not represent the correct virgin creep function. Virgin creep curves are represented much more accurately by affine creep curves than by parallel ones. If similarity of the creep curves is adopted for the formulation of the creep function, it is to be represented in the form of a product of age and duration (Equation (9.1)). If this type of creep function is used, a numerical solution technique is required, the accuracy of which depends on how accurately the test data are represented by the creep function for a particular concrete, and on the number of steps used in the numerical solution. The step-by-step procedure is actually completely general in that it can deal with any creep function and any prescribed history of stress or strain.

For the purpose of analysis, the total time is divided into a number of time steps the length of which should increase with time. The notation used for the step-by-step numerical analysis is defined in Figure 9.6. For best results under continuously varying stress the time intervals $\Delta t_j$ should be chosen such that their lengths are approximately equal on a log-time plot.

Three different formulations of the step-by-step method are reported in the literature.

According to the method used by Ghali *et al.*,[19] the stress increment $\Delta\sigma_j$ is assumed to be applied at the middle of the interval (time $t_j$). The elastic strain component is calculated at that time with $E(t_j)$, and creep is determined from time $t_j$ on. At the end of the interval (time $t_{j+1}$) the strain increment is:

$$\Delta\varepsilon_j(t_{j+1}) = \Delta\sigma_j\Phi(t_{j+1}, t_j) = \frac{\Delta\sigma_j}{E(t_j)}[1+\phi(t_{j+1}, t_j)] \qquad (9.16)$$

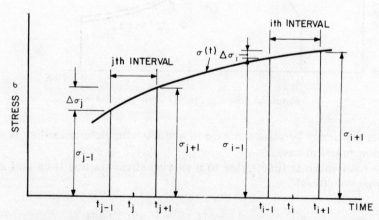

**Figure 9.6** Definition of the time intervals for the step-by-step method

and at the end of the $i$th interval at time $t_{i+1}(>t_{j+1})$

$$\Delta \varepsilon_j (t_{j+1}) = \Delta \sigma_j \Phi(t_{i+1}, t_j) = \frac{\Delta \sigma_j}{E(t_j)} [1 + \phi(t_{i+1}, t_j)] \qquad (9.17)$$

The second method, used by Bazant and Najjar,[31] corresponds to the trapezoidal rule. If relatively large time steps are used Simpson's rule may be used to obtain accurate results.[32]

The total strain at the end of the $i$th interval is the sum of the strains due to stress increments $\Delta \sigma_j$ applied during all the previous increments[19] (cf. Equation (9.17)):

$$\varepsilon(t_{i+1}) = \sum_{j=1}^{i} \Delta \varepsilon_j (t_{i+1}) = \sum_{j=1}^{i} \frac{\Delta \sigma_j}{E(t_j)} [1 + \phi(t_{i+1}, t_j)] \qquad (9.18)$$

This equation can be written in abbreviated form

$$\varepsilon_{i+1} = \sum_{j=1}^{i} \Delta \varepsilon_j = \sum_{j=1}^{i} \frac{\Delta \sigma_j}{E_j} (1 + \phi_{ij}) \qquad (9.19)$$

Using Equation (9.19) to predict the stress for a prescribed strain history, we put $\Delta \sigma_j = (\sigma_{j+1} - \sigma_{j-1})$ and separate the stress increment occurring during the $i$th interval from the summation term, to obtain

$$\sum_{j=1}^{i} \Delta \varepsilon_j = (\sigma_{i+1} - \sigma_{i-1}) \frac{1}{E_i} (1 + \phi_{ii}) + \sum_{j=1}^{i-1} (\sigma_{j+1} - \sigma_{j-1}) \frac{1}{E_j} (1 + \phi_{ij}) \qquad (9.20)$$

Rearranging this equation, we find an expression for the stress at the end of the $i$th interval, provided the stress at the beginning of the same interval is known:

$$\sigma_{i+1} = \sigma_{i-1} + \frac{E_i}{1 + \phi_{ii}} \left( \sum_{j=1}^{i} \Delta \varepsilon_j - \sum_{j=1}^{i-1} (\sigma_{j+1} - \sigma_{j-1}) \frac{1}{E_j} (1 + \phi_{ij}) \right) \qquad (9.21)$$

The term $\Delta \varepsilon_j$ can include any type of strain change during the $j$th interval, such as strain imposed by compatibility conditions, by external stress, shrinkage, or swelling. Temperature strains can also be included, providing it is remembered that creep characteristics of concrete are a function of temperature. For pure relaxation (i.e. without shrinkage), the term $\Delta \varepsilon_j$ is equal to the elastic imposed strain during the first interval ($\Delta t_1 = 0$) and zero for all subsequent intervals.

The accuracy of the results increases with the number of steps but reasonable results are obtained with as few as five steps so that hand computation is possible.[19]

### 9.3.7   The Trost–Bazant method (TB method)

A practical method for computing directly the strain under a varying stress, or stress under a constant or varying strain has been developed by Trost in 1967[20] and later improved by Bazant[21] who calls his method the 'age adjusted modulus method'. This introduces the concept of an 'aging' coefficient.

To explain the aging coefficient concept we express the total strain resulting from an initial stress applied at age $t_0$, $\sigma(t_0) = \sigma_0$, and the subsequent continuously varying stress in the form

$$\varepsilon(t) = \frac{\sigma_0}{E(t_0)}[1 + \phi(t, t_0)] + \int_{t_0}^{t} \frac{1 + \phi(t, t')}{E(t')} \frac{\partial \sigma(t')}{\partial t'} \, dt' \qquad (9.22)$$

Evaluating the integral of this equation for an assumed variation of stress with time and expressing the change of stress in the form

$$\Delta\sigma(t) = \int_{t_0}^{t} \frac{\partial \sigma(t')}{\partial t'} \, dt' = \sigma(t) - \sigma_0 \qquad (9.23)$$

where $\sigma(t)$ is the total stress at time $t$, we can rewrite Equation (9.15):

$$\varepsilon(t) = \frac{\sigma_0}{E(t_0)}[1 + \phi(t, t_0)] + \frac{\sigma(t) - \sigma_0}{E(t_0)}[1 + \chi(t, t_0)\phi(t, t_0)] \qquad (9.24)$$

In this equation, $\chi(t, t_0)$ is the aging coefficient which accounts for the effect of aging on the ultimate value of creep for stress increments or decrements occurring gradually after the application of the original load.

Trost established the value of the aging coefficient $\chi$ by assuming the variation in strain to be affine to the creep-time function. Later Bazant[21] formulated the method more rigorously, proposing that the aging coefficient be determined from the relaxation functions established by a step-by-step procedure from creep functions (e.g. Equation (9.21)). The aging coefficient can be extracted from Equation (9.24) if the stress $\sigma(t)$ is known. Realizing that for relaxation, $\varepsilon(t)$ is constant and equal to $\sigma_0/E(t_0)$ we can solve for $\chi(t, t_0)$:

$$\chi(t, t_0) = \frac{\sigma_0}{\sigma_0 - \sigma(t)} - \frac{1}{\phi(t, t_0)} \qquad (9.25)$$

If a unit strain is applied at age $t_0$, we obtain

$$\chi(t, t_0) = \frac{E(t_0)}{E(t_0) - R(t, t_0)} - \frac{1}{\phi(t, t_0)} \qquad (9.26)$$

The aging coefficient $\chi$ can be defined for any creep function, however, for practical creep analysis there is a need to establish the values of $\chi$ for different ages at loading, different times under load, and for different

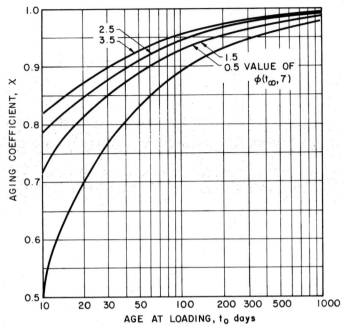

**Figure 9.7** Aging coefficient χ as a function of the age at loading, $t_0$, based on the ACI creep function for structural concrete

member dimensions. Bazant[21] used the ACI creep functions for structural concrete and mass concrete[3] to define numerical values of the aging coefficient (Figure 9.7).

The aging coefficients based on the 1978 CEB-FIP creep functions have recently been established by Favre *et al.*[22] and by Neville and Dilger.[23] Samples of the charts of Neville and Dilger[23] are given in Figures 9.8 and

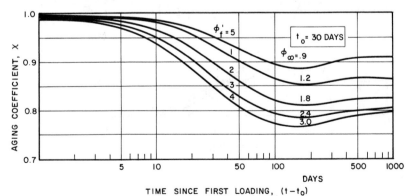

**Figure 9.8** Aging coefficient $\chi(t, t_0)$ as a function of time under load for $t_0 = 30$ days and for all theoretical member thicknesses

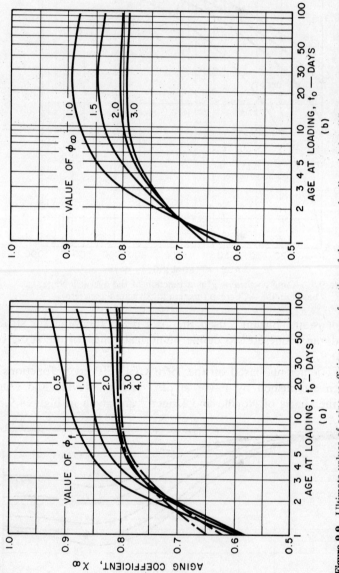

**Figure 9.9**  Ultimate values of aging coefficient χ as a function of the age at loading: (a) for different values of the flow coefficient $\phi_f$; (b) for different values of the ultimate creep coefficient $\phi_\infty$ (based on $E(t_0)$).

9.9. It is interesting to note that the aging coefficients vary very little for different member thicknesses for load durations up to 1000 days so that only one curve is needed for particular values of $t_0$ and $\phi_f$ and all member thicknesses (Figure 9.8). The final values for an effective thickness $h =$ 0.20 m are presented in Figure 9.9. Similar graphs are needed for other values of $h$. It is also interesting to note that, theoretically, the aging coefficient must be equal to 0.5 shortly after the initial application of the load if a continuous creep function is used.[32] This can be derived easily from Equation (9.16) by considering only one interval after the application of the initial load. However, if the CEB-FIP creep formulation is used which contains a more or less instantaneous irreversible creep term $\beta_a$, the value of $\chi$ is approximately 1.0 shortly after loading.

A comparison of aging coefficients extracted from experimental data of Bastgen[24] with those according to Trost,[20] Bazant,[21] Schade[32] and Neville and Dilger,[23] the latter based on the 1978 CEB-FIP creep function, is shown in Figure 9.10.

The aging coefficients vary widely for load duration less than 100 days but are rather close to the experimental values for $(t - t_0) > 100$ days. From this figure it is obvious that the aging coefficients established on the basis of the CEB-FIP creep function best represent the experimental trend. The fact that values of the aging coefficient $\chi$ differ significantly during the early period under load is not very significant because the aging coefficients are associated with small creep coefficients so that the product of the two factors

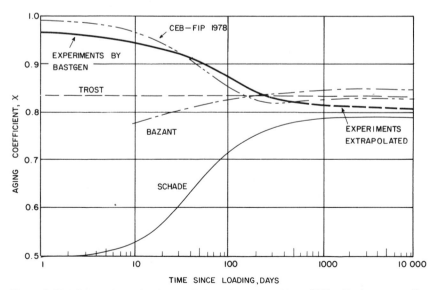

**Figure 9.10** Comparison of aging coefficients from experiments[24,34] with those according to the CEB creep function[23] to Trost[20], to Bazant[21], and to Schade[32]

does not significantly affect the total value of the term $[1 + \chi\phi(t, t_0)]$ which is associated with the time-dependent stress.

Some further remarks should be made regarding the aging coefficients and the relaxation function from which they were derived. Under certain conditions, the 1964 and 1970 CEB-FIP creep functions yield nonsensical results. As shown by Haas,[33] the relaxation function obtained with the 1970 CEB-FIP creep function for concrete members with an effective thickness $h_0 > 0.20$ m and loaded at a relatively early age, ends up by giving tensile stresses even though the initial stress was compression. The same was found by El-Shafey[35] for the ACI creep function for $t_0 < 5$ days. No tensile stresses develop under any condition when the 1978 CEB-FIP creep function is used.

From these observations and the discussions of the different types of creep functions, it must be concluded that at the present time no method can be considered to be exact. This is so because there are certain non-linearities in the behaviour of concrete, particularly shortly after loading, which cannot be represented by linear creep models. In addition, creep is not strictly linearly related to stress, which further complicates the situation.

Recently, Bazant and Kim[36] proposed a simple method to approximate the relaxation function $R(t, t')$ for a given creep function $\Phi(t, t')$ so that the aging coefficient $\chi(t, t')$ can be readily determined from Equation (9.26). This method does not require tabulation of any coefficient and works well for all conceivable time shapes of the concrete creep function. The following empirical formula represents the approximate relationship between the relaxation function and the creep function:

$$R(t, t') = \frac{1 - \Delta_0}{\Phi(t, t')} - \frac{0.115}{\Phi(t, t-1)}\left(\frac{\Phi(t' + \theta, t')}{\Phi(t, t-\theta)} - 1\right) \qquad (9.27)$$

where

$t'$ = age at loading (days)
$t$ = time since casting of the concrete
$\Delta_0$ = correction factor
$\theta = \frac{1}{2}(t - t')$

For a particular age at loading the variable $t'$ is replaced by $t_0$. The coefficient $\Delta_0$ introduces a relatively minor age-independent correction factor which is generally less than 0.02 and may be neglected, i.e. $\Delta_0 = 0$. More accurately $\Delta_0 \approx 0.008$.

## 9.4 REDISTRIBUTION OF STRESSES IN UNCRACKED SECTIONS

Having established the general principles of the different methods of creep analysis, we apply the ID method to a simple column and discuss the analysis of cross sections using the 'creep-transformed section modulus' which is based on the TB method.

### 9.4.1 ID method

The redistribution of stresses in a concentrically loaded column is investigated. At time $t_0$, the stress in concrete due to a normal force $N_0$ is

$$\sigma(t_0) = \frac{N_0}{A_c + n_0 A_s} = \frac{N_0}{A_c(1 + \rho n_0)} \tag{9.28}$$

At time $t$, the equilibrium conditions are expressed by

$$d\sigma_s(t) = -\frac{1}{\rho} d\sigma(t) \tag{9.29}$$

and compatibility requires that

$$d\varepsilon(t) = d\varepsilon_s(t) \tag{9.30}$$

The strain increment in the concrete developing during time interval $dt$ is expressed by Equation (9.9b). Putting $d\varepsilon_{sh} = (\varepsilon_{sh\infty}/\phi_{f\infty}) d\phi_f$ in Equation (9.9a) we can write Equation (9.30) in the form

$$\frac{\sigma(t)}{E_{28}} d\phi_f + \frac{1 + \phi_d}{E_{28}} d\sigma + \frac{\varepsilon_{sh\infty}}{\phi_{f\infty}} d\phi_f = \frac{d\sigma_s(t)}{E_s} \tag{9.31}$$

With $n_0 = E_s/E_{28}$ we obtain after rearranging:

$$\frac{d\sigma(t)}{d\phi_f} + \frac{\rho n_0}{1 + \rho n_0(1 + \phi_d)} \left( \sigma(t) + \frac{E_s \varepsilon_{sh\infty}}{\phi_{f\infty}} \right) = 0 \tag{9.32}$$

Solution of this differential equation is obtained with the boundary conditions at $t = t_0$:

$$\phi_f(t_0, t_0) = 0 \quad \text{and} \quad \bar{\sigma}(t_0) = \sigma(t_0) \frac{1 + n_0 \rho}{1 + \rho n_0[1 + \phi_d(t, t_0)]}$$

(i.e. assuming $\phi_d(t, t_0)$ to occur instantaneously). For brevity we put $\phi_d(t, t_0) = \phi_d$ and obtain

$$\sigma(t) - \bar{\sigma}(t_0) = -\left( \bar{\sigma}(t_0) + \frac{\varepsilon_{sh\infty} E_{28}}{\phi_{f\infty}} \right)$$
$$\times \left[ 1 - \exp \left( -\frac{\rho n_0}{1 + \rho n_0(1 + \phi_d)} \phi_f(t, t_0) \right) \right] \tag{9.33a}$$

Introducing the notation $\alpha = \rho n_0/(1 + \rho n_0)$ we find:

$$\sigma(t) - \bar{\sigma}(t_0) = -\left( \bar{\sigma}(t_0) + \frac{\varepsilon_{sh\infty} E_{28}}{\phi_{f\infty}} \right) \left[ 1 - \exp \left( \frac{\alpha \phi_f(t, t_0)}{1 + \alpha \phi_d} \right) \right] \tag{9.33b}$$

The steel stress is calculated from equilibrium considerations requiring that $\sigma(t)A_c + \sigma_s(t)A_s = N_0$:

$$\sigma_s(t) = [\sigma(t_0)(1 + \rho n_0) - \sigma(t)]\frac{1}{\rho} \tag{9.34}$$

so that the change in stress

$$\Delta\sigma_s(t) = \sigma_s(t) - \sigma_s(t_0) \tag{9.35}$$

Expressing the change in stress in the steel in terms of the initial concrete stress $\sigma(t_0)$ we find after some mathematical manipulation the rather lengthy expression

$$\Delta\sigma_s(t) = n_0\left\{\left[\frac{1}{\alpha} - \frac{1-\alpha}{\alpha(1+\phi_d)}\exp\left(-\frac{\alpha\phi_f}{1+\alpha\phi_d}\right)\right] - 1\right\}\sigma(t_0)$$
$$+ \frac{1-\alpha}{\alpha\phi_f}\left[1 - \exp\left(\frac{\alpha\phi_f}{1+\phi_d}\right)\right]\varepsilon_{sh}(t, t_0)E_s \tag{9.36}$$

To simplify calculations, Rüsch et al.[14,37] introduced the multipliers:

$$\xi_g = \frac{1}{\alpha} - \frac{1-\alpha}{\alpha(1+\phi_d)}\exp\left(-\frac{\alpha\phi_f}{1+\alpha\phi_d}\right) \tag{9.37}$$

and

$$\xi_s = \frac{1}{\alpha\phi_f}\left[1 - \exp\left(\frac{\alpha\phi_f}{1+\alpha\phi_d}\right)\right] \tag{9.38}$$

With these multipliers we write Equation (9.36) in the simpler form

$$\Delta\sigma_s(t) = n_0(\xi_g - 1)\sigma(t_0) + \xi_s(1-\alpha)\varepsilon_{sh}(t, t_0)E_s \tag{9.39}$$

Considering a prestressed member with the tendons eccentric by $y_1$ we can express the loss of prestress in the form:[14,37]

$$\Delta\sigma_{ps}(t) = n_0\left((\xi_g - 1)\sigma(t_0) + (\xi_p - 1)\frac{1-\alpha}{\alpha}\sigma_p(t_0) + \xi_s(1-\alpha)\varepsilon_{sh}(t, t_0)E_c\right) \tag{9.40}$$

Here

$$\xi_p = \frac{1}{1+\phi_d}\exp\left(-\frac{\alpha\phi_f}{1+\alpha\phi_d}\right) \tag{9.41}$$

In case of an eccentric cable, the stiffness coefficient $\alpha$ is defined by the relation

$$\alpha = n_0\frac{A_s}{A_c'}[1 + (y_1'/r')^2] \tag{9.42}$$

where $A_c'$, $r'$, and $y_1'$ are, respectively, the area, the radius of gyration, and

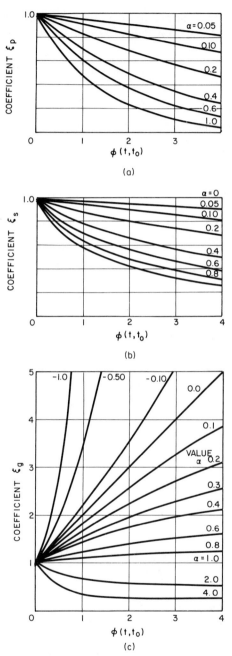

**Figure 9.11** Coefficients $\xi_p$, $\xi_s$ and $\xi_q$ as a function of the creep coefficient $\phi(t, t_0)$ for different values of the stiffness coefficient $\alpha$.

the eccentricity of the prestressing steel in the cross section transformed with $n_0 = E_s/E(t_0)$. The stiffness coefficient $\alpha$ can be defined in a more general way as the ratio

$$\alpha = \frac{\delta_{c1,1}}{\delta_{c1,1} + \delta_{s1,1}} \tag{9.43}$$

where

$\delta_{c1,1}$ = displacement due to a unit force applied to concrete
$\delta_{s1,1}$ = displacement due to a unit force applied to steel

The three coefficients can be described as follows[26]

$\xi_g$ = multiplier of the elastic solution in the case of permanent load applied to the concrete structure after the introduction of additional elastic restraints (Figure 9.11(c))
$\xi_p$ = multiplier of the elastic solution in the case of constant imposed deformation upon additional restraint (Figure 9.11(a))
$\xi_s$ = multiplier of the elastic solution in the case of gradually imposed stress-independent deformations related by affinity to the creep coefficient (Figure 9.11(a))

The three multipliers are depicted in Figure 9.11 as a function of the creep coefficient $\phi(t, t_0)$ for different values of $\alpha$ (from Reference 26). It is evident from this simple case that the numerical effort to compute the change in steel stress is quite considerable if Figure 9.11 is not available.

If more than one layer of steel is present in a cross section the ID method leads to coupled differential equations which are difficult to solve. To circumvent the solution of the coupled differential equations, Busemann's[27] creep fibre method is used.

### 9.4.2 TB method using creep-transformed section properties

Equations for time-dependent stresses in steel and concrete can easily be derived on the basis of equilibrium and compatibility of strains in a section. For a single layer of steel this leads to the well-known equation for loss of prestress:[20,23]

$$\Delta\sigma_{sp}(t) = \frac{n_0\sigma_0\phi(t, t_0) + E_s\varepsilon_{sh}(t, t_0) + \sigma_r'(t)}{1 + \rho n_0(1 + y_1^2/r^2)[1 + \chi\phi(t, t_0)]} \tag{9.44}$$

where

$n_0 = E_s/E(t_0)$ at age of loading
$\sigma_0$ = concrete stress at the level of the steel due to dead load and prestressing
$\sigma_r'(t)$ = reduced relaxation (to be discussed later)
$r$ = radius of gyration of net concrete section
$y_1$ = distance between centroid of net concrete section and steel
$\chi$ = aging coefficient

For $n$ layers of steel (prestressed and/or non-prestressed), a set of $n$ equations with $n$ unknowns has to be solved.

A more elegant solution is obtained using the creep-transformed section properties.[28]

To arrive at the time-dependent stresses and strains, the forces in the steel corresponding to the unrestrained creep, to free shrinkage and to reduced intrinsic relaxation are applied to the creep-transformed cross section in which the steel is included with the modular ratio $n^* = n_0[1 + \chi\phi(t, t_0)]$. For reasons of internal equilibrium the forces change signs when applied to the concrete. The concrete stresses resulting from this analysis are due to all time-dependent effects, and the corresponding steel stresses (obtained with the modular ratio $n^*$) are added to the stresses due to unrestrained creep, free shrinkage, and (reduced) relaxation, to obtain the time-dependent steel stress. The method is entirely general and rigorous and can be applied to any cross section (even a composite one) containing any number of layers of non-prestressed or prestressed steel.

The procedure is now explained in detail for the simple case of a prestressed beam with one layer of prestressing steel.

The steel stress corresponding to unrestrained creep, free shrinkage, and intrinsic relaxation is obtained from the relationship

$$\sigma_s^* = n_0\sigma_0\phi(t, t_0) + \varepsilon_{sh}(t, t_0)E_s + \sigma_r'(t) \tag{9.45}$$

The corresponding normal force is found by multiplying this stress by the steel area $A_s$:

$$N_s^* = A_s\sigma_s^* \tag{9.46}$$

The eccentric normal force $N^*$ generates a bending moment

$$M_s^* = A_s\sigma_s^* y_1^* \tag{9.47}$$

In these equations the term not previously defined is

$y_1^* =$ distance of centroid of steel from the centroid of the creep-transformed cross section

It is interesting to note that the terms on the right-hand side of Equation (9.45) are those of the numerator of the equation for prestress loss, Equation (9.44).

The concrete stresses corresponding to the forces $N_s^*$ and $M_s^*$,

$$\Delta\sigma(t) = -\left(\frac{N_s^*}{A_c^*} + \frac{M_s^*}{I_c^*} y^*\right) \tag{9.48}$$

are the actual time-dependent stresses in the concrete. In this equation:

$A_c^* =$ cross-sectional area of the concrete cross section in which the steel is transformed with $n^* = n_0[1 + \chi\phi(t, t_0)]$

$I_c^* =$ moment of inertia of the concrete cross section in which the steel is transformed with $n^* = n_0[1 + \chi\phi(t, t_0)]$

The steel stress obtained from the relationship

$$\Delta\sigma_s^*(t) = -\left(\frac{N_s^*}{A_c^*} + \frac{M_s^*}{I_c^*} y_1^*\right)n^*$$ (9.49)

is to be added to the stress $\sigma_s^*$, expressed by Equation (9.45), to obtain the time-dependent change in stress. Thus,

$$\Delta\sigma_s(t) = \sigma_s^*(t) + \Delta\sigma_s^*(t)$$ (9.50a)

$$= [n_0\sigma_0\phi(t, t_0 + \varepsilon_{sh}(t, t_0)E_s + \sigma_r'(t)] + \Delta\sigma_s^*(t)$$ (9.50b)

For more than one layer of steel, the steel $\sigma_s^*(t)$, has to be found for each individual layer, and the normal force and bending moment due to the stresses in all layers have to be determined. For $m$ layers:

$$N_s^* = \sum_{i=1}^{m} \sigma_{s,i}^*(t)A_{s,i}$$ (9.51)

and

$$M_s^* = \sum_{i=1}^{m} \sigma_{s,i}^*(t)y_i^*A_{s,i}$$ (9.52)

### 9.4.3 Deformations

The time-dependent deformations are calculated by multiplying the initial deformations by the creep coefficient $\phi(t, t_0)$, adding the deformations due to free shrinkage, and then deducting the deformations due to the moments $N_s^*$ and $M_s^*$ which are:

$$\Delta\varepsilon^*(t) = -\frac{N_s^*}{A_c^*E_c^*} = -\frac{N_s^*[1 + \chi\phi(t, t_0)]}{A_c^*E_0}$$ (9.53)

and

$$\Delta\psi^*(t) = -\frac{M_s^*}{I_c^*E_c^*} = -\frac{M_s^*[1 + \chi\phi(t, t_0)]}{I_c^*E_0}$$ (9.54)

The total axial strain at the level of the centroid of the elastic (transformed) section is then given by the expression

$$\varepsilon(t) = \frac{N_0}{A_c'E_c}[1 + \phi(t, t_0)] + \varepsilon_{sh}(t, t_0) - \frac{N_s^*}{A_c^*E^*} - \frac{M_s^*}{I_c^*E^*} y_c^*$$ (9.55)

where $y_c^*$ is the distance between the centroid of the elastic (transformed) section and that of the creep-transformed section. The total curvature is expressed by

$$\psi(t) = \frac{M_0}{I_c'E_0}[1 + \phi(t, t_0)] - \frac{M_s^*}{I_c^*E_0}[1 + \chi\phi(t, t_0)]$$ (9.56)

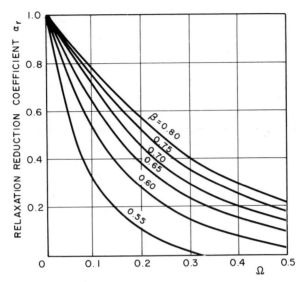

**Figure 9.12** Relaxation reduction coefficient versus $\Omega$ for different values of $\beta$

The section properties with a prime refer to the section transformed with $n_0$, and those with an asterisk to those transformed with $n^* = n_0(1 + \chi\phi)$.

In the above equations, the reduced relaxation $\sigma_r'$ and not the intrinsic relaxation $\sigma_r$ is used in order to include the reduction of the relaxation losses due to creep and shrinkage. The interrelationship between losses due to creep and shrinkage of the concrete and relaxation of the steel can be taken into account accurately by the procedure developed by Tadros *et al.*.[29,30] Based on a step-by-step numerical procedure, a chart has been developed (Figure 9.12) which gives the relaxation reduction coefficient $\alpha_r$ as a function of the ratio

$$\Omega = \frac{\text{loss due to creep and shrinkage}}{\text{initial prestress}} = \frac{\Delta\sigma_{ps(cp+sh)}}{\sigma_{ps0}}$$

for different ratios

$$\beta = \frac{\text{initial prestress}}{\text{ultimate strength}} = \frac{\sigma_{ps0}}{f_{pu}}$$

The reduced relaxation is expressed by

$$\sigma_r'(t) = \alpha_r\sigma_r(t) \qquad (9.57)$$

Comparing this approach with that recommended by CEB[5] it is found that CEB is conservative in all cases.

The creep-transformed section method is particularly elegant in solving the complex problem of time-dependent stresses and deformations in a composite member. Refer to Figure 9.13. The initial elastic strain $\varepsilon_1$ of the precast section (subscript 1) due to girder weight (moment $M^{(1)}$) and prestressing, will increase owing to unrestrained creep and free shrinkage from the beginning of the composite action (time $t_1$) till time $t$ by $\{\varepsilon_1^{(1)}[\phi_1(t, t_0) - \phi_1(t_1, t_0)] + \varepsilon_{sh1}(t, t_1)\}$. At the level of the centroid of the cast-in-place deck (fibre 2) this increase is $\{\varepsilon_{1,2}^{(1)}[\phi_1(t, t_0) - \phi_1(t_1, t_0)] + \varepsilon_{sh1}(t, t_1)\}$. During the same time the deck shrinks by an amount $\varepsilon_{sh2}(t, t_1)$ where the ages $t$ and $t_1$ are counted from the moment at which the composite action begins, which is normally one or two days after casting of the deck concrete.

If the weight of the slab is carried by the precast girder (unshored construction) the time-dependent strain in fibre 2 is increased by $\varepsilon_{1,2}^{(2)}\phi_1(t, t_1)$ where $\varepsilon_{1,2}^{(2)}$ is the elastic strain in fibre 2 due to the moment $M^{(2)}$ (caused by the slab weight) in the precast girder. Moments due to superimposed loads applied after the commencement of the composite action are considered separately.

With these strains we now calculate the steel stresses $\sigma_s^*$, for each fibre containing steel, and the corresponding normal forces and bending moments in the way just described. Relaxation of the steel is included by adding the reduced relaxation $\sigma_r'(t)$ to the stresses of the prestressed layer(s), if any. In addition to the steel forces, the deck generates a normal force and a bending moment. The normal force corresponds to the difference between the free shrinkage of the deck, $\varepsilon_{sh2}(t, t_1)$, and the strain in fibre 2 that develops after time $t_1$ due to the forces acting on the girder:

$$\Delta\varepsilon_2^*(t, t_1) = \varepsilon_{1,2}^{(1)}[\phi_1(t, t_0) - \phi_1(t_1, t_0)] + \varepsilon_{1,2}^{(2)}\phi_1(t, t_1) + \varepsilon_{sh1}(t, t_1) - \varepsilon_{sh2}(t, t_1) \tag{9.58}$$

The force in the deck corresponding to this difference in strains is

$$N_c^* = \Delta\varepsilon_2^*(t, t_1)E_2^*(t_1)A_{c2} \tag{9.59}$$

where $E_2^*(t_1) = E_2(t_1)/[1 + \chi_2\phi_2(t, t_1)]$ is the age adjusted modulus of deck concrete. The moment about the centroid of the creep-transformed section is $N_c^* y_c^*$.† In addition to this moment, a small moment is caused in the concrete deck by the time-dependent curvature that develops in the precast girder after the beginning of the composite action. This moment is found from the relationship

$$M_{c2}^* = \Delta\psi(t, t_1)I_{c2}E_2^*(t_1)$$
$$= \left(\frac{M^{(1)}}{I_c'E_1(t_0)}[\phi_1(t, t_0) - \phi_1(t_1, t_0)] + \frac{M^{(2)}}{I_c'E_1(t_1)}\phi_1(t, t_1)\right)I_{c2}E_2^*(t_1) \tag{9.60}$$

† $y_c^*$ is defined in Figure 9.13.

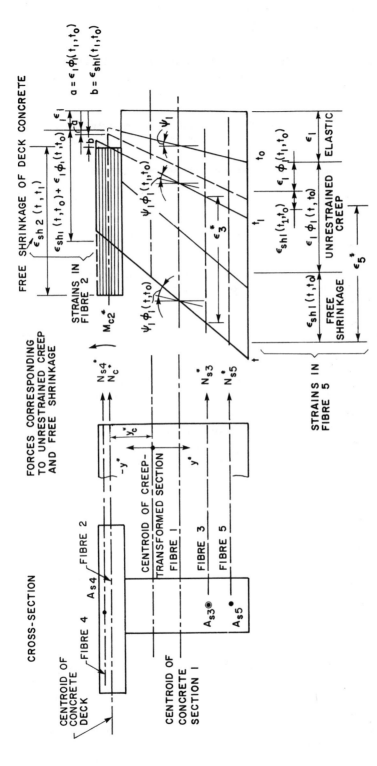

**Figure 9.13** Strains in composite girder due to initial forces on girder, unrestrained creep, and free shrinkage: $t_0$ is age at prestressing of the precast girder; $t_1$ is age at casting of the deck and cast-in-place deck. Subscripts 1 and 2 refer respectively to precast girder and cast-in-place deck

where $I_{c2}$ is the moment of inertia of the concrete deck. The total moment generated by the deck concrete is thus

$$M_c^* = N_c^* y_c^* + M_{c2}^* \tag{9.61}$$

The total normal forces and bending moments caused by the steel and the concrete deck are applied to the creep-transformed section in which the deck is included with $E_{c2}^*/E_{c1}^*$ its area.

The time-dependent deformations (strains, curvatures, and deflections) are obtained by adding to those due to unrestrained creep and shrinkage the values due to the forces $N^*$ and $M^*$, calculated with the age-adjusted effective modulus $E_{c1}^* = E_1(t_1)/[1 + \chi\phi(t, t_1)]$. More details and numerical examples are given in Neville and Dilger[23] and Dilger.[28]

## 9.5 ANALYSIS OF STATICALLY INDETERMINATE STRUCTURES

Nearly all concrete structures are built in stages. This means that the structural analysis performed for the structure as a whole gives only approximate values of the forces present in such a structure. For most structures such an analysis is satisfactory, but there are cases in which the forces shortly after a certain stage of construction differ substantially from those obtained from an analysis of the structure as a whole. In this case, and under conditions of changing support conditions, substantial time-dependent forces are induced, and these may alter radically the elastic moment configuration. The structures in which time-dependent effects are most pronounced are those made continuous at a later stage, or those undergoing differential settlement.

### 9.5.1 ID method

The time-dependent forces (reaction or moment) resulting from creep and shrinkage can be expressed in a way similar to that discussed in context with the redistribution of stresses in a section. For a structural system with one redundant we find:[14,37]

$$X_1(t) = \xi_g X_{1,g} + \xi_p X_{1,p0} + \xi_s X_{1,s\infty}' / \phi_f \tag{9.62}$$

In this equation:

$X_1(t) =$ value of the redundant at time $t$

$X_{1,g} =$ elastic value of the redundant at coordinate 1 due to permanent external loads

$X_{1,p} =$ elastic value of the redundant due to imposed force (e.g. prestressing or instantaneous settlement)

$X_{1,sh}' =$ fictitious elastic value of the redundant due to shrinkage without relief by creep.

With this equation, simple cases can be solved very easily, but if different material properties are present in a continuous structure, the equations become rather complex and no solutions are known which can deal with the general case of a continuous composite member which can consider the effects of prestressed and non-prestressed steel.

### 9.5.2 TB method

The analysis of the time-dependent forces in continuous structures due to creep and shrinkage is conceptually easy to understand if the aging coefficient is used. However, it should be emphasized that the use of the aging coefficient is applicable only under the following conditions:[20,21,23,39]

(a) the support conditions are changed suddenly as in the case of an instantaneous differential settlement or in stage construction where additional forces are applied suddenly (e.g. by prestressing); and

(b) the support conditions or the imposed deformations change at the same rate as creep, as in the case of shrinkage or differential settlement developing approximately at the same rate as creep.

If the changes occur considerably faster or slower than creep, the aging coefficient cannot be used. The case of slow or rapid differential settlement is to be treated by a different approach.

When the condition (a) or (b) is satisfied we can find the time-dependent forces in a continuous (uncracked) concrete structure by expanding the well-known elastic equations of Müller–Breslau to include the time-dependent effects. To do this we determine the *time-dependent* displacements resulting from:

(a) statistically determinate forces due to sustained loads (including prestressing);

(b) known statically indeterminate (redundant) forces due to sustained loads (including prestressing);

(c) unknown statically indeterminate time-dependent forces.

Due to creep, the displacement at coordinate $i$ due to external loads in the released structure is equal to

$$\delta_i(t) = \delta_i(t_0)\phi(t, t_0) \tag{9.63}$$

where

$\delta_i(t_0)$ = elastic displacement at coordinate $i$ due to a sustained load applied at age $t_0$ to the released structure

$\phi(t, t_0)$ = creep coefficient at time $t$ for a load applied at age $t_0$.

Due to the constant redundant force $X_i(t_0)$, we have a time-dependent displacement at coordinate $i$

$$X_i(t_0)\,\delta_{ii}(t) = X_i(t_0)\,\delta_{ii}\phi(t, t_0) \tag{9.64}$$

and at coordinate $j$

$$X_i(t_0)\,\delta_{ij}(t) = X_j(t_0)\,\delta_{ij}\phi(t,\,t_0) \tag{9.65}$$

A time-dependent force $\Delta X_i(t)$ (which is zero at age $t_0$) developing at coordinate $i$ results in the following time-dependent displacements:

at coordinate $i$

$$\Delta X_i(t)\,\delta_{ii}(t) = \Delta X_i(t)\,\delta_{ii}[1 + \chi\phi(t,\,t_0)] \tag{9.66}$$

and at coordinate $j$

$$\Delta X_i(t)\,\delta_{ij}(t) = \Delta X_i(t)\,\delta_{ij}[1 + \chi\phi(t,\,t_0)] \tag{9.67}$$

In the above equations, $\delta_{ii}$ and $\delta_{ij}$ are flexibility coefficients, i.e. displacements at coordinate $i$ or $j$ due to a force $X_i = 1$, and $\chi$ is the aging coefficient, which is introduced here because the force $\Delta X_i(t)$ develops gradually with time. The argument of the aging coefficient is always the same as that of the creep coefficient with which it is associated. The final total force $X_i(t)$ is obtained by adding the initial force $X_i(t_0)$ and the time-dependent force $\Delta X_i(t)$, namely

$$X_i(t) = X_i(t_0) + \Delta X_i(t) \tag{9.68}$$

The above equations do not include the effect of reinforcement on the time-dependent displacements. This, however, does not normally have a pronounced effect on the time-dependent forces because both the displacements due to the initial forces and those due to the time-dependent forces are affected by the presence of steel in much the same way. The effect of the steel can be included accurately by using the properties of the creep-transformed section.

Many examples using the aging coefficient are available in the literature.[20,21,23] We will only look at one case, namely a composite beam made continuous by a cast-in-place joint cast at the same time at the deck.

This composite beam can be analysed using the properties of the creep-transformed section. The properties of the creep-transformed section are used in combination with the age-adjusted effective modulus so that the displacements at coordinates $i$ and $j$ become (cf. Equations (9.66) and (9.67)):

$$\Delta X_i(t)\,\delta_{ii}(t) = \Delta X_i(t)\int \frac{M_{ui}^2}{E_{c1}^* I_c^*}\,\mathrm{d}x = \Delta X(t)\int \frac{M_{ui}^2}{E_1 I_c^*}[1 + \chi_1\phi_1(t,\,t_0)]\,\mathrm{d}x \tag{9.69}$$

and

$$\Delta X_i(t)\,\delta_{ij}(t) = \Delta X_i(t)\int \frac{M_{ui}M_{uj}}{E_c^* I_c^*}\,\mathrm{d}x$$

$$= \Delta X_i(t)\int \frac{M_{ui}M_{uj}}{E_1 I_c^*}[1 + \chi_1\phi_1(t,\,t_0)]\,\mathrm{d}x \tag{9.70}$$

Introducing the notation

$$\delta^*_{ii} = \int \frac{M^2_{ui}}{E_1 I^*_c} \, dx \qquad (9.71)$$

and

$$\delta^*_{ij} = \int \frac{M_{ui}M_{uj}}{E_1 I^*} \, dx \qquad (9.72)$$

we can write Equations (9.69) and (9.70) in the form

$$\Delta X_i(t) \, \delta_{ii}(t) = \Delta X_i(t) \, \delta^*_{ij}[1 + \chi_1 \phi_1(t, t_0)] \qquad (9.73)$$

and

$$\Delta X_i(t) \, \delta_{ij}(t) = \Delta X_i(t) \, \delta^*_{ij}[1 + \chi_1 \phi_1(t, t_0)] \qquad (9.74)$$

These equations are similar to Equations (9.66) and (9.67) except that $I^*_c$ instead of $I_c$ is used in defining the flexibility coefficients. The time-dependent redundant forces are found from the usual compatibility conditions. For a two-span beam, we obtain:

$$\delta_1(t) + \Delta X_1(t) \, \delta^*_{11}[1 + \chi_1 \phi_1(t, t_0)] \qquad (9.75)$$

Rearranging:

$$\Delta M(t) = \Delta X_1(t) = -\frac{\delta_1(t)}{\delta^*_{11}} \frac{1}{1 + \chi_1 \phi_1(t, t_0)} \qquad (9.76)$$

For continuous composite structures this approach offers the possibility to analyse the time-dependent effects directly once the displacement $\delta_1(t)$ is known. $\delta_1(t)$ is defined by the integral

$$\delta_1(t) = \int \psi(t) M_{u1} \, dx \qquad (9.77)$$

where $\psi(t)$ is expressed by Equation (9.56). The term $I^*_c$ includes the concrete deck (with an appropriate value $n^*_2$ for the deck concrete) and any prestressed and non-prestressed steel.

## 9.6 TIME-DEPENDENT ANALYSIS OF COMPLEX STRUCTURES

For complex structures the expressions for time-dependent effects become so involved that solutions are possible only by using the computer. Segmental construction (precast or cast-in-place), multi-stage construction involving multiple layers of concrete and steel and structures with creep and shrinkage properties varying through the section fall into this category. The principles involved in such analysis are, however, the same as discussed previously. A general analysis involves the following:

(1) Adoption of an appropriate constitutive equation capable of modelling the instantaneous and time-dependent properties of the different layers of concrete as a function of time, temperature, humidity, etc..
(2) Compatibility analysis for individual cross sections, based on the assumption that plane sections remain plane.
(3) Analysis of the structure as a whole observing the boundary conditions of the structure.

A step-by-step procedure similar to that discussed before is used for the analysis in which the time is divided into discrete intervals. The stresses and deformations at the end of each interval are calculated by the displacement method of structural analysis, using the stress increments that occurred in the preceding intervals.

In such an analysis, the temperature and humidity fields are, of course, assumed to have reached a steady state. If this is not so, a transient thermal and/or hygral diffusion analysis has to be carried out to establish the temperature and/or moisture conditions at any point in the structure during the time interval in question.[40,41,42] Such analyses are not considered here.

Before discussing the analytical procedure in more detail, a few remarks should be made about the modelling of the material behaviour.

For the analysis of concrete structures the constitutive relations of three materials: concrete, prestressing steel, and non-prestressed steel, are required. The latter is considered to be a linear elastic material obeying Hooke's law. Because of creep the first two materials have time-dependent stress–strain relations. The constitutive equation for concrete under constant temperature and humidity may be of simple form given by ACI Committee 209,[3] or of the form given in Reference 43 expressing the 1978 CEB-FIP creep and shrinkage function in a simplified way. More accurate representations of any constitutive relation is possible by using exponential series expressions. Such expressions are used by many researchers (e.g. Argyris *et al.*,[44] Pister *et al.*,[45] Khalil *et al.*[46]).

The time-dependent properties of the prestressing steel are available in the form of relaxation curves from tests in which the steel tendon is stressed between two fixed points and the reduction in stress is recorded. Empirical equations are available expressing the intrinsic relaxation $\sigma_{r0}$ in terms of the initial stress, $\sigma_{ps0}$, and the temperature $T$.[38,47]

### 9.6.1 Incremental analysis

The incremental analysis involves a pseudo-elastic analysis[48,49] for each interval $i$ using an effective modulus $\bar{E}_i$, for the concrete to be defined shortly, and the free concrete strains developing during that interval. These strains have to be made compatible with those in the various layers of steel.

The strain increment in the concrete, without restraint by the steel, at a point in the concrete cross section, occurring during the interval $i$ (see Figure 9.6) is the difference between the strain at the end and at the beginning of interval $i$ (cf. Equations (9.16) and (9.19)):

$$\Delta \varepsilon_i = \frac{\Delta \sigma_i}{E_i}(1 + \phi_{ii}) + \sum_{j=1}^{i-1} \frac{\Delta \sigma_j}{E_j}(\phi_{ij} - \phi_{i-1,j}) + \Delta \varepsilon_{\text{sh},i} \qquad (9.78)$$

Defining the 'effective' modulus of elasticity for interval $i$ as

$$\bar{E}_i = E_i/(1 + \phi_{ii}) \qquad (9.79)$$

and the 'initial' strain as

$$\Delta \bar{\varepsilon}_i = \sum_{j=1}^{i-1} \frac{\Delta \sigma_j}{E_j}(\phi_{ij} - \phi_{i-1,j}) + \Delta \varepsilon_{\text{sh},i} \qquad (9.80)$$

Equation (9.78) becomes

$$\Delta \varepsilon_i = \frac{\Delta \sigma_i}{\bar{E}_i} + \Delta \bar{\varepsilon}_i \qquad (9.81)$$

Both the effective modulus $\bar{E}_i$ and the 'initial' strain $\Delta \bar{\varepsilon}_i$ are independent of the stresses introduced during interval $i$. The term 'initial' (often used in context with thermal stress analysis) is not to be confused with the instantaneous elastic deformation.

Equation (9.81) reduces the analysis of the time-dependent stresses to a pseudo-elastic analysis during interval $i$.

With the assumption that plane sections remain plane, the time-dependent stresses become linearly distributed over the cross section. It is, therefore, possible to write the increments of axial strain $\Delta \varepsilon_0$ and curvature $\Delta \psi$ of a concrete section during interval $i$ in terms of the axial force and bending moment increments $\Delta N_{c,i}$ and $\Delta M_{c,i}$ as follows:

$$\Delta \varepsilon_{0,i} = \frac{\Delta N_{c,i}}{A_c \bar{E}_i} + \Delta \bar{\varepsilon}_{0,i} \qquad (9.82)$$

and

$$\Delta \psi_i = \frac{\Delta M_{c,i}}{I_c \bar{E}_{c,i}} + \Delta \bar{\psi}_i \qquad (9.83)$$

where $A_c$ and $I_c$ are the area and the moment of inertia of the concrete section. Similarly, the incremental strain in the prestressing steel during the $i$th interval is

$$\Delta \varepsilon_{\text{ps},i} = \frac{\Delta N_{\text{ps},i}}{A_{\text{ps}} E_{\text{ps}}} + \Delta \bar{\varepsilon}_{\text{ps},i} \qquad (9.84)$$

where $A_{ps}$ is the area of the prestressing steel and $\Delta\bar{\varepsilon}_{ps,i}$ is the 'initial' strain increment, equal to $-\Delta f_{r,i}/E_{ps}$. The incremental strain is a function of the initial stress in the prestressing steel and the tendon shortening prior to interval $i$, both of which are known when performing the analysis for interval $i$.

Finally, the incremental strain in the non-prestressed steel is

$$\Delta\varepsilon_{n,i} = \frac{\Delta N_{s,i}}{A_s E_s} \tag{9.85}$$

For linear structures, the analysis for the effects for the initial strains $\Delta\bar{\varepsilon}$ and $\Delta\bar{\psi}$ can be performed by the standard displacement method and for complex structures by the finite element method in a way similar to that used for solving problems of thermal stress analysis.

The analysis of interval $i$ yields increments of stresses and displacements occurring during that interval. These are to be added to the increments occurring during all preceding intervals to give the total values. The analysis continues till the end of the last interval.

Examples of time-dependent analyses of a frame bridge built from long precast beams and leg elements, of a segmental bridge and of a cable-stayed bridge have been reported in the literature.[48,49,50,51] A comprehensive description of the time-dependent analysis by computer for composite structures is given in Schade and Haas.[52] Complex three-dimensional structures have been analysed by the program SMART as reported in Argyris *et al.*.[53]

## 9.7 SUMMARY

The principles behind the better-known methods of creep analysis are reviewed and compared. The improved Dischinger method, the use of rheological models, and the aging coefficient introduced by Trost and Bazant are methods which are suited for creep analysis of structures. The use of the aging coefficient seems to be superior to the other methods because the analysis can be reduced to simple quasi-elastic solutions by using the age-adjusted effective modulus of concrete $E_c^* = E_0/[1 + \chi\phi(t, t_0)]$ or a modular ratio $n^* = n_0[1 + \chi\phi(t, t_0)]$. The main advantage is, perhaps, that well-known concepts such as the transformed cross section and the familiar elasticity equations of Müller–Breslau can be used with only minor modifications for the solution of time-dependent problems of concrete structures.

For complex structures, a step-by-step scheme is discussed in which the time-dependent analysis is performed as a series of quasi-elastic steps with an effective modulus of elasticity for the time interval considered, in combination with the 'initial' strain method well known for the analysis of temperature effects in structures.

# REFERENCES

1. Comité Européen du Béton (1968), *Recommendations for an International Code of Practice for Reinforced Concrete*, American Concrete Institute, Detroit and Cement and Concrete Association, London, 156 pp.
2. CEB-FIP (1970), 'International recommendations for the design and construction of concrete structures—principles and recommendations', *Comité Européen du Béton—Federation Internationale de la Precontrainte, FIP 6th Congr., Prague, June 1970*, Cement and Concrete Association, London, 80 pp.
3. ACI Committee 209 (1971), 'Prediction of creep, shrinkage, and temperature effects in concrete structure', *Designing for Effects of Creep, Shrinkage and Temperature in Concrete Structures*, **SP-27**, American Concrete Institute, Detroit, pp. 21–93.
4. Bažant, Z. P., and Osman, E. (1976), 'Double power law for basic creep of concrete', *Mater. Struct.*, **9**, No. 49, 3–11.
5. CEP-FIP (1978), 'Model Code for Concrete Structures'.
6. McHenry, D. (1943), 'A new aspect of creep in concrete and its application to design', *Proc. ASTM*, **43**, 1069–86.
7. Maslov, G. N., 'Thermal stress state in concrete masses with account to creep of concrete, (in Russian).
8. Glanville, W. H. (1930), 'Studies in reinforced concrete, III—creep or flow of concrete under load', *Building Research Tech. Pap. No. 12, Department of Scientific and Industrial Research, London*.
9. Whitney, C. S., (1982), 'Plain and reinforced concrete arches', *J. Am. Concr. Inst.* **28**, 479–519.
10. Dischinger, F. (1937), 'Untersuchungen über die Knicksicherheit, die elastische Verformung und das Kriechen des Betons bei Bogenbrücken', *Der Bauingenieur* **18**, No. 33/34, 487–520; No. 35/36, 539–52; No. 39/40, 595–621.
11. England, G. L., and Illston, J. M. (1965), 'Methods of computing stress in concrete from a history of measured strain', *Civil Engineering, London*, April, May and June.
12. Glucklich, J., and Ishai, O. (1960–61), 'Rheological behaviour of hardened cement paste under low stress', *J. Am. Concr. Inst.*, **57**, 947–64.
13. Nielsen, L. F. (1970), 'Kriechen und Relaxation des Betons', *Beton. Stahlbeton.* **65**, 272–5.
14. Rüsch, H., Jungwirth, D. and Hilsdorf, H. (1973), 'Kritische Sichtung der Einflüsse von Kriechen und Schwinden des Betons auf das Vehalten der Tragwerke', *Beton. Stahlbeton.*, **68**, Nos. 3, 4, 5.
15. Bažant, Z. P., and Najjar, L. T. (1973), 'Comparison of approximate linear methods for concrete creep', *J. Struct. Div. ASCE*, **99**, No. ST9, 1851–74.
16. England, G. L., and Jordaan, I. J. (1975), 'Time-dependent and steady-state stresses in concrete structures with steel reinforcement at normal and raised temperature', *Mag. Concr. Res.*, **27**, No. 92, 131–42.
17. Jordaan, I. J., England, G. L., and Khalifa, M. A. (1977), 'Creep of concrete, a consistent engineering approach', *J. Struct. Div. ASCE*, **103**, No. ST3, 475–91.
18. Trost, H., Cordes, H., and Abele, G. (1978), 'Kriech- und Relaxationsversuche an sehr altem Beton', *Dtsch. Ausschuss Stahlbeton, Schriftenr. Heft* 295, 1–27.
19. Ghali, A., Neville, A. M., and Jha, P. C. (1967), 'Effect of elastic and creep recoveries of concrete on loss of prestress', *J. Am. Concr. Inst.*, **64**, 802–10.
20. Trost, H. (1967), 'Auswirkungen des Superpositionsprinzips auf Kriech- und Relaxations-probleme bei Beton und Spannbeton, *Beton. Stahlbeton.*, **62**, No. 10, 230–8; No. 11, 261–9.

21. Bažant, Z. P. (1972), 'Prediction of concrete creep effects using age-adjusted effective modulus method', *J. Am. Concr. Inst.* **69**, No. 4, 212–7.
22. Favre, R., Koprna, M., and Radojicic, A. (1980), *Effects différés, fissuration et déformations des structures en béton*, Ed. Georgi, Saint-Saphorin, Switzerland, 220 pp.
23. Neville, A. M., and Dilger, W. H., (1982), *Creep of Plain and Structural Concrete*, Longman, London.
24. Bastgen, K. (1979), 'Zum Spannungs–Dehnungs–Zeit–Verhalten von Beton, Relaxation, Kriechen und deren Wechselwirkung', *Dissertation*, Technical University Aachen.
25. Nielsen, L. F. (1982), 'The improved Dischinger method as related to other methods and practical applicability', *Symp. on the Design for Creep and Shrinkage in Concrete Structures, ACI Spec. Publ. No. 77.*
26. *CEB Bull. d'Information, No.* 136, 190 pp., Paris, (July 1980).
27. Busemann, R. (1950), 'Kriechberechnung von Verbundträgern unter Benutzung von zwei Kriechfasern', *Der Bauingenieur*, **25**, No. 11, 418–20.
28. Dilger, W. H. (1982), 'Creep analysis of prestressed concrete members using creep-transformed section properties' *J. Prestress. Concr. Inst.* **27**, No. 1, Jan.–Feb.
29. Tadros, M. K., Ghali A., and Dilger, W. (1975), 'Time-dependent prestress loss and deflection in prestressed concrete members', *J. Prestress. Concr. Inst.*, **20**, No. 3, May–June.
30. Tadros, M. K., Ghali A., and Dilger, W. (1977), 'Effect of non-prestressed steel on prestress loss and deflection', *J. Prestress. Concr. Inst.*, **22**, No. 2, March–April.
31. Bažant, Z. P., and Najjar, L. T. (1973), 'Comparison of approximate linear methods for concrete creep', *J. Struct. Div. ASCE*, **99**, No. ST9, 1851–74.
32. Schade, D. (1977), 'Alterungsbeiwerte für das Kreichen von Beton nach den Spannbetonrichtlinien', *Beton. Stahlbeton.*, **5**, 113–17.
33. Haas, W. (1974), 'Comparison of stress–strain laws for the time-dependent behaviour of concrete', *Proc. RILEM-CISM Symp., Udine*, September 1974.
34. Ross, A. D. (1957), 'Creep of concrete under variable stress', *J. Am. Concr. Inst.*, March, 739–58.
35. El-Shafey, O. A. B. (1979), 'Time-dependent effects in structural concrete members', *PhD Thesis*, The University of Calgary, Canada.
36. Bažant, Z. P., and Kim, S. S. (1979), 'Approximate relaxation functions for concrete creep', *J. Struct. Div. ASCE.*, **105**, No. ST12, December.
37. Rüsch, H., and Jungwirth, D. (1976), *Berücksichtigung der Einflüsse von Schwinden und Kriechen auf das Verhalten der Tragwerke*, Werner Verlag, Düsseldorf.
38. Papsdorf, W., and Schwier, F. (1968), *Creep and Relaxation of Steel Wire, Particularly at High Temperature*, Stahl und Eisen.
39. Dilger, W. H. (1981), 'Creep analysis of concrete structures', *Symp. on the Design for Creep and Shrinkage in Concrete Structures, ACI Spec. Publ.*. No. 77.
40. Kiessl, K., and Gertis, K. (1977), 'Nichtisothermer Feuchtetransport in dickwandigen Betonteilen von Reaktordruckbehältern', *Dtsch. Ausschuss Stahlbeton, Schriftenr. Heft* 280, 1–19.
41. Maes, M. A. (1980), 'Effects of environmental and material characteristics on behaviour of concrete strucures', *MSc Thesis*, The Univesity of Calgary.
42. Jordaan, I. J., and Khalifa, M. M. A. (1978), 'Time-dependent stresses in heated concrete strucures', *Douglas McHenry Int. Symp. on Concrete and Concrete Structures, ACI Spec. Publ.*, SP55, 321–346.

43. CEB-Bulletin No. 94 (1973), *Manuel de Calcul,* 'Effets structuraux du fluage et des déformations différées', August.
44. Argyris, J. H., Pister, K. S., and William, K. J. (1976), 'Thermo-mechanical creep of aging concrete—a unified approach', *Int. Assoc. Bridge Struct. Eng. Mem.,* **36-I,** 23–57.
45. Pister, K. S., Argyris, J. H., and William, K. J. (1978), 'Creep and shrinkage of aging concrete', *Douglas McHenry Int. Symp. on Concrete and Concrete Structures, ACI Spec. Publ.,* SP55.
46. Khalil, M. S., Dilger, W. H., and Ghali, A. (1982), 'Design considerations dictated by creep and shrinkage in prestressed concrete beams', *Symp. on the Design for Creep and Shrinkage in Concrete Structures, ACI Spec. Publ. No.* 77.
47. PCI Committee on Prestress Loss (1975), 'Recommendations for estimating prestress losses', *J. Prestress. Concr. Inst.,* **20,** No. 4, 43–75.
48. Tadros, M. K., Ghali, A., and Dilger, W. H. (1979), 'Long-term stresses and deformations of segmental bridges', *J. Prestress. Concr. Inst.,* **24,** No. 4, 66–87.
49. Tadros, M. K., Ghali, A., and Dilger, W. H. (1977), 'Time-dependent analysis of composite frames', *J. Struct. Div. ASCE,* **103,** No. ST4, Proc. Paper 12893, 871–84.
50. Danon, J. R., and Gamble, W. L. (1977), 'Time-dependent deformations and losses in concrete bridges built by the cantilever method', *Rep. No.* UILI-ENG-77-2002, *Civil Engineering Studies, University of Illinois at Urbana-Champaign, Urbana, Illinois,* 169 pp.
51. Khalil, M. (1979), 'Time-dependent analysis of concrete structures', *PhD Thesis,* Department of Civil Engineering, The University of Calgary, Canada.
52. Schade, D., and Haas, W. (1975), 'Elektronische Berechnung der Auswirkungen von Kriechen und Schwinden bei abschnittsweise hergestellten Verbundstabwerken', *Dtsch. Ausschuss Stahlbeton, Schriftenr. Heft.* 244.
53. Argyris, J. H., Szimmat, J., and William, K. J. (1977), 'Zur Konvertierung von SMART I', *Dtsch. Ausschuss Stahlbeton, Schriftenr. Heft.* 279.

*Chapter 10*

# Observations on Structures

*H. G. Russell, B. L. Meyers, and M. A. Daye*

## 10.1  INTRODUCTION

The creep and shrinkage behaviour of concrete and concrete structures has been investigated since the early 1900s.[1,2] Investigations have progressed from the observation of structures and empirical descriptions of the observed behaviour, to electron microscope studies of the components of cement and theoretical hypothesis describing the observed behaviour. Although significant progress has been made toward understanding the creep and shrinkage behaviour of cement and concrete, neither an all inclusive empirical relationship nor a complete theoretical formulation have been developed to describe these phenomena.

For these reasons, it is important that measurements on structures be continued and carefully reported in the literature. It is equally important that when such observations are made, the measurement program be planned with specific objectives in mind. Such objectives could include:

(1)   Verify predicted behaviour and/or empirical prediction method.
(2)   Determine structural adequacy.
(3)   Study structural behaviour.
(4)   Create data bank to check future theoretical formulations.
(5)   General interest.

It can be seen that the type of structural observations that are required for each of the above vary in accuracy and complexity. It should be further noted that in any observation program significant variability, beyond that caused by the measurement techniques, is introduced. Consider for example, objective (1), above. The following must be evaluated when such a structural observation program is developed:

(1)   *Variability of the prediction method.* Recent data presented at the International Symposium on Fundamental Research on Creep and Shrinkage of Concrete in Lausanne, Switzerland, in 1980, indicate that presently available Level 2, as defined by Hilsdorf and Muller,[44] prediction methods for creep properties have a coefficient of variation of approximately 25%.

This variability can be reduced if creep properties are estimated from experimental data obtained from the concrete used in the structure.

(2) *Variability of structural analysis method.* Analysis methods can range from simple elastic beam analysis to complex finite element procedures. The accuracy to which such methods predict stress and deformations vary significantly.

(3) *Variability of material properties.* It can be assumed that the general variability (coefficient of variation) of material properties, i.e. strength and elastic modulus, are approximately 10% or less.

(4) *Variability of loads.* Prior to evaluating such variability, the investigator should realize that although structures are designed to withstand highly variable loads (i.e. wind, earthquake, etc.), the loads actually experienced by a structure usually are fairly well defined (i.e. dead load, permanent live load, etc.). Although these latter loads are well defined, they are often significantly less than full design loads, and therefore, result in lower stresses and deformations than assumed in design.

(5) Finally, the effect of seasonal and daily temperature and humidity variations on measured response should be considered, since creep and shrinkage behaviour is significantly affected by such variation.

Therefore, variation of as much as ±50% from predicted deformations can be expected from even the most carefully planned and executed testing program.

## 10.2  GENERAL GUIDELINES FOR STRUCTURAL OBSERVATIONS

Although it is difficult to generalize concerning the type of programme that should be developed for a particular field measurement project, the following general guidelines should be considered:

(1) The measurement system should be compatible with the environment of the structure, and the duration of the test programs.

(2) Expected deformations should be evaluated under service load conditions (i.e. loads experienced by the structure during the measurement period).

(3) Expected deformations and stresses should be calculated using a stress–strain relationship that approximates actual conditions (i.e. in most cases, strength procedures should not be used, and all load factors and $\phi$ factors should be 1.0)

(4) If possible, companion laboratory creep specimens should be cast from concrete used to fabricate the structure under investigation. These speci-

mens should be tested under a sustained stress in the range of that experienced by the structure under service loads. Environmental conditions should be some predetermined standard, preferably the expected averages of site temperature and humidity during the test period. Age of loading should be consistent with age at which structure is loaded.

(5) Unloaded companion specimens should be tested in the same environment as (4), above, to determine the unrestrained shrinkage behaviour of the concrete.

(6) In appropriate situations, strain measurements should be taken perpendicular to the direction of principal strain to determined shrinkage strains in the structure. This is particularly useful in columns. However, the effect of Poisson's ratio and different percentages of reinforcement should be taken into account.

(7) Structural measurements should be taken at about the same time each day in order to minimize environmental variations. Ideally, measurements should be taken shortly before sunrise or on overcast days if the effect of sunshine is critical.

(8) The expected variability of the planned measurements should be evaluated.

## 10.3 FIELD MEASUREMENTS

In the sections that follow, actual measurements taken on various structures throughout the world will be presented. These observation programmes were undertaken for various reasons. High-rise concrete buildings were investigated to study structural behaviour. Bridge structures were investigated to determine structural adequacy. The data and programmes discussed in the following sections represent a small percentage of those available. No attempt was made to be all inclusive. For the most part, only programmes that have not been widely disseminated in the literature are discussed. The authors have, therefore, left out some significant work; for example, the work on bridges in France by the Ministry of Transportation,[45] in Portugal, on dams,[46,47] and the United States by the US Corps of Engineers,[48] among others,[49] have not been discussed in the pages that follow. It is hoped that researchers in this field will add to the data bank by widely disseminating their work.

## 10.4 BUILDINGS

Although the majority of published research on creep and shrinkage[1,2] has been concerned with calculation methods and laboratory measurements, some observations on actual buildings have been reported. In these observations, the total measured deformations have generally resulted from a combination of instantaneous shortening, creep, shrinkage, temperature, and relative humidity effects. It has been important to separate the different effects to obtain a proper interpretation of data.

Field data have also been influenced by the construction schedule and rate of loading. In many cases, the researcher did not control these effects and only interpreted their influence after the data had been obtained. To obtain field data on an actual building over a long period of time, complete cooperation of owner, architect, engineer, contractor, and subcontractors is needed.[3]

The following sections of this chapter summarize some of the information on observations of buildings from different countries.

### 10.4.1  Australia

Four experimental flat plate structures were erected at the Commonwealth Scientific and Industrial Research Organization, Division of Building Research, Australia, in the 1960s. The structures were all erected and loaded in an outdoor environment. Bay sizes were $9 \times 12$ ft ($2.7 \times 3.7$ m). Lightweight concrete was used. The first two structures were conventionally reinforced flat plates. The second two were post-tensioned flat plates. Details of these structures are given in a series of publications.[4-13]

Structure no. 1[4,6,8] was 3 bays × 3 bays with a $3\frac{1}{2}$ in (89 mm) thick concrete slab supported on 16 steel tubular columns. The unusual feature of this structure was the large span-to-depth ratio of 41. In addition to measurements of initial deflection when the formwork was released, long-term deflections under dead load were measured for a period of 200 days.[6] Further deflection under live load was measured for an additional 50 days.

Structure no. 2[7] was 3 bays × 2 bays and had a 4 in (102 mm) thick slab. It was reinforced with a wire mesh and was tested to destruction under vertical loads.

Structure no. 3[9,10] was a 3 in (76 mm) thick, 3 bays × 2 bays structure that was post-tensioned with uniform prestress in both directions.

Structure no. 4[11] was similar to no. 3 except the bays were 2 × 4 and draped post-tensioning tendons were used. The draped tendons were used to eliminate bending moments and deflections at predicted service loads and to reduce cracking and deflection that were observed in structure no. 3. Results from structure no. 4 have been described by Brotchie and Beresford.[13]

The test programme on structure no. 4 involved three parts. Measurements were taken during prestressing, under long-term loading, and during a test to failure. During the long-term loading, measurements of concrete strains were made at 108 locations using a mechanical strain gauge. Deflections were also measured optically. The sustained uniform load used for long-term measurements was applied between ages of 10 and 30 days. The load remained in place until an age of 950 days.

Initially, an upward deflection of slab structure no. 4 occurred. Before the end of the first year, the deflection had returned to zero. Afterwards, downward deflection occurred. Longitudinal tendon stresses decreased by approximately 7% and transverse tendon stresses decreased by 8% during sustained loading. Long-term movements were small and shrinkage was less than expected.

Long-term deflections of five reinforced concrete flat slabs and flat plates have been reported by Taylor and Heiman.[14] Their programme was undertaken so that structural designers could more accurately assess long-term deflection performance of reinforced concrete flat slabs and plates. Measurements on some structures were made for up to nine years. It was found that measured values were up to five times greater than calculated values. They attributed this to possible higher creep and shrinkage of the ready-mix concretes used in the Sydney area. Shrinkage tests in the laboratory confirmed that the concretes possessed higher shrinkage than comparable European concretes.

### 10.4.2 United Kingdom

The work of the Cement and Concrete Association[15,16] has been aimed at measurements on commonly used structures. Consequently, they have measured strains on prestressed ribbed slabs in a hospital prototype building,[15] and on similar reinforced ribbed slabs in a telephone exchange extension.[16] The structures were subjected to the normal construction loads and to the climatic conditions of the British Isles.

In the prestressed ribbed slab floor, two ribs were instrumented at three depths in three sections. A total of 18 gauges were applied. Each rib was designed to have a uniform prestress on it. Paralleling the measurements on the ribs, measurements of shrinkage were made on site stored prisms. Elastic modulus, creep, and shrinkage were also measured on prisms stored in the laboratory. Data measured on the prisms were used to calculate deformations in the structure. Excellent agreement was obtained. The data illustrated the need to consider the influence of relative humidity for each geographic location when calculating long-term deformations.

The long-term deformations of reinforced concrete columns in two structures have been measured by Swamy.[17,18] The columns were instrumented to measure strains, temperature, and moisture changes. Results confirmed observations[19] that considerable load transfer to the steel reinforcement takes place in reinforced concrete columns. In one of the structures, steel stresses were in excess of the permissible design values.

Swamy reported a major proportion of the loads carried by the columns resulted from dead loads and only 10 to 20% of the total design loads constituted the live load portion. An assessment of the strain components showed that in the lower columns of a building, instantaneous deformations accounted for the highest proportion of the total shortening. In the upper columns of the buildings, shrinkage strain contributed the greatest component. This is consistent with other observations.[19,20]

A 428 ft (130 m) high reinforced concrete chimney was instrumented[21] to measure vertical deformations at two different heights above ground level. Vertical strains were measured at heights of 49 ft (15 m) and 197 ft (60 m) above ground level. Horizontal strains were also measured at the 197 ft (60 m) level. Measured vertical strains after 1200 days were reported to be 475 and 340 millionths at the 197 ft (60 m) and 49 ft (15 m) levels, respectively.

### 10.4.3  United States

In recent years, the Portland Cement Association has undertaken a series of projects to measure long-term deformations in high-rise buildings. All projects have included field instrumentation on actual structures, laboratory tests on samples of concrete used in the instrumented structures, and correlation between calculated and measured deformations. The three instrumented structures were Lake Point Tower, Water Tower Place, and River Plaza.

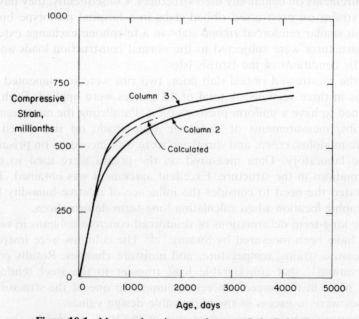

**Figure 10.1**  Measured strains on columns at Lake Point Tower

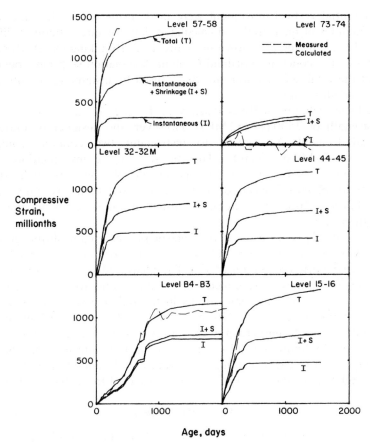

**Figure 10.2**  Calculated and measured vertical shortening of columns at
Water Tower Place

Instrumentation of Lake Point Tower began in 1966 and continued
through 1978.[19,22,23] The purpose of the investigation was to measure
deformations in the columns of an unusually tall structure during and after
construction. At the time of construction, Lake Point Tower with a height of
645 ft (197 m) was the world's tallest concrete building. A sample of the
data obtained over a 12-year period is shown in Figure 10.1, together with a
comparison of the calculated deformations.[19]

Water Tower Place is a 76-story reinforced concrete building with a
height of 859 ft (262 m). Measurements of column shortening were made at
six levels in the building.[20,22,23] The instrumented levels were selected to
measure shortening in each of five different concrete strengths used in the
columns. In addition to measurements of column shortening, measurements

of differential vertical movement were made using precise leveling techniques. Samples of measured and calculated data for one column at different
levels of the building are shown in Figure 10.2. By taking measurements on
samples of the concrete, Portland Cement Association staff have been able
to split the total deformation into that caused by instantaneous shortening,
creep, and shrinkage.

The lower columns of River Plaza were designed for a concrete compressive strength of 7500 psi (52 MPa). However, the designer permitted the
Portland Cement Association to instrument two columns containing
11,000 psi (76 MPa) concrete. The purpose of the investigation was to
identify problems involved in producing and placing very high strength
concrete in columns. Measured strains on the two columns were not significantly different from measurements obtained on columns having lower
strength concrete.[22,24]

Results of a $6\frac{1}{2}$ year observation of elastic plus creep and shrinkage strains
in a number of columns and walls in a second story of a 38-story building
have been reported by Fintel *et al.*[25] The measured strains were compared
with the analytically determined values and found to be in acceptable
agreement. Measured inelastic strains in the building showed good agreement with predicted values. However, the effect of relative humidity and
temperature on drying creep were not considered due to the lack of
information. Preliminary results of this investigation indicated that the creep
of members subjected to varying temperature and humidity may be considerably higher than the creep of columns protected from the weather.

## 10.5  BRIDGES

Although bridges in other countries[26-29] have been instrumented, field
measurements of creep and shrinkage on concrete bridges in the United
States of America are limited. However, some programmes have been
reported and results are summarized in this section. The three parts of this
section consider three different types of concrete bridges. These are reinforced concrete, prestressed concrete with solid sections, and prestressed
box girder bridges. In addition, prestressed box girder bridges in Czechoslovakia are briefly reviewed.

### 10.5.1  Reinforced concrete

In a paper published in 1971, Pauw[30] reported deflection measurements on
a five-span continuously reinforced box girder bridge located in Missouri.
Deflection measurements were made by means of permanently installed
inserts. Analysis of field observations revealed that elevations were significantly affected by temperature fluctuations, humidity fluctuations, and
differential expansion and contraction of the bent columns.

Observations were made over a period of 438 days following removal of the forms and shores. No significant changes in elevation could be detected after eight months. Pauw[30] concluded that ultimate sustained load deflections may be calculated on the basis of some reasonable reduced modulus value.

## 10.5.2 Prestressed concrete: solid sections

The first major use of prestressed concrete in the United States occurred with construction of Walnut Lane Bridge in Philadelphia in 1950. One of the first prestressed concrete structures designed by the Missouri State Highway Department was instrumented to determine behaviour during construction and in service.[31] Two post-tensioned concrete girders were instrumented to measure strains and deflections. Measurements were made before, during, and after the post-tensioning operations, before and after all major construction operations, and at appropriate time intervals following completion of the structure. A companion test programme was undertaken to measure physical properties of representative samples of the materials used in construction of the bridge. Measured deformations for 540 days after post-tensioning have been reported.[31]

Beginning in 1965, Gamble began a series of investigations into the long-term behaviour of prestressed concrete highway bridges in-service in the State of Illinois.[32-37] Three structures were instrumented to measure strains, camber, and temperature. The first structure was a four-span continuous composite prestressed concrete highway bridge.[32] Girders were cast in June, 1967 and camber measurements were made over a period of 12 years.

The second structure was a three-span bridge[33] on which measurements were taken for slightly more than two years.

The third structure[34] was a three-span bridge built in 1972. Readings were taken for about three years after the first girders were cast.

Instrumented bridges were standard designs for the time of their construction. They were built without special features related to their inclusion in an experimental programme. Objectives of all three programmes were to obtain better information to aid the designer in accounting for prestress loss and prediction of camber changes.

As part of the same programme, analytical tools[35,36] were developed to help interpret the data. Extensive parametric studies were carried out to develop a relatively complete understanding of the factors influencing camber change, loss of prestress, development of moments at interior supports, and interactions between the time dependent strains in deck and girder concretes. Major emphasis of the analytical work was development of procedures whereby measured creep and shrinkage data from concrete used in a particular structure could be used as input data in the analysis. This led to a relative precise prediction for a specific structure with specific material properties.

To provide information on properties of concretes used in the concrete bridges, samples were obtained for measurements of creep and shrinkage deformations. These programs demonstrated that there is an important difference between the creep and shrinkage strains of concrete kept outdoors, as in a bridge structure and the same concrete kept in a constant environment of the laboratory.

### 10.5.3 Prestressed concrete: box girders

With the introduction of post-tensioned segmental cantilever bridge construction into the United States in the early 1970s, a significant growth in the long-span concrete bridge market occurred. At least three major structures have been instrumented for measurement of time-dependent strains.

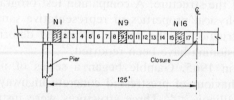

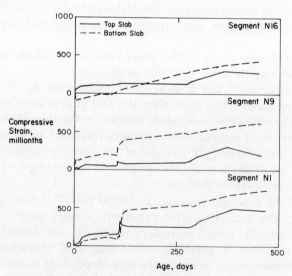

**Figure 10.3**  Measured longitudinal strains on Kishwaukee
River Bridge

The bridges are Denny Creek Bridge,[22,38] Kishwaukee River Bridge,[22,23] and Turkey Run State Park.[39]

Denny Creek Bridge located near Snoqualmie Pass in the State of Wahington was built using a three-stage cast-in-place method. In the first stage, the bottom slab and webs were cast and post-tensioned. In the second stage, the top slab between the webs was cast and post-tensioned. In the third stage, the cantilever deck on the outside of the webs was cast and post-tensioned. Instrumentation was installed in one span of the bridge with the objective of determining the strain distribution caused by creep and shrinkage effects.[22,38]

Kishwaukee River Bridge,[22,23] located in Illinois, consists of twin, single-cell prestressed concrete box girders. Each bridge has five spans measuring 170 ft (52 m) for the end spans and 250 ft (76 m) for three interior spans. Total length of each structure is approximately 1170 ft (357 m). The bridge is a constant depth precast concrete post-tensioned box girder built by the cantilever method. Longitudinal strains were measured on three segments of a main span cantilever.

Sample data obtained on Kishwaukee Bridge are shown in Figure 10.3. Coinciding with field measurements, laboratory tests were conducted on cylinders made from the same concrete used to cast the instrumented structures. Compressive strength, elastic modulus, coefficient of thermal expansion, creep, and shrinkage at various ages were measured. Based on the laboratory tests and detailed construction records, prestress losses and camber changes were calculated.

### 10.5.4 Prestressed concrete bridges in Czechoslovakia

Of special note in the area of field observation of structures is work performed in Czechoslovakia and reported by Dr T. Jávor.[43] Measurements were taken for significant periods of time, in excess of five years, on box girder cantilever and framework bridges. Data were gathered at various locations along the span of the bridges and at various points within the box cross sections. In addition, companion prismatic specimens were tested in the laboratory as well as *in situ* near the bridge or placed in the hollow cross sections of the bridges. The calculated results were compared to classical Dischinger theory with excellent results.

### 10.6 Nuclear Power Plant Prestressed Concrete Containment Structures

During the lifeime of a prestressed concrete containment structure, the effect of time-dependent concrete deformation and steel relaxation, on the stress in the prestressed tendons must be measured and reported to the Nuclear Regulatory Commission (for structures in the United States). Such

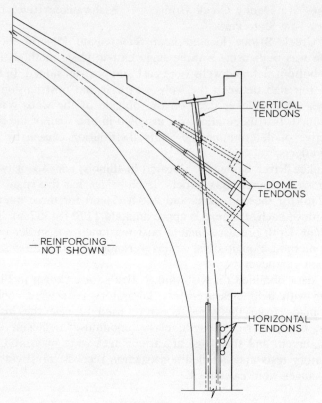

**Figure 10.4** Typical tendon location, prestressed concrete containment structure

measurements are taken at predetermined intervals, usually one, three, and five years after initial stressing, then at five-year intervals, until the plant is shutdown. The measurements are not sophistcated, and involve determining the forces at the tendon anchorages, by releasing the stress (lift off) on a limited number of tendons randomly distributed throughout the structure. It should be noted that the purpose of these measurements is not to determine creep and shrinkage but rather to evaluate structural adequacy by comparing prestressing forces to predicted and allowable levels. A typical containment structure is shown in Figure 10.4.

The prestressing used in the structures reported on in this section are 90 wire, 240 ksi ultimate strength unbonded tendons, resulting in a concrete compressive stress of 1530 psi. Nominal wire relaxation is 8% and concrete strength is 5000 psi. Total concrete deformation can be estimated from the loss in stress measured during tendon lift off. Further, strain due to concrete time-dependent deformation, including creep and shrinkage, can be approximately estimated by subtracting measured relaxation behaviour of the

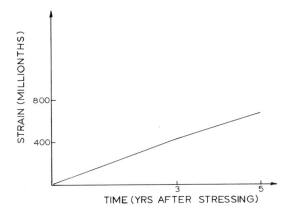

**Figure 10.5** Strain, vertical tendon

tendons used. This calculation does not consider the interaction of concrete time-dependent strain and steel relaxation.

Figures 10.5, 10.6, 10.7, and 10.8 show the results of estimates of concrete strain using the method described above for three types of tendons on two different containment structures[40,41] over a five-year period. Figures 10.5 and 10.6 show results from a vertical and horizontal tendon on the Calvert Cliffs Nuclear Power Plant in Calvert County, Maryland.[40] Figures 10.7 and 10.8 show results from dome tendons from Calvert Cliffs and the Turkey Point Nuclear Power Plant,[41] near Homestead, Florida. Refer to Figure 10.4 for the location of the various tendon types. Both of the structures described here are first generation power plant designs. Since their design and construction, containment size, tendon size, and distribution have changed. However, the behaviour shown in Figures 10.5, 10.6, 10.7, and 10.8 is typical of nuclear power plant structures.

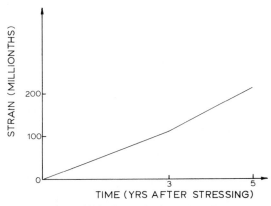

**Figure 10.6** Strain, horizontal tendon

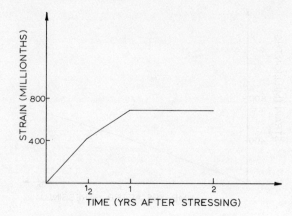

**Figure 10.7**  Strain, dome tendon

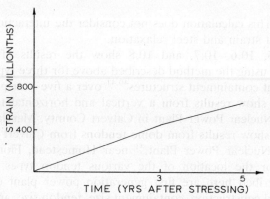

**Figure 10.8**  Strain, dome tendon

It can be observed that in all cases, except the vertical tendon, the estimated concrete strains are approximately the same, and close to concrete strain measured on a sealed $6 \times 12$ in concrete cylinder.[42] Such a typical strain curve for concrete used in nuclear containment structures is shown in Figure 10.9.

It can be concluded that for non-accident loads, i.e. dead and permanent live loads, the evaluation of the creep and shrinkage behaviour of nuclear containment structures can be performed using rather simple methods such as those described by ACI 209[2], if measured time-dependent strains on sealed $6 \times 12$ in concrete cylinders are available.

## 10.7  CONCLUSIONS

Some of the more important field measurements of long-term concrete behaviour have been discussed briefly and some general guidelines to be

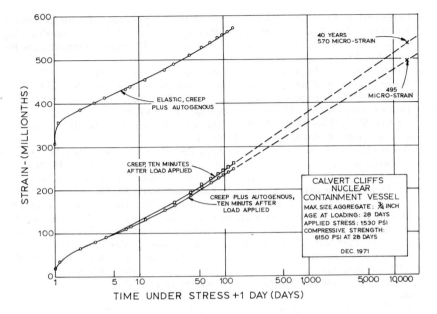

**Figure 10.9** Typical time-dependent strain

used when planning such a field programme presented. The authors would again like to encourage future workers in this field to report their data in the literature.

As previously indicated, most of the published data on creep and shrinkage were obtained in the laboratory.[1,2] These data are extremely important in the development of the, as yet unobtainable, all inclusive empirical relationship or the complete theoretical formulation for the time-dependent deformation of concrete. However, while we continue to strive for complete knowledge, more and more complicated and creep-sensitive structures are being designed and built every day as witnessed by 859 (262 m.) concrete high-rise buildings like Water Tower Place, and long-span prestressed concrete bridges. In order to adequately and safely construct them, accurate field observations are required.

## REFERENCES

1. ACI Committee 209 (1967), 'Shrinkage and creep in concrete', *ACI Bibliogr. No. 7*, American Concrete Institute, Detroit.
2. ACI Committee 209 (1972), 'Shrinkage and creep in concrete', *ACI Bibliogr. No. 10*, American Concrete Institute, Detroit.

356 *Creep and Shrinkage in Concrete Structures*

3. Beresford, F. D. (1970), 'Measurement of time-dependent behavior in concrete buildings', 4th *Australian Building Research Congr., Sydney, Australia, Aug. 1970.* Also *Commonwealth Scientific and Industrial Research Organization Reprint No. 554.*
4. Anon (1961), 'Experimental lightweight flat plate structure, Part I—Measurements and observations during construction', *Construct. Rev.,* 34, No. 1.
5. Anon (1961), 'Experimental lightweight flat plate structure, Part II—Deformation due to self weight', *Construct. Rev.* 34, No. 3.
6. Anon (1961), 'Experimental lightweight flat plate structure, Part III—Long term deformations', *Construct. Rev.,* 34, No. 4.
7. Anon (1962), 'Experimental lightweight flat plate structure, Part IV—Design and erection of structure with concrete columns', *Construct. Rev.,* 35, No. 2.
8. Beresford, F. D. (1962), 'Experimental lightweight flat plate structure, Part V—Deformation under lateral load', *Construct. Rev.,* 35, No. 12.
9. Lewis, R. K. (1963), 'Experimental lightweight flat plate structure, Part VI—Design and erection of post-tensioned flat plate', *Construct. Rev.,* 36, No. 3.
10. Beresford, F. D., and Blakey, F. A. (1963), 'Experimental lightweight flat plate structure, Part VII—A test to destruction', *Construct. Rev.,* 36, No. 6.
11. Gamble, W. L. (1964), 'Experimental lightweight flat plate structure, Part VIII—Test to failure of prestressed slab', *Construct. Rev.,* 37. No. 10.
12. Blakey, F. A. (1963), 'Australian experiments with flat plates', *J. Am. Concr. Inst.,* 60, No. 4.
13. Brotchie, J. F., and Beresford, F. D. (1967), 'Experimental study of prestressed concrete flat plate structure', *Civ. Eng. Trans. Inst. Eng., Australia,* CE9, No. 2, October.
14. Taylor, P. J., and Heiman, J. L. (1977), 'Long-term deflection of reinforced concrete flat slabs and plates', *J. Am. Concr. Inst.,* 74, No. 11.
15. Parrott, L. J. (1976), 'Long-term deformation of concrete in a prestressed concrete floor', *Proc. Conf. on Performance of Building Structures, Glasgow, Scotland, April 1976.*
16. Parrott, L. J. (1977), 'A study of some long-term strains measured in two concrete structures', *Int. Symp. on Testing In-Situ of Concrete Structures, Budapest, Hungary, Sept. 1977.*
17. Swamy, R. N., and Arumugasaamy, P. (1978), 'Deformations in service of reinforced concrete columns', *Douglas McHenry Int. Symp. on Concrete and Concrete Structures, ACI Spec. Publ.* SP55, Paper SP55-15.
18. Swamy, R. N., and Potter, M. M. A. (1976), 'Long-term movements in an eight-story reinforced concrete structure', *Proc. Conf. on Performance of Building Structures, Glasgow, Scotland, April. 1976.*
19. Pfeifer D. W., Magura, D. D., Russell, H. G., and Corley, W. G. (1971), 'Time-dependent deformations in 70-story structure', *Symp. on Designing for Effects of Creep, Shrinkage, and Temperature in Concrete Structures, ACI Spec. Publ.* SP-27, American Concrete Institute, Detroit.
20. Russell, H. G., and Corley, W. G. (1978), 'Time-dependent behavior of columns in Water Tower Place', *Douglas McHenry Int. Symp. on Concrete and Concrete Structures, ACI Spec. Publ.* SP55, Paper SP55-14. Also *Portland Cement Association, Res. Dev. Bull.* RD052–01B.
21. Reith, R. D., Kohn, R., and Watson, P. (1976), 'Full-scale measurements of a reinforced concrete chimney', *Proc. Conf. on Performance of Building Structures, Glasgow, Scotland, April 1976.*
22. Russell, H. G. (1980), 'Field instrumentation of concrete structures', *Full-Scale*

*Load Testing of Structures*, ASTM STP-702, American Society for Testing and Materials.
23. Russell, H. G., and Shiu, K. N. (1980), 'Creep and shrinkage behavior of tall buildings and long span bridges', *Symp. on Fundamental Research on Creep and Shrinkage of Concrete, Lausanne, Switzerland, 15–17 Sept. 1980.*
24. 'High-strength concrete in Chicago high-rise buildings' (1977), *Task Force Report No. 5, Chicago Committee on High-Rise Buildings, Chicago, Ill.*, Feb.
25. Fintel, M., and Khan, F. R. (1971), 'Effects of column creep and shrinkage in tall structures—analysis for differential shortening of columns and field observation of structures', *Symp. on Designing for Effects of Creep, Shrinkage, and Temperature in Concrete Structures, ACI Spec. Publ.* SP-27, American Concrete Institute, Detroit.
26. Tyler, R. G. (1976), 'Creep, shrinkage and elastic strain in concrete bridges in the United Kingdom, 1963–71', *Mag. Concr. Res.*, **28**, No. 95, June.
27. Finsterwalder, U. (1958), 'Die Ergebnisse von Kriech- und Schwind messungen an Spann betonbauten' ('Results of creep and shrinkage measurements on prestressed concrete structures'), *Beton. Stahlbeton.*, **53**, No. 5.
28. Roelfstra, P. E. (1980), 'Computerized structural analyses applied to large span bridges', *Symp. Fundamental Research on Creep and Shrinkage of Concrete, Lausanne, Switzerland, 15–17 Sept. 1980.*
29. Haas, W. (1980), 'Numerical analysis of creep and shrinkage in concrete structures', *Symp. on Fundamental Research on Creep and Shrinkage of Concrete, Lausanne, Switzerland, 15–17 Sept. 1980.*
30. Pauw, A. (1971), 'Time dependent deflections of a box girder bridge', *Symp. on Designing for Effects of Creep, Shrinkage, and Temperature in Concrete Structures, ACI Spec. Publ.* SP-27, American Concrete Institute, Detroit.
31. Pauw, A., and Breen, J. E. (1961), 'Field testing of two prestressed concrete girders', *Highway res. Board Bull.* 307, 'Prestressed concrete structures—creep, shrinkage, deflection studies', *Natl. Acad. Sci.—Natl. Res. Counc. Publ. No. 937.*
32. Gamble, W. L. (1970), 'Field investigation of a continuous composite prestressed I-beam highway bridge located in Jefferson County, Illinois', *Civ. Eng. Stud., Struct. Res. Ser. No. 360*, University of Illinois at Urbana-Champaign, Urbana, Ill.
33. Houdeshell, D. M., Anderson, T. C., and Gamble, W. L. (1972), 'Field investigation of a prestressed concrete highway bridge located in Douglas County, Illinois', *Civ. Eng. Stud., Struct. Res. Ser. No. 375*, University of Illinois at Urbana-Champaign, Urbana, Ill.
34. Gamble, W. L. (1979), 'Long-term behavior of a prestressed I-girder highway bridge in Champaign County, Illinois', *Civ. Eng. Stud., Struct. Res. Ser. No. 470*, University of Illinois at Urbana-Champaign, Urbana, Ill.
35. Mossiossian, V., and Gamble, W. L. (1972), 'Time-dependent behavior of noncomposite and composite prestressed concrete structures under field and laboratory conditions', *Civ. Eng. Stud., Struct. Res. Ser. No. 384*, University of Illinois at Urbana-Champaign, Urbana, Ill.
36. Hernandez, H. D., and Gamble, W. L. (1972), 'Time-dependent prestress losses in pre-tensioned concrete construction', *Civ. Eng. Stud., Struct. Res. Ser. No. 417*, University of Illinois at Urbana-Champaign, Urbana, Ill.
37. Gamble, W. L. (1980), 'Final summary report, field investigation of prestressed reinforced concrete highway bridges', *Civ. Eng. Stud., Struct. Res. Ser. No. 479*, University of Illinois at Urbana-Champaign, Urbana, Ill.

358     *Creep and Shrinkage in Concrete Structures*

38. Corley, W. G. (1980), 'Instrumentation of Denny Creek Bridge', reported in 'Highlights of the Denny Creek Symposium', *Concr. Int. Des. Constr.*, **2**, 11.
39. Holman, R. J. (1977), 'Development of an instrumentation program for studying behavior of a segmental concrete box girder bridge', *Joint Highway Res. Project*, JHRP-77-4, Purdue University, Indiana, March.
40. Baltimore Gas and Electric Corporation, Calvert Cliffs Nuclear Power Plant-Unit #1, 'Containment structural post-tensioning system', *1 and 3 Years Surveillance Rep*, Bechtel Power Corporation, Gaithersburg, Maryland, July, 1975 and May, 1977.
41. Turkey Point Nuclear Power Plant-Unit #3, 'Containment structural post-tensioning system', *Six Months, 1, 2, and 5 Years Surveillance Reps*, Bechtel Power Corporation, Gaithersburg, Maryland.
42. 'Concrete properties of Calvert Cliffs nuclear containment vessel' (1971), *Prog. Rep. No. 1, University of California, Berkeley.*
43. Jávor, T. (1980), 'Creep observations of prestressed concrete bridges', *Res. Inst. Civ. Eng., Bratislava, Czechoslovakia.*
44. Hilsdorf, H. K., and Müller, H. S. (1979), 'Comparison of methods to predict time-dependent strains of concrete', *Institut fur Beustofftechnologie, Universitat Karlsruhe.*
45. 'Les Ouvrages D'Art' (1978), *Bull. Liaison Labo. Ponts Chaussées*, Ministère des Transports, Paris, France.
46. Rocha, M., Serafim, J. L., Da Silveira, A. F., and Guerreiro, M. Q. (1958), 'Observations of concrete dams. Results obtained in Cabril Dam', *Lab. Nac. Eng. Civ., Tech. Pap. No. 129, Lisboa.*
47. Rocha, M. and Da Silveira, A. F. (1965), 'Assessment of observation techniques used in Portuguese concrete dams', *Mem. No. 254, Lab. Nac. Eng. Civ.*
48. ACI Committee 224 (1972), *Control of Cracking in Concrete Structures*, American Concrete Institute, Detroit.
49. 'Testing In-Situ of Concrete Structures' (French 'Essais *in-situ* des structures en beton') (1977) *RILEM, Int. Symp. on Testing of Concrete Structures In-Situ*, Budapest, Hungary, 1977.

# Index